THE THERMIONIC VACUUM TUBE

AND ITS APPLICATIONS

THE THERMIONIC VACUUM TUBE

AND ITS APPLICATIONS

BY

H. J. VAN DER BIJL, M.A., Ph.D.

*M.Am.I.E.E., M.I.R.E., Mem. Am. Phys. Soc., Scientific & Technical
Adviser, Dept. of Mines & Industries, Union of South Africa,
Late Research Physicist, American Tel. & Tel. Co.
and Western Electric Co., New York*

FIRST EDITION

NINTH IMPRESSION

McGRAW-HILL BOOK COMPANY, INC.

NEW YORK: 370 SEVENTH AVENUE
LONDON: 6 & 8 BOUVERIE ST., E. C. 4

1920

PREFACE

In a comparatively short time the applications of Thermionics have grown to a considerable extent, and are now not only of great value in engineering fields, but are also penetrating more and more into university and college laboratories. It is difficult for those who are interested in the subject, but who have not had the opportunity or the time to follow its development closely, to abstract from the literature, which has become quite voluminous, the principles of operation of thermionic vacuum tubes. This and the popularity which the remarkable ability of these tubes to perform a great variety of functions has gained for them, have created a need for a book describing in a connected manner the more important phenomena exhibited by the passage of electrons through high vacua.

In this work I have endeavored to set forth the principles of operation of thermionic vacuum tubes, and to coordinate the phenomena encountered in a study of this field. Such a procedure is sure to result in a more valuable book than a detailed description without proper coordination of the many investigations that have been published on this subject.

I have tried to make the treatment sufficiently elementary to meet the demands that will necessarily be made on a book of this kind. This is especially the case with the first few chapters, which must be regarded as very elementary and are mainly intended for those who are interested in the applications of thermionic tubes but are not sufficiently acquainted with the properties and behavior of electrons to understand the operation of these tubes.

I wish to express my indebtedness to several of my colleagues who have read parts or all of the manuscript. In this connection I wish to mention especially Mr. C. A. Richmond and Dr. P. I. Wold.

H. J. v. d. B.

CONTENTS

CHAPTER VII

THE THERMIONIC AMPLIFIER

CHAPTER VIII

THE VACUUM TUBE AS AN OSCILLATION GENERATOR

CHAPTER IX

Modulation and Detection of Currents with the Thermionic Tube

CHAPTER X

Miscellaneous Applications of the Thermionic Tube

INTRODUCTION

THE achievements in the art of intelligence communication which we have witnessed in the past seven or eight years are the result of an extensive series of investigations that stretch over many decades. It is easy to see the factors that directly influenced that part of the work that has been conducted in recent years to develop our systems of telephone and telegraph communication. But we should not forget that this work rests on a foundation laid by such pioneers as Maxwell, Hertz, H. A. Lorentz, J. J. Thomson and O. W. Richardson—men who conducted their research without any monetary motive and with little or no thought to any possible future commercial application. Through their efforts we came into possession of the electromagnetic theory, which has taught us much about the undulatory propagation of energy, and the electron theory, which enabled us to explain many a baffling phenomenon. It is a glowing compliment to scientific research that just those two theories which seemed so abstruse and so speculative that hardly anybody believed in them should have grown into such valuable commercial assets. The electron theory in particular has been of invaluable assistance in the development of the audion or three-electrode thermionic tube, the device which forms the nucleus of the research and development work that was carried on in efforts to improve our means of intelligence communication.

The audion tube consists of an evacuated vessel containing a filament which can be heated by passing a current through it, an anode which is usually in the form of a plate or pair of plates or a cylinder, and a third electrode, usually in the form of a wire grating placed between the filament and the anode. The hot filament emits electrons, and these are drawn to the anode or plate under the influence of a potential difference between filament and plate such as to maintain the plate positive with respect to the filament. The third electrode, commonly referred to as the grid, functions as the controlling electrode and is for the purpose of controlling the flow of electrons from filament to anode. By applying potential variations to the grid the electron current flowing from filament to plate can be varied.

It has been known for many years that hot bodies possess the property of imparting a charge to conductors placed in their neighborhood. This phenomenon was investigated by Elster and Geitel in the eighties of the last century. They found that if a metallic filament was placed in a glass vessel and heated to incandescence a plate placed close to the filament inside the vessel acquired a positive charge, but when the vessel was exhausted the charge on the plate became negative. About the same time Edison noticed that if a plate be inserted in a carbon filament incandescent lamp, a current flowed between the plate and the filament when the plate was connected to the positive end of the filament but not when the plate was connected to the negative end. The direction in which the current appeared to flow was from plate to filament when the former was positive with respect to the filament. J. J. Thomson showed in 1899 that the current flowing through the space between filament and plate was carried by electrons. He did this by measuring the ratio of the charge to the mass of the particles that appeared to convey the current, and found a value from which it was to be concluded that these particles were electrons. The mechanism of the emission of the electrons from the hot filament was explained by O. W. Richardson in 1901. Richardson made use of a view that had previously been suggested to explain metallic conduction, namely, that the free electrons in a metal possess kinetic energy like the molecules of a gas. He furthermore assumed that there is a force at the surface of the filament which tends to hold the electrons within the filament, and in order to escape from it the electrons have to do a certain amount of work, depending on the value of this force. At ordinary temperatures the energy of the electrons in the substance is not sufficient to enable them to overcome this surface force, but if the temperature of the filament be raised sufficiently high the energy of the electrons increases enough to enable some of them to overcome the surface force and escape. According to Richardson, therefore, the electrons escape from hot bodies solely in virtue of their kinetic energy. A large number of experiments have been made by various investigators on this phenomenon. These experiments not only verified Richardson's theory, but also gave results that have an important bearing on the operation of the thermionic vacuum tube. The study of the emission of electrons from a hot filament and their transport to the anode or " plate " involves a large number of problems, the solution of which is based on the electron theory. The most important developments of the thermionic vacuum tube were carried out by men who were familiar with this theory, and indeed their knowledge of its fundamental principles contributed in a large measure to the rapidity with which the thermionic tube was nursed from the stage almost of a scientific toy to a very important commercial device. For these reasons the first few chapters of this book are devoted to an elementary discussion of the properties of electrons and the phenomena encountered in the conduction of electricity by dislodged charges. The discussion of these phenomena, however, has to be elementary and concise in consideration of the great amount of material dealing directly with the applications of the vacuum tube. In choosing the material for these earlier chapters I was guided mainly by my own experiences in my work on the development of this type of device.

When Richardson in 1901 gave an explanation of the mechanism of the emission of electrons from hot bodies he made an important contribution to science, but there was at that time no thought of the practical value that this theory was destined to have. It was in 1905 that J. A. Fleming conceived the idea of using a thermionic valve, that is, an evacuated bulb containing a cold anode and a hot filament as a rectifier for the detection of electromagnetic waves. At about the same time Wehnelt, who had previously carried out investigations on the emission of electrons from hot bodies and had produced the oxide-coated filament which bears his name, and which gives off electrons much more readily than metallic filaments, also suggested using a hot cathode device as a rectifier and described experiments that he had conducted with such a tube. Since electrons are emitted from the hot filament but not from the cold anode, such a device has a unilateral conductivity, and can therefore be used to rectify alternating currents. Wehnelt's idea in using a hot cathode rectifier was to obtain large currents for small potential differences between the electrodes. If, for example, a glow discharge is passed through a partially evacuated tube containing cold electrodes the space between the electrodes contains both electrons and positive ions. The electrons move to the anode while the positive ions move towards the cathode. Since the positive ions are large and heavy compared with the electrons they move more slowly than the electrons under the influence of the same field. The space in the neighborhood of the cathode therefore contains more positive ions than electrons, and this causes the establishment of a positive space charge near the cathode. This positive space charge is responsible for a large part of the voltage drop in such tubes. By using a hot cathode which spontaneously emits electrons Wehnelt could neutralize the space charge due to the positive ions, because the electrons coming from the cathode combine with the positive ions to form neutral gas molecules. By this means therefore the voltage drop in the tube was greatly decreased.

If the tube be evacuated to such an extent that there are practically no collisions between the electrons and the residual gas molecules, the space between cathode and anode contains practically only electrons, and therefore there is only a negative space charge between the electrodes of the tube. We then have what is commonly known as a thermionic valve. The characteristics of this device are discussed in Chapter IV, while the influence of gases in the bulb on the characteristics of the valve is dealt with in Chapter V.

The tubes discussed in the following pages operate under conditions that are characterized by the absence of gas. To maintain this condition it is necessary to insure that the gas pressure is at all times so low that the mean free path of the electrons in the residual gas is large compared with the distance between the electrodes. This requires not only that the gases in the space be removed to a sufficient extent, but also that the electrodes and walls of the vessel be sufficiently freed of occluded gases by heating these parts during the process of evacuation. This is necessary because the bombardment of the anode by the electrons and the heating current in the filament during the operation of the tube would result in a rise in temperature of the electrodes and walls, thus causing the liberation of occluded gases and a consequent impairment of the vacuum. The extent to which the electrodes and walls of the tube must be denuded of gases during evacuation depends on the power dissipated in the device during operation. If we are concerned only with electrodes that can be heated by passing a current through them we can adopt the practice that lamp manufacturers have been following for the last twenty or thirty years, viz., passing a current through the electrodes to raise them to a higher temperature than they attain during operation, and by baking the bulbs in ovens to as high a temperature as the glass can stand. But thermionic devices of the most commonly employed types also contain electrodes (anode and grid) which cannot be heated by passing a current through them. These electrodes can be heated during evacuation by electron bombardment; that is, by passing a thermionic current from cathode to anode through the tube. The amount of this current and the voltages applied should be higher than the values used when the tube is subsequently put in operation. Although these tubes operate under the condition that gas has no influence on the discharge, the operation of the tube will always be better understood if the effects of small traces of gases are known. These effects are therefore discussed in Chapter V.

The thermionic valve (by this we mean a two-electrode thermionic tube) is still used at the present time as a rectifier of alternating current, and as such it is a valuable instrument, and is capable of performing useful functions. Its operation and some of its uses are discussed in Chapter VI.

As a detector of electromagnetic waves, the valve has no commercial value. The device which is used for detecting purposes is the three-electrode tube which in addition to the anode and hot cathode, also contains a grid to control the flow of electrons from cathode to anode. The grid was inserted by de Forest in 1907, who called the device the " Audion " and it is the insertion of the grid which has made the thermionic tube of such great value. Since the flow of electrons from filament to anode or plate can be varied by applying potential variations to the grid, the circuit in which this tube is used consists of two branches: the output circuit connecting the filament to the plate through a current or power indicating device and the input circuit connecting the filament to the grid through the secondary of a transformer or other means of supplying potential variations to the grid.

The audion was first used as a radio detector but was subsequently found to be capable of performing a number of other important functions. In fact, the insertion of the grid into

the valve resulted in a device of tremendous potentialities—
one that can justly be placed in the same category with such
fundamental devices as the steam engine, the dynamo and the
telephone.

Since a mere variation in the potential of the grid produces a
variation in the plate current, it could reasonably be expected
that more power would be developed in the output circuit than the
power expended in the input circuit, and this has actually been
found to be the case. The operation of the tube as an amplifier
is simpler than its operation as a detector. I have therefore found
it best first to discuss the manner in which the tube operates as an
amplifier, reserving its operation as detector for discussion in a
later chapter. Chapter VII not only describes the amplifier and
the manner in which it operates and the circuits that can be used,
but also discusses the characteristics of the three-electrode tube.

Since the power in the output circuit of the audion tube is
greater than the power expended in the input, it is possible to
increase the degree of amplification by feeding back part of the
energy in the output to the input. If the proportion of the
energy thus returned to the input circuit is large enough and the
phase relations of the currents in the output and input circuits are
right, the tube can be made to produce sustained oscillations.
What is usually done is to connect the tube in an oscillation circuit
having the desired capacity and inductance, and then couple the
output circuit to the input in such a way that current variations
in the output circuit cause potential variations to be impressed
on the grid. There are a great variety of circuits whereby this can
be accomplished. To make a three-electrode tube produce sustained oscillations is an extremely simple matter, but to make it
operate in the most efficient way as an oscillation generator requires
a knowledge of the various factors that influence its operation
as such. These matters are discussed in Chapter VIII.

When the tube is used as an amplifier or oscillation generator
it is desirable that the characteristic representing the relation
between the plate current and the potential of the plate or grid
be as nearly linear as possible. On account of the negative space
charge of the electrons in the space between filament and anode
the characteristic is not linear, but convex towards the axis of
voltage. When the applied voltage becomes sufficiently high to
attract the electrons over to the plate as fast as they are emitted

from the filament, the characteristic curve becomes concave towards the voltage axis and finally becomes nearly horizontal, thus giving the saturation current. In order to obtain the best operation of the tube as an amplifier, it is necessary to straighten out the characteristic of the plate circuit. Means whereby this can be done are discussed in Chapter VII. The curvature of the characteristic causes the shape of the current wave in the output to be distorted, and is therefore a very undesirable feature of the audion as an amplifier. On the other hand, the fact that the characteristic is curved simply increases the number of uses to which the audion can be put. Its ability to detect electromagnetic waves lies in the curvature of its characteristic. This also makes it possible to use the tube for modulating high frequency waves for the purpose of radio or carrier telephony and telegraphy. The processes involved in detection and modulation are identical, and these are therefore treated together in Chapter IX.

In Chapter X are described a few miscellaneous applications of the thermionic tube. This list is intended to exemplify the manner of applying the principles of the tube and does not make any pretense at being complete. It is believed that the number of such applications is destined to increase considerably and that the tube will become of increasing importance not only in engineering practice, but also in university and college laboratories.

A large number of names have been used to designate the three-electrode type of thermionic tube, such as audion, pliotron, triode, thermionic valve, etc.—an impressive array of names which certainly attests the importance of this device. To forestall any possible confusion in the mind of the uninitiated reader I may say that these names all apply to one and the same thing, namely the audion or three-electrode tube discussed in the following pages.

The major portion of the development of the audion has taken place in the past eight years, but while the number of applications of the tube increases almost daily, we must frankly admit that as far as the tube itself is concerned, it was developed to a full grown and powerful instrument as early as 1914. The rapid development of this device, both in the United States and Europe and the popularity it has gained, are due to a number of factors that have concurred to place it in the foreground. One obvious reason, of course, is its ability to perform such a large number of important functions. No wonder that it has been referred to as the " versatile talking bottle."

Another factor which stimulated its development in the United States was the pressing necessity for a satisfactory system of telephone communication over long distances—a necessity which resulted from the recognition of the telephone as a very important factor in the industrial and commercial development of the country and the fact that the industries of the country are scattered over extensive regions. It was evident to those skilled in the art that the *sine qua non* of such a system of telephone communication is a device that will amplify telephone currents without impairing the quality of the transmitted speech more than can be tolerated in commercial service. When the audion made its appearance, telephone amplifiers or repeaters had already been developed, one of which, the mechanical repeater, still gives satisfactory service on some of the long distance lines. But the potentialities of the audion were immediately recognized by the leading telephone engineers when it came into their hands in 1912. As the result of a far-sighted policy based on the recognition of the influence of scientific research on industrial development, the fate of this device was placed in the hands of a number of well-trained research physicists and engineers. Its development proceeded so rapidly that the summer of 1914 saw the three-electrode tube as the repeater on a commercial system of telephone communication connecting New York with San Francisco. In order to use these tubes as repeaters on such a long telephone line, it stands to reason that the tubes must be carefully designed and constructed to have not only characteristics of definite predetermined value, but characteristics that remain constant over long periods of time and differ but little from one tube to another. This required a full understanding of the operation of the device, something which was secured in the comparatively short time only as the result of a very intensive and well organized series of investigations during the years 1912 to 1914. These investigations were carried out by the engineers of the American Telephone and Telegraph Company, and the Western Electric Company. About the same time the engineers of the General Electric Company made a study of the characteristics of thermionic tubes and enriched the world with valuable information regarding them.

The research and development work on the tube during these

years resulted in a large number of other uses to which it might be applied. Its possibilities in the radio field were recognized, and its application to this field resulted in March, 1915, in the successful transmission of speech by radio from Montauk Point, New York, to Wilmington, Delaware. These experiments which were undertaken by the American Telephone and Telegraph Company, and the Western Electric Company, were continued with the cooperation of the U. S. Navy Department and resulted, in the fall of 1915, in the successful transmission of speech from Arlington, Va., to Paris and Honolulu, a distance in the latter case of 5000 miles.

The need for satisfactory systems of intelligence communication in the war zone and the success of the audion tube resulted in an industrious development of this device and its applications in Europe. Most of the work that was done by the British, the French, the Germans, etc., is only coming to light now after peace has been declared. Military necessity forbade the exchange of scientific information that can reasonably be expected in times of peace. This has caused a great deal of duplication of work, and makes it extremely difficult to give recognition to individuals for important contributions. However, the matter of crediting individual investigators is insignificant in comparison with the great benefit that the world has derived from a general recognition of the value of scientific research and the many-sided and intensive investigations to which this very important and even more promising device has been subjected.

THE
THERMIONIC VACUUM TUBE AND ITS APPLICATIONS

CHAPTER I

PROPERTIES OF ELECTRONS

1. Electron and Corpuscle. The word electron was introduced by Prof. G. Johnstone Stoney in 1891 to denote the " natural unit of electricity," that is, the quantity of electricity which was found to be invariably carried by an atom of any univalent element (such as hydrogen) in electrolysis. Stoney did not imply that the electron was a small particle of something that carried a certain charge—the picture often formed of the electron; according to his definition the electron is simply a unit of charge and that is all. Corpuscle, on the other hand, is the name given by Sir J. J. Thomson to the carriers of electricity shot off from the cathodes in vacuum tubes. The researches of Thomson on the discharge through vacuum tubes showed that this corpuscle has a negative charge equal to one electron.

The two terms, electron and corpuscle, are frequently used indiscriminately. For this there is some justification, provided we understand, as is usually done, by the term electron the natural unit of negative electricity, and do not extend its meaning to include both positive and negative units as was originally intended by Stoney. With this restriction of the word electron there is no difference between the electron and Thomson's corpuscle. The corpuscle was found to have a charge equal to the electron and a mass which is $\frac{1}{1845}$ of that of the hydrogen atom. But the indiscriminate use of the terms electron and corpuscle is unfortunate, because it robs us of a name for the natural unit of charge, irrespective of whether it is negative or positive. This objection is usually overcome by referring to the unit of positive charge as the *positive electron*. This quantity, which has the same absolute value as the (negative) electron, is found always to be associated with a mass about 1800 times that associated with the electron. If the mass of the electron should be found to be entirely electromagnetic, then there would be no difference between it and the ultimate unit of negative charge.

In attempting to form a mental picture of the electron it is best not to associate in the mind the idea of a small particle having definite size and mass, in the ordinary sense, because the size and mass of an electron are things about which we must speak somewhat reservedly. The electron manifests itself only in virtue of the electric and magnetic fields created by its presence in the surrounding medium, but whether or not its mass is entirely electromagnetic is as yet an open question. Furthermore, the theory of Abraham and Lorentz, for example, leads to the conception of two masses for the electron, namely, the so-called transverse and longitudinal masses. As regards the size of the electron, although an estimate has been made of what might be considered its effective radius by the simple process of integrating the energy of the field, due to the slow-moving electron, it is not unlikely that this represents only one size obtained under one particular set of conditions. A discussion of these matters is beyond the scope of this book, and we must content ourselves with only the most elementary explanation of some of the properties of electrons, and only insofar as they are needed for understanding the more important phenomena encountered in vacuum tubes.

2. Lines of force and tubes of force. When Faraday, almost a hundred years ago, found himself in a position to repudiate the hypothesis of the action of forces at a distance, he had formed a conception of the nature of force action between electrically charged bodies which forms the basis of conceptions upon which modern physics is built. He pointed out that the electrical energy lies in the medium between charged bodies and not in the bodies themselves. This conception, which is the parent of the electromagnetic theory, has also been of valuable service in the development of the electron theory. The motion of the electron through the medium, the ether, is somewhat analogous to the motion of a sphere through a liquid; the moving sphere sets the surrounding liquid in motion with a velocity proportional to its own, so that to move the sphere the surrounding liquid must also be set in motion, with the result that the sphere behaves as if its mass were increased. The analogy is not complete because while the mass of the sphere is apparently increased in virtue of its being surrounded by the liquid, the mass of the electron is apparently entirely due to the condition of the surrounding medium. The analogy would be more nearly complete if the sphere were supposed to be practically weightless, such as a tennis ball kept completely immersed in a liquid, say by means of a string tied to the bottom of the liquid container. If we imagine that the string could move freely parallel to itself, then the inertia of the ball when moved in a plane perpendicular to the direction of the string will be due almost entirely to the motion of the liquid surrounding the ball. Such is the case when an electron moves in a plane. There is, however, this important difference: the moving electron does not drag the ether (that is, the surrounding medium) with it, as the ball drags the liquid in which it is immersed; what is dragged along is a modification of the ether. To get an understanding of this it is necessary to introduce the conception of lines and tubes of force.

The conception of lines of force was introduced by Faraday to form a mental picture of the processes going on in the electric field. To him these lines were not mere mathematical abstractions. He ascribed to them properties that gave them a real physical significance. They terminate on opposite charges, are always in a state of tension, tending to shorten themselves, and are mutually repellent. The direction of a line of force at any point gives the direction of the field at that point. With the help of these properties of lines of force it is possible to obtain an idea of the distribution of the intensity of the field surrounding electrically charged bodies.

The idea of *tubes of force* has been introduced to make the method of Faraday metrical rather than merely descriptive. A tube of force is obtained by drawing a number of lines of force through the boundary of any small closed curve. The lines then form a tubular surface which, it can be proved, will never be cut by any lines of force, and the extremities of which enclose equal and opposite charges. By properly choosing the area of the surface enclosed by the curve through which the lines are drawn the extremities of the tube can be made to enclose **unit charge**. Such a unit tube is called a Faraday tube. Maxwell and J. J. Thomson have made an exhaustive study of these tubes of force and expressed their properties in mathematical terms. The result that interests us here is that a tube of force behaves as though it had inertia, so that in order to move a tube work must be done. This explains why a charge behaves as if it had mass.[1]

3. Field of the "Stationary Electron." The electric field surrounding a point charge equal to one electron far removed from other charges may be represented by lines of force in the manner shown in Fig. 1. If this charge is at rest the field is a symmetrical

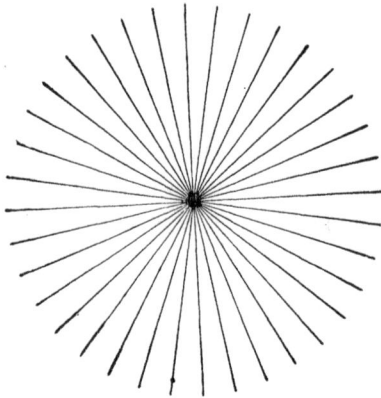

.Fig. 1

electrostatic field, the lines of force being distributed uniformly in radial fashion. This is what may be called the field of the stationary electron. Whether the electron actually consists of a small charge located at O, or whether the modification of the ether, as represented by Fig. 1, itself constitutes what is known as the electron, is as yet an unanswered question. For our present purposes it makes no difference which of the two views is the correct one, for the following reasons: It is known from elementary electrostatics that a charge which is uniformly distributed over the

---

[1] It must be remarked that the conception of tubes of forces is used here merely to aid in understanding the phenomena. Whether or not tubes of force, or even the ether, possess any physical significance is a question. Modern developments seem to indicate that this question must be answered in the negative.

surface of a sphere acts as if the whole charge were concentrated at the center of the sphere. Hence, if the isolated electron is only a symmetrical radial field with lines of force converging uniformly to a mathematical point, we can still look upon the field as being due to a charge concentrated at the point, or uniformly distributed over the surface of a small sphere whose center is at the point. We can therefore treat the electron as though it had definite size. When we say that the electronic charge is e we mean that the electron constitutes an electric field equivalent to that which would be obtained if a charge equal to e were concentrated at the center of the field. The intensity E of the field at a distance r from the center of the field is then $\frac{e}{r^2}$. The number of Faraday tubes passing through unit area at right angles to the direction of the tubes can be expressed by

$$D = \frac{E}{4\pi} = \frac{e}{4\pi r^2}. \quad \cdot \quad \cdot \quad \cdot \quad \cdot \quad \cdot \quad \cdot \quad (1)$$

This quantity is often referred to as the electric displacement. Maxwell's displacement current is nothing else but the time variation of the density of the Faraday tubes.

4. Field of the "Moving Electron." Whenever an electron is in motion it is accompanied by a magnetic field. This field is created by the motion of the Faraday tubes invariably associated with the electron. J. J. Thomson has shown that if θ be the angle between the direction of a Faraday tube and the direction in which it is moving at a point P with a velocity v, the motion of the tube produces a magnetic force at P equal to $4\pi v \sin \theta$. The direction of the magnetic force is at right angles to the tube and the direction in which it is moving. Hence, if a charge at O (Fig. 2) were moving in the direction OQ the lines of magnetic force would be circles having their centers along OQ. Here we have a magnetic field produced by the motion of a charge. But an electric current always produces a magnetic field. Hence a moving charge and a current have the same effect. It can be shown that a charge e moving with a velocity v is equivalent to a current of magnitude ve, and if there are n charges the current is nev. This result, though seemingly obvious, is very important. We shall have occasion to make use of this result in the study of the discharge through vacuum tubes. A circuit in which a vacuum tube is inserted consists of two parts: one, the ordinary metallic circuit, and the other, the space between the cathode and the anode in the tube. In the tube we have to deal with the motion of electrons through space; this constitutes the so-called "space current." Problems encountered in this type of circuit are not so easily handled as those dealing with ordinary metallic circuits, because explanations of electrical phenomena in the latter are based on Ohm's law, whereas circuits in which vacuum tubes are included do not obey Ohm's law. The extent of the deviation from Ohm's law in such circuits depends upon the nature of the discharge through the tube and the relative values of the impedance of the tube and that of the metallic circuit.

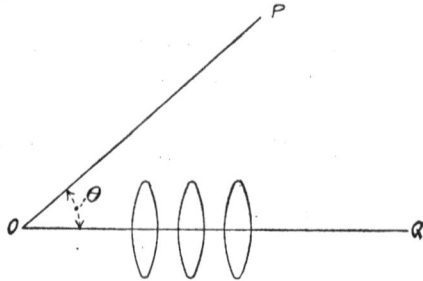

FIG. 2.

5. Mass of the Electron. Let us consider the case of an electron moving with uniform speed through the ether in the direction OQ (Fig. 2). If the velocity with which the electron moves is small compared with that of light the tubes of force will be uniformly distributed as in the case of the stationary electron. Hence the displacement or density of the Faraday tubes at P is, as we have seen above, equal to $\dfrac{e}{4\pi r^2}$, and therefore the magnetic force produced at the point P by the moving electron is

$$H = \frac{ev \sin \theta}{r^2}. \qquad \qquad \text{(1a)}$$

It is known from elementary physics that the energy in unit volume of a magnetic field at a point where the magnetic intensity is H is $\dfrac{H^2}{8\pi}$. Hence the energy of the field at the point P is

$$\frac{e^2 v^2 \sin^2 \theta}{8\pi r^4}.$$

Integrating this over the whole space from infinity up to a small distance a from O, the total energy of the field is found to be $\dfrac{e^2v^2}{3a}$. Now, the kinetic energy of mass m moving with a velocity v is $\frac{1}{2}mv^2$. If the body has a charge e we have to add to this energy $\dfrac{e^2v^2}{3a}$, so that the total energy of the system is

$$\frac{1}{2}\left(m'+\frac{2e^2}{3a}\right)v^2, \quad \ldots \ldots \quad (2)$$

and it therefore appears that the mass of the moving charged body is $m'+\dfrac{2e^2}{3a}$ instead of only m', the mass of the uncharged body. The second term represents the electromagnetic mass. The quantity a is what we may term the radius of the electron. This, however, does not necessarily mean that the electron is a well-defined sphere of radius a; all it means is that where an electron manifests itself the modification of the ether is such as would exist if a charge e were uniformly distributed over the surface of a sphere of radius equal to a. The quantity a merely represents one of the limits of integration arbitrarily assumed in summing the total magnetic energy in the whole space through which the electron moves.

In deriving the above expression for the energy of the moving electron, it was assumed that the field of the moving electron is the same as that of the stationary electron. This is, however, only the case if the electron moves slowly, because when a Faraday tube is moved it tends to set itself at right angles to the direction of motion. The tubes constituting the electron therefore tend to crowd together in a plane perpendicular to the direction of motion of the electron. The result is an increase in the inertia or mass of the electron, because more work must be done to move a Faraday tube parallel to itself than along its own direction, just as it is harder to move a log of wood in the water parallel to itself than to move it endwise. This increase in the mass of the electron only becomes appreciable when it moves with a speed greater than about one-tenth that of light; for speeds less than this the expression (2) can be taken to give the mass of the electron to a first approximation, but for higher speeds the determination of the mass becomes more complicated. The mass of the electron is measured by the ratio of the force to the acceleration to which it gives rise. According to the theory of Abraham and Lorentz the electron has two masses: the longitudinal mass, when it is accelerated in the direction of motion, and the transverse mass, when it is accelerated perpendicular to the direction of motion of the electron. If m represents the mass of the slow-moving electron, then the longitudinal and transverse masses m_1 and m_2 are given by

$$\frac{m_1}{m} = \frac{1}{\left(1 - \dfrac{v^2}{c^2}\right)^{3/2}}; \quad \frac{m_2}{m} = \frac{1}{\left(1 - \dfrac{v^2}{c^2}\right)^{1/2}};$$

where c is the speed of light. It is seen that as the speed of the electron approaches that of light, its electromagnetic mass tends to become infinitely large. The transverse mass of the high-speed electron for various speeds has been determined by Kaufmann and Bucherer.[1] Their experiments verify the above expression for the transverse mass. From this it would seem that the mass of the electron is entirely electromagnetic. Later developments of the Theory of Relativity have rendered this conclusion somewhat questionable, so that there does not seem to be definite experimental evidence to indicate that the electronic mass is entirely electromagnetic.[2]

On the assumption that the mass is entirely electromagnetic equation (2) would give the following expression for the simple mass of the slow-moving electron.

$$m = \frac{2e^2}{3a}. \qquad \ldots \ldots \ldots \quad (3)$$

If the known values of e and m are inserted in this expression we find a value for a which is 2×10^{-13} cm. This effective radius of the slow-moving electron is therefore only about one fifty-thousandth of the radius of the hydrogen atom.

[1] W. KAUFMANN, Göott. Nachr. Math.-Phys. Kl., p. 143, 1901; p. 291, 1902, p. 90, 1903. For later experiments of KAUFMANN, Ann. d. Phys., Vol. 19, p. 487, 1906. A. H. BUCHERER, Phys. Zeitschr., Vol. 9, p. 755, 1908.

[2] For a discussion of this and allied questions, the reader might refer to H. A. LORENTZ, " The Theory of Electrons " and L. SILBERSTEIN, " The Theory of Relativity."

6. Effect of Electric Field on the Motion of an Electron.
To find the effect of an electric field on an electron is a comparatively simple matter as long as the field is uniform. Suppose we
have two infinitely large parallel plates OY and QR (Fig. 3) with
a potential difference V between them,
and let the electron be projected with
a velocity v_0 from the point O in the
direction OQ. On account of the
electric field between the plates the
velocity of the electron will be continually increased on its way to Q.
The kinetic energy of the electron at
the moment of its leaving O is $\frac{1}{2}mv_0^2$;
if its velocity on reaching Q is v its
kinetic energy at Q will be $\frac{1}{2}mv^2$. The
difference in these values of the kinetic

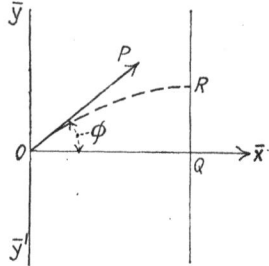

Fig. 3.

energy is due to the applied electric field. If E is the intensity
of the field at a point between the plates, the force acting on the
electron is Ee and the work done on it is $e\int Eds$, where ds is
an element of the path along OQ. But $Eds = dV$, the potential
difference between the ends of the element of path ds.
Hence

$$e\int Eds = Ve = \tfrac{1}{2}mv^2 - \tfrac{1}{2}mv_0^2. \tag{4}$$

Now, suppose the electron be projected with a velocity v_0
in the direction OP, where OP makes an angle ϕ with the direction
OQ of the field. Since the only force acting is the electric force
Xe, we have for the equations of motion:

$$\left. \begin{array}{l} Y = 0 \\ Xe = m\dfrac{d^2x}{dt^2} \end{array} \right\} \quad . \quad . \quad . \quad . \quad . \quad . \quad . \tag{5}$$

where m is the mass of the electron, e its charge and X the intensity
of the field which is supposed to be constant and uniform between
the plates. The Y component of the initial velocity is $v_0 \sin \phi$,
so that

$$y = v_0 t \sin \phi. \quad . \quad . \quad . \quad . \quad . \quad . \quad . \tag{6}$$

Integrating (5) and inserting the value of t from (6) we get:

$$x = \frac{Xe}{2m} \frac{y^2}{v_0^2 \sin^2 \phi} + y \cot \phi. \quad \cdots \quad \cdots \quad (7)$$

This equation gives the point R at which the electron will strike the plate QR when the distance x between the plates and the intensity of the field are known. If the electron starts from O in the direction OY, ϕ is 90° and equation (7) becomes

$$x = \frac{Xey^2}{2mv_0^2}. \quad \cdots \quad \cdots \quad \cdots \quad (8)$$

This equation enables us to calculate the deviation x of an electron from its path by an electric field perpendicular to the original direction of motion of the electron.

So far we have assumed that the lines of force between the plates are straight. If this is not the case the motion of the electron is not easily determined. The case in which the electron moves from a straight wire to a plate is one in which the field is not uniform. Such cases are frequently met with in the study of discharge through vacuum tubes, and the problems involved become so difficult that the desired result is often more easily determined empirically. Such is, for example the case with the three-electrode thermionic amplifier. The classification of cases dealing with electric fields that can be represented by straight lines of force and which can be handled mathematically is a purely geometrical matter. Such fields are obtained with the following structures: (a) Both electrodes are infinitely large parallel plates; (b) one electrode is an infinitely long cylinder and the other an infinitely long wire in the axis of the cylinder; (c) both electrodes are infinitely long co-axial cylinders; (d) one electrode is a sphere and the other a point in the center of the sphere; (e) both electrodes are concentric spheres. It will be recognized that in all these cases the lines of electric force are straight.

7. Effect of Magnetic Field on the Motion of an Electron. Now, instead of an electric field let us apply a magnetic field to the moving electron. As was shown above an electron moving with a velocity v is equivalent to an electric current $i = ve$. We can therefore directly apply the well-known law connecting the mechanical force F exerted by a magnetic field of intensity H on a current i; namely, $F = \mu H i$, where μ is the permeability of the medium. Since we are considering the motion of an electron through space we can put $\mu = 1$, so that the force on the electron is

$$F = Hev. \qquad \ldots \ldots \ldots \quad (9)$$

This force is at every instant at right angles to both the direction of motion of the electron and that of the magnetic field. Thus, referring to Fig. 3, if the electron starts in the direction OX and the magnetic field be perpendicular to the plane of the paper and directed downwards, the electron will be deviated from OX in the direction OY'. Now, when the force acting on a body is always at right angles to its direction of motion the body must describe a circular path, the force being given by

$$F = \frac{mv^2}{r}. \qquad \ldots \ldots \ldots \quad (10)$$

where r is the radius of curvature. Hence, we get from (9) and (10):

$$r = \frac{mv}{He}. \qquad \ldots \ldots \ldots \quad (11)$$

This equation shows how strong the magnetic field must be to make the electron travel in a circle of any desired radius.

Equation (11) expresses an interesting and useful result. We shall mention a few of its applications here. We saw above that a moving electron creates a magnetic field whose lines of force are circles having their centers along the path of the electron. Now consider two electrons moving side by side in the same direction. Obviously the magnetic field produced by each must exert a mechanical force on the other in the sense explained above. A consideration of the directions of these mechanical forces will show that the two moving electrons tend to attract each other. This result is not contrary to the fundamental law of electrostatics that like charges are repellent. The mutual attraction exerted by the electrons is due only to their motion and increases with their velocity. If they moved in opposite directions they would repel each other. It follows from this that an electron stream in a vacuum tube would tend to shrink together if the velocity with which the electrons move become sufficiently high, the shrinkage increasing with the velocity with which the electrons comprising the stream move. In ordinary cases the velocity of the electrons in a vacuum tube is so small (of the order of a million centimeters per second) that the shrinkage of the electron stream due to the reduction in the mutual electrostatic repulsion is inappreciable.

If the electron source in a vacuum tube is a hot cathode the electrons are emitted from it in all directions: the electron stream will therefore generally spread out as the distance from the cathode increases. This spreading can be prevented by means of a magnetic field applied in a suitable way to the stream. In Fig. 4 let C be the hot cathode, P an anode and A an electrode with an aperture in its center. Let A and P be connected and a potential difference applied between them and the cathode. Of the electrons moving away from the cathode some go to A and some shoot through the aperture in A and pass on to P. Between C

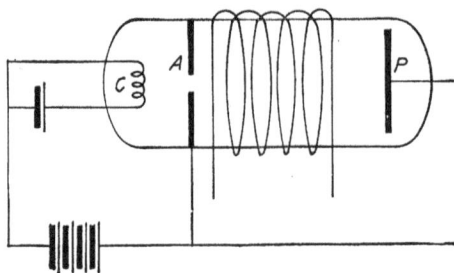
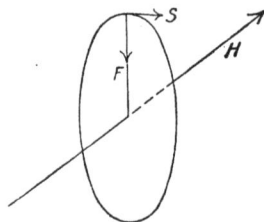

Fig. 4.   Fig. 5.

and A their velocity will be continually increased by the electric field existing between C and A, but after passing A they will continue to move with the same velocity which they had on reaching A, since A and P are at the same potential. If now a magnetic field in the direction AP be applied by means of a coil as shown in the figure, it will be seen, by applying the above laws, that the electrons will travel along a helical path, the diameter of which decreases as the strength of the magnetic field is increased. In Fig. 5, H represents the direction of the magnetic field, F the direction of the force on the electron, and S the path of the electron, which is at right angles to F and H. The motion due to the force F, when added to the primary motion in the direction of H, which the electron has when passing through the aperture in A of Fig. 4, results in the electron describing a helical path. If the magnetic field be made sufficiently strong the diameter of the helix can be made so narrow that the electrons practically travel in a straight line along the axis of the tube.

The study of the motion of an electron in a magnetic field has been successfully applied to the determination of the mass of the electron. Referring to equation (8) it is seen that if we know the velocity v_0 with which an electron moves and determine experimentally the extent y to which the electron stream is deflected by an electric field X whose direction is at right angles to the direction of motion of the electron, we can calculate the value of $\frac{e}{m}$. In order to obtain the velocity all that is necessary is to apply a magnetic field in such a way as to counterbalance the deflection of the electron stream caused by the electric field. Then the magnetic force given by equation (9) must be equal to the electric force eX, Hence

$$v_0 = \frac{X}{H}. \qquad \cdots \cdots \cdots \quad (12)$$

Millikan has accurately determined the value e of the electronic charge.[1] Hence, knowing e and $\frac{e}{m}$ we can obtain the mass m of the electron. This value has been found to be 9.01×10^{-28} grm.

8. The Accelerated Electron. Radiation. We have seen that an electron possesses inertia. From this it follows that in order to accelerate an electron work must be done on it and if it is retarded in its motion it must give up part of its kinetic energy. If the inertia of an electron is wholly electromagnetic the work done in accelerating it is work done on lines of force. Suppose a charge with its connecting lines of force moves through space with a uniform velocity. If this charge is suddenly retarded the ends of the lines of force terminating on it will be, so to speak, jerked backwards. In accordance with the properties of lines of force this kink created at the end of the line will not be transmitted along the whole line instantaneously but will be propagated along it with a finite velocity—the velocity of light. These kinks in the lines are the seat of that part of the energy which the electron gives up when retarded. It can be shown that the electric and magnetic forces associated with a kinked line are more intense than those associated with a straight line. In

[1] R. A. MILLIKAN, "The Electron," University of Chicago Press, 1917.

the latter case the electric force at a distance r from the center of the electron is $\frac{e}{r^2}$ electrostatic units, or $\frac{ec}{r^2}$ electromagnetic units, c being the velocity of light, and the magnetic force that given by equation (1a). If, however, an electron be retarded the electric and magnetic forces E and H at a point distant r from the center of the electron at the moment the kink passes through that point are:

$$\left. \begin{array}{l} E = \dfrac{ef \sin\theta}{r} \\[3mm] H = \dfrac{ef \sin\theta}{rc} \end{array} \right\} , \quad \cdot \ \cdot \ \cdot \ \cdot \ \cdot \ \cdot \ \cdot \quad (13)$$

where f is the acceleration and θ the angle between the r and the direction of motion of the electron. E and H are at right angles to each other and to the direction of propagation of the kink in the line. The energy radiated by the electron is therefore radiated as electromagnetic energy with the speed of light. The total amount of energy radiated by the electron is $\frac{2}{3}\frac{e^2f^2}{c}$.

If now such an electromagnetic disturbance passes over an electron moving with uniform velocity the electric and magnetic fields associated with it will be modified by the intense fields in the disturbances and this modification is propagated to the center of the moving electron along the lines of force constituting it. The result is a change in the motion of the electron. It is seen therefore that the energy of a moving electron can be transformed into radiation energy and vice versa, the transformation always taking place when the electron is retarded or accelerated. This result is an important agency in the production of dislodged electrons, that is, electrons in such a state that they can be readily utilized for purposes of discharge in vacuum tubes. An electron which is bound to an atom of a gas or vapor, or to a substance, can be dislodged by passing an electromagnetic disturbance in the form of light or X-rays over it, in which case the energy imparted to the electron may be so great that it can overcome the forces that bind it to the atom or substance.

A bound electron can also be dislodged by arresting the motion of a high-speed electron in its neighborhood. In this case the kinetic energy of the moving electron is first transformed into energy of radiation, part of which is in turn transferred to the bound electron.

9. Relation between Space Charge and Potential Distribution. In dealing with the conduction of electricity by dislodged electrons or positive ions, it is necessary to consider the effect exerted by their presence on the potential distribution between the electrodes. The difference between the number of electrons and positive ions in unit volume, multiplied by the charge per ion, is usually referred to as the space charge or volume density of electrification.

If, in the space between two electrodes, there are n_0 positive ions, and n electrons, and if the charge on the positive ion and the electron be e_0 and e, then the distribution of potential between the electrodes can be expressed by

$$\frac{\partial^2 V}{\partial x^2} + \frac{\partial^2 V}{\partial y^2} + \frac{\partial^2 V}{\partial z^2} = 4\pi(ne - n_0 e_0), \quad \ldots \quad (14)$$

where V is the potential at a point having the coordinates x, y and z. This equation is known as *Poisson's equation*. It has been used extensively in investigations dealing with the conduction of electricity through gases and high vacua.

In applying this equation to the case in which the charges are contained between two infinitely large parallel plates, between which a potential difference is applied, the lines of force are straight and everywhere perpendicular to the plates, so that the equipotential surfaces are planes parallel to the plates. The last two terms on the left-hand side of equation (14) therefore vanish and we get

$$\frac{d^2 V}{dx^2} = 4\pi(ne - n_0 e_0) = 4\pi\rho, \quad \ldots \ldots \quad (15)$$

where ρ is the volume density of electrification or space charge.

If there are no free charges between the plates, or if the total positive charge in every volume element is equal to the total negative charge, we have $ne - n_0 n_0 = 0$, and $\frac{dV}{dx} = $ constant. We then have the simple case of infinitely large parallel plates at different potentials, but with no charges between them, in which the potential at different points is a linear function of the distance x from one plate. For the case of high vacuum tubes in which the current is carried almost exclusively by electrons, $n_0 = 0$ and $\rho = ne$.

CHAPTER II

DISLODGMENT OF ELECTRONS FROM ATOMS OF VAPORS AND GASES. IONIZATION

10. Occurrence of Electrons. In this and the following chapter will be discussed the conditions in which electrons normally exist and the means whereby they can be brought into such a state that they are readily available for discharge in vacuum tubes.

Since all charged bodies attract, or are attracted by, oppositely charged or uncharged bodies, it is to be expected that there are comparatively few electrons floating around free in nature. By far the larger number of electrons exist as the building stones of which all matter is built up, and they are held in this condition by very strong forces. These forces are due to the positive electrons in the nuclei of the atoms. Such atomic systems, consisting of electrons and positive electrons, are electrically neutral and are not affected by an electric field. The fact that the conductivity of a gas or vapor is very small is an indication that there can only be very few electrons in the gas or vapor that are not bound in electrically neutral systems. In the case of conducting solids the number of electrons that are free, or that can readily be made free by the application of an electric field, is comparatively large, so that such solids are said to be good conductors of electricity. But such electrons cannot be said to be dislodged. They are only available for discharge through conductors and not for discharge between conductors separated by a gaseous or vacuous medium. For the latter purpose they must be dislodged not only from the atoms in the substance but also from the substance itself.

11. Ionization. The process of the production of dislodged electrons is known as *ionization*. In all cases this process involves overcoming the forces that hold the electrons in the atoms or in the substance. In accordance with the properties of electrons described in the preceding chapter it will be evident that this can be done in three ways, viz.: (a) by means of the impact of electrons or positive ions on the atoms or substance; (b) by means of electromagnetic radiation: (c) by means of heat. The first of these three processes gives rise to the phenomenon of delta rays, or secondary electron emission, the second gives rise to the so-called photoelectric effect, and the third forms the basis of the subject of thermionics. The present chapter will be devoted to a discussion of the phenomena accompanying the dislodgment of electrons from the atoms of vapors and gases, the problems of the dislodgment of electrons from solid substances being reserved for the next chapter.

12. Constitution of the Atom. Although very little is known about the exact nature of the processes going on in an atom when its equilibrium is disturbed, there are nevertheless a certain number of experimentally determined facts giving rise to theories that successfully account for many of the phenomena encountered in the ionization of atoms.

There is very little doubt but that the atom consists of a number of electrons grouped around a number of positive electrons. The absolute value of the positive electronic charge is the same as the electron, but its mass is 1845 times as great. The positive electrons in an atom form the atomic nucleus, while the electrons are separated from the nucleus by distances that are large compared with their size. For our present purposes it does not matter how these mutually repellent positive electrons are held together in the nucleus; it is likely that they are held together by electrons so that the nucleus really consists of a group of positive and negative electrons having a resultant positive charge equal to the sum of the electronic charges outside the nucleus. There is reason to believe that the electrons grouped around the positive nucelus revolve in nearly circular orbits round the nucleus. If the electrons did not revolve, the force of attraction between them and the nucleus would cause them to drop into the nucleus. On the other hand, it follows from ordinary mechanics that when an electron revolves round a positive nucleus it must be constantly accelerated and must therefore be constantly radiating energy. In such case a simple system like the hydrogen atom, which consists of a single positive and one negative electron, would radiate all of its energy in such a short time that it really could not exist at all. An attempt to overcome this difficulty has been made by Bohr by making an assumption which frankly repudiates Newtonian mechanics for atomic systems. Bohr assumes that although the electron revolves round the positive nucleus, it does not radiate any of its energy as long as it remains at the same distance from the nucleus, energy being only radiated or absorbed when this distance is decreased or increased, and this happens only when the distance is changed by definite amounts so that the energy is radiated or absorbed in definite quanta. Bohr's atom has been successful in explaining and predicting a number of phenomena, but although there is an element of truth in it, it is still far from the whole truth. We shall therefore not enter into any further discussion of it. Suffice it to say that the necessity for introducing such assumptions as Bohr's, and the assumption of energy radiation by definite quanta, which was originated by Planck, seems to indicate that in dealing with atomic systems we can apply the Newtonian system of mechanics only when the atoms are in a steady but not in a varying state. Since Newtonian mechanics was built up on an experimental basis of large-scale phenomena, one would not necessarily expect it to give an explanation of atomic phenomena.

But apart from the question of the behavior of the electrons in the atom, recent experiments have given conclusive evidence that the atom consists of a number of electrons held together in some configuration by a heavy positive nucleus. The total charge of the nucleus is equal to the sum of the charges of the electrons, so that the atom is electrically neutral. The total positive charge, or the number of electrons, determines the chemical nature of the atom. Starting with the lightest known element, hydrogen, all the elements with a few slight deviations are obtained, in the order of their atomic weights, by successively adding one electron and the equivalent positive charge.

The process of ionization consists in the detachment of one or more electrons from the atom, thus leaving the atom positively charged. An atom from which one or more electrons have been removed is known as a positive ion. If the atoms of a gas be ionized and a potential difference be applied between two plates immersed in the gas, the positive ions will move, under the influence of the electric field, to the negative plate and the electrons to the positive plate. If the pressure of the gas is not too low and the speed of the ions or electrons not too high, the electrons that have been detached from the atoms will attract other neutral atoms and thus form negative ions, and these will move to the positive plate. A negative ion is therefore an atom which has more electrons than are necessary to balance the charge due to the positive nucleus.

In order to ionize an atom the forces that hold the electrons to the nucleus must be overcome. These forces depend partly on the distance between the electron and the positive nucleus. Considering the case of an atomic system, consisting of a single positive and one negative electron, the mass of the former being very much larger than that of the latter, the electron can revolve round the positive electron, or escape from it, according as its kinetic energy is smaller or greater than its potential energy, and in the formation of the system a certain amount of work is done by the electrical forces until this equality is attained. This energy of formation therefore gives a measure of the work which must be done to remove the electron from the nucleus, and will be greater the smaller the distance between the electron and the nucleus. Secondly, the work necessary to remove the electron depends on the number of electrons grouped round the nucleus, and on the configuration of the system. It can be seen in a general way that if there are a number of electrons grouped, for example, in a ring round the nucleus, the repulsion exerted by the other electrons would make it easier to remove an electron than would be the case of a system consisting of only one electron and one positive electron. Now, the way in which the electrons are grouped depends upon the number of electrons in the atom. If they are grouped in rings the heaviest atom that can have all its electrons in one ring only is that which contains eight electrons, namely, the oxygen atom. Heavier atoms would then have their electrons arranged in two rings, still heavier in three rings, and so on. It can therefore be seen in a general way that it would require a smaller expenditure of energy to detach an electron from an oxygen atom than from an atom of hydrogen.

13. Radiation from Atoms caused by Bombardment of Electrons. Let us now look into the process of ionization in greater detail. Suppose we have a tube containing mercury vapor and two electrodes, A and B, one of which is a source of electrons. Let a potential difference be applied between the electrodes so that the electrons are driven from the one to the other, say, from A to B. On their way these electrons will collide with some of the atoms of the vapor, the velocity with which they collide increasing on their way in virtue of the electric field between the electrodes. If the electrons start with zero velocity from the electrode A, their velocity v after having dropped through a potential difference V is given by $Ve = \frac{1}{2}mv^2$, where e is the electronic charge. Now, it has been found[1] that as long as the electrons strike the atoms of the vapor with a velocity which is less than that corresponding to a drop through a certain definite voltage, which, in the case of mercury, is about 5 volts, they are reflected from the atoms without any loss of energy. The impact is therefore elastic. If, however, the electrons strike the atoms with a velocity greater than this value, they lose part or all of their energy, and at the same time the atom radiates energy in the form of monochromatic light. The frequency ν of the light radiated is given by the following relation:

$$Ve = h\nu \quad . \quad . \quad . \quad . \quad . \quad . \quad . \quad (1)$$

where V is the voltage through which the electron has dropped, and h is Planck's constant of action. The product $h\nu$ has the dimensions of energy. The above equation expresses one of the most important relations of modern physics. It was not derived from the impact experiments of Franck and Hertz just mentioned; these experiments give only one of the experimental verifications of the relation. It was originally deduced by Einstein on the basis of Planck's quantum theory of radiation. Einstein's equation will be more fully discussed when we come to consider the photo-electric effect.

The emission of light in the form of monochromatic radiation is due to the electron in the atom not acquiring sufficient energy from the colliding electron to get out of reach of the forces of attraction of the nucleus, and it consequently drops back to its original position, thus giving up the energy, in the form of monochromatic radiation, which it has acquired from the colliding electron. The frequency of the light emitted is characteristic of the atom and is referred to as *characteristic radiation*. All known atoms have a large number of characteristic frequencies.

[1] FRANCK AND HERTZ, Verh. d. D. Phys. Ges., Vol. 16, pp. 457 and 512, 1914.

These frequencies form the line spectra observed in the discharge through gases and vapors.

The blue glow observed in vacuum tubes that are not well evacuated is due to the impact of electrons on the molecules of the residual gas and is the resultant of a large number of characteristic frequencies emitted when the electrons in the atoms that are displaced by the colliding electrons drop back to their original positions of equilibrium. Whenever the blue glow appears some of the electrons in the atoms are displaced beyond the forces of attraction of the atoms and take part in the current convection through the tube. This process of completely detaching electrons from the atoms by means of colliding electrons is known as *ionization by collision*. The blue glow in vacuum tubes is therefore always an indication that ionization by collision takes place.

As was explained in the first chapter, an electron can also acquire energy from a light wave passing over it. And since the energy in a wave and the energy of an electron are related as shown by equation (1), it follows that a wave of frequency ν can produce the same effects explained above that are produced by an electron that has dropped through a voltage V, where V and ν are related by equation (1).

14. Ionization Voltage and Convergence Frequency. The least energy with which an electron must collide with an atom in order to completely detach an electron from the atom of any gas or vapor is known as the ionization energy of the gas or vapor. This amount of energy is usually expressed in terms of the voltage through which the electron drops before it collides with the atom, and is then referred to as the *ionization voltage*. The ionization voltage is the ionization energy divided by the charge of the electron.

When an electron in the atom is displaced to such an extent that the force of attraction of the parent atom just manages to pull it back to its original position within the atom, a characteristic radiation is emitted whose frequency is known as the *convergence frequency*. It is the shortest wave length that can be emitted by the most loosely-bound electrons in the normal atom. Applying equation (1) we can, if we know the convergence frequency, compute the ionization voltage. This relation is important because it is often easier to obtain the ionization voltage by measuring the convergence frequency from observation of the line spectra of the gas or vapor than by measuring the ionization voltage itself.   Direct determinations have been made of the ionization voltage of various gases and vapors, but there is reason to believe that most of the values obtained are not reliable.   The following values of the ionization voltage, computed from the convergence frequencies, give an idea of the order of magnitude of this important quantity.

Substance.	Ionization Voltage Volts.
Mercury vapor	10.4
Zinc vapor	9.24
Magnesium vapor	9.13
Calcium vapor	9.96
Helium	29.00
Hydrogen	13.6

Since an electron can ionize a gas atom after having dropped through the ionization voltage of the gas, it follows that in a vacuum tube which contains some residual gas, ionization always takes place if the voltage applied between cathode and anode exceeds the ionization voltage.   Now, it is impossible completely to remove the last traces of gases and vapors from a vacuum tube. Hence, in thermionic tubes operating on applied voltages greater than those given in the above table some ionization by collision always takes place, although the amount of ionization in well-evacuated tubes may not be large enough to cause an appreciable effect on the operation of the tube.   A discharge which is carried entirely by electrons is a pure electron discharge.   When the term is applied to the discharge through vacuum tubes, as actually realized in practice, it does not necessarily mean a discharge which is carried entirely by electrons, but one in which the number of positive ions formed by collision ionization is so small as not to have any appreciable influence on the operation of the tube.

CHAPTER III

DISLODGMENT OF ELECTRONS FROM SOLID SUBSTANCES

15. Free Electrons. If a substance contains electrons that are not bound to atoms to form electrically neutral systems, the substance must be conducting, because if it were placed in an electric field the free electrons would move in the direction of the field and thus establish a current in the substance. In order to account for the conductivity of metallic substances the assumption has been made that metals contain a large number of free electrons. This assumption has been questioned. On the other hand, the fact that a substance conducts electricity indicates that it must enable electrons to pass freely through it under the application of an electric field. It is possible that the conductivity of metals is due to the frequency of collision of the atoms of the metal with each other. When two atoms collide there is a chance that an electron originally belonging to one of the atoms comes so well within the field of force of the other atom that it is attracted with equal forces by both atoms, so that the resultant force on the electron is very small. The electron would therefore be essentially free at that moment and if the metal were placed in an electric field by applying a potential difference between its ends, the electron would move in the direction of the field. This would, of course, leave one of the atoms positively charged, but its loss would immediately be compensated for by electrons coming from the source of potential difference. On account of the frequency of collisions of the atoms in the metal a large number of electrons can thus be set free and made to move along the lines of force of an applied field. In the case of gases where the atoms are relatively far apart, the chance of this happening is very small, so that a gas is not a good conductor of electricity.

16. Force that Holds Electrons in Substance. The electrons and atoms of a solid substance, like the atoms of a gas, possess kinetic energy and are in a constant state of motion. Now, if the electrons in a substance possess kinetic energy the question arises why do they not escape from the substance. The answer to this question is an assumption that there exists at the surface of the substance a force which tends to keep the electrons in the substance. Being an assumption it is necessarily a very unsatisfactory answer. But there is a good reason to believe that this assumption, which was made by O. W. Richardson in 1901, is one which did not lead us astray. In fact recent developments regarding the structure of the atom lead us to believe that such a force which tends to hold electrons in a substance must necessarily exist at the surface of the substance, and Richardson's assumption can be explained in a manner which is entirely consistent with our physical conceptions. In order to escape from the surface of the substance an electron must do work in overcoming the force which tends to hold it in the substance, and this amount of work it does at the expense of its own kinetic energy. For all known substances the work which an electron must do to escape, and the amount of kinetic energy possessed by the electrons in the substances are of such order of magnitude that only very few electrons manage to escape at ordinary temperatures. By far the larger number of them would have to expend more energy than they possess, so that they are held within the substance.

The work which an electron must do to escape from the surface of a substance is sometimes referred to as the " electron evaporation constant." Generally it is expressed in terms of equivalent volts. The evaporation constant w and the equivalent voltage ϕ are connected by the relation

$$\varphi = \frac{w}{e}, \quad . \quad . \quad . \quad . \quad . \quad . \quad . \quad (1)$$

where e is the electron charge. We shall in the following refer to ϕ as the *electron affinity*. This quantity is the most important constant in thermionics. It determines the thermionic current that can be obtained from any particular type of cathode at any desired temperature, and is characteristic of the substance used as cathode. The smaller ϕ is, the larger is the thermionic current that can be obtained. It is desirable that cathodes be used in thermionic tubes for which ϕ is as small as possible, because the power that must be dissipated in the heating of the cathode to obtain a definite thermionic current decreases as ϕ is decreased. This implies economy of operation as well as increased life of the tube, because of the lower temperature at which the cathode can be operated. The coated type of cathode (Wehnelt cathode) is an example of a cathode that has been so treated as to obtain a low value of the electron affinity.

To obtain a better insight into the nature of the electron affinity, let us consider an evacuated enclosure divided into two parts A and B by a surface S. Let A and B each contain a number of electrons and suppose that the electrons in B can pass freely through the surface into A, but in passing from A to B they must do a certain amount of work w. The electrons in A and B possess kinetic energy like the molecules of a gas; they will therefore move about at random, continually passing through the surface S in either direction. The steady state will be reached when as many electrons pass per second from A to B as from B to A. When this state is attained there will be more electrons in A than in B. The relation between these numbers is given by Boltzman's equation:

$$n = N\epsilon^{-\frac{w}{kT}} \quad \text{or} \quad w = kT \log \frac{N}{n}, \quad \ldots \ldots \quad (2)$$

where N and n denote the number of electrons per unit volume in A and B respectively, T the temperature of the system, w the work which an electron must do to move from A to B and k the gas constant per electron, sometimes referred to as Boltzmann's constant. This constant k is two-thirds of the average kinetic energy which an electron possesses at a temperature of $1°$ absolute and is equal to 1.36×10^{-16} erg.

We can now replace the part A within our enclosure by a metal, so that the surface S is the surface of the metal. The replacement is entirely in accordance with our fundamental assumption that in escaping from the metal an electron must do a certain definite amount of work. The number of electrons immediately outside the surface of the metal will then be related to the number inside the metal by the above equation (2). It is seen then that the number n that escape from the metal depend upon the two factors T and w, and that this number increases as the temperature of the system is increased or as w is decreased. Moreover, it is seen that, since both T and w appear in the exponent of ϵ, a small change in either of them causes a considerable change in the number of electrons that escape from the metal. If, on the other hand, both T and w be kept constant, the number which escapes increases as the number of electrons N in the metal is increased. This increase is, however, not nearly as effective as that occasioned by a decrease in w. These considerations show the importance of the constant w. Its influence on the phenomena encountered in the thermionic discharges goes even further than this, as we shall now proceed to show.

17. Contact Electromotive Force. The advent of galvanic electricity was the discovery by Volta, over a hundred years ago, that when two pieces of different metals were placed in contact and then separated they acquire electric charges. If the two pieces of different metals were placed in an electrolyte and joined by a wire outside the electrolyte a current is established in the circuit so formed. The nature of this force which drives electricity round the circuit and is known as contact electromotive force was never understood until recently. It was only in the last two decades that it was shown conclusively, mainly by O. W. Richardson and P. Debije, that the contact electromotive force is an intrinsic property of metals and is determined by the electron evaporation constant w. The connection between the contact E.M.F. and the evaporation constant w can be gathered from the following. Since an electron must do work in escaping through the surface of a metal, it follows that two points, one outside of the surface and the other inside, must be at different potentials. This difference of potential is given by equation (1). Now suppose we have two slabs of different material, such as copper and zinc. Let them be connected by a copper wire as shown in Fig. 6, and let the number of electrons per unit volume in the copper and zinc be, respectively, N_1 and N_2. Since the two pieces of metal are metallically connected all points in the metallic circuit must be at the same potential, except for a very small potential difference which occurs at the junction AB of the two metals. Let the circuit be grounded, so that it can be considered to be at zero potential. Let the work which an electron must do to escape from the copper slab be w_1, so that in moving from the metal to a point just outside the surface its potential changes from zero to a value V_1, say. Let the corresponding values for the zinc slab be w_2

and V_2, and let us see how much work must be done in moving an electron from a point in the copper through the space to a point inside the zinc. In moving through the surface of the copper an amount of work, w_1, must be done by the electron. In moving from here to a point just outside the zinc surface it does an amount of work equal to $(V_1 - V_2)e$, where e is the charge of the electron, and in moving through the surface of the zinc slab the work done is $-w_2$. The total amount of work done is therefore

$$w_1 + (V_1 - V_2)e - w_2.$$

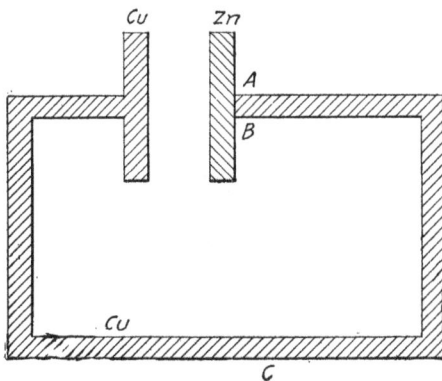

FIG. 6.

Now, since the number of electrons per unit volume in the copper and the zinc is N_1 and N_2, respectively, we have by equation (2):

$$V_1 - V_2 = \frac{kT}{e} \log \frac{N_1}{N_2} + \frac{w_2 - w_1}{e}. \quad . \quad . \quad . \quad (3)$$

The term $\frac{kT}{e} \log \frac{N_1}{N_2}$ gives rise to the small E.M.F. set up at the junction AB of the metals and accounts for the Peltier effect. It is so small in comparison with the other terms in equation (3) that it can be neglected. Hence, referring to equation (1), it follows that

$$V_1 - V_2 = \phi_1 - \phi_2. \quad . \quad . \quad . \quad . \quad (4)$$

The difference $V_1 - V_2$ is called the contact potential difference between the two metals, and, as is seen from equation (4), it is equal to the difference between the electron affinities of the metals.

18. Measurement of Contact E.M.F. That this difference of potential actually exists between the two metals can be shown in the following way: Suppose the circuit be cut at a place C, and the two ends connected to the quadrants of a quadrant electrometer. Let us connect one of the plates (say, the copper plate) and the corresponding pair of quadrants to ground, while the other side of the system remains insulated. The system constituted by the plates Cu and Zn, and the quadrants, has a definite electrostatic capacity depending on the distance between the plates. If the plates first be placed close together and then jerked apart, the capacity of the system will change, and if a potential difference exists between the plates the electrometer will show a deflection. If the potential difference is P and the deflection d_1, the sensitivity of the measuring system is given by

$$S = \frac{d_1}{P}.$$

Now, instead of directly grounding the copper plate, let it be connected to ground through a battery which maintains it at a constant potential V, so that the potential difference between the two plates is $P+V$. If the plates are now placed the same distance apart as they were initially in the first operation, and then again pulled apart to the same extent as before, the electrometer will show a different deflection d_2, and the sensitivity of the measuring system is now given by

$$S = \frac{d_2}{P+V}.$$

Equating these two expressions for the sensitivity, the contact potential difference P between the two plates is:

$$P = V\frac{d_1}{d_2 - d_1}.$$

We shall see later that the contact potential difference, and hence also the electron affinity, depends very much on the nature of the surface of the substance. It can be modified very appreciably by gas occluded in the surface. The notable effect of gas has led some to believe that the contact potential difference is not an intrinsic property of metals but is occasioned entirely by the presence of gas. There is, however, no doubt but that contact potential difference must exist between metals in the best obtainable vacuum, and that it is determined by the electron affinity. A film of gas on the surface of the substance can increase or decrease the electron affinity and so change the contact potential difference between it and another substance. It will be shown later that this also produces a change in the thermionic current obtainable from the substance. Thus, when hydrogen is occluded in the surface of platinum the work which an electron must do to escape from the surface of the platinum is decreased, while oxygen occluded in the surface of calcium increases it.

The following table gives the electron affinities for a number of substances,[1] expressed in volts:

Tungsten	4.52	Zinc	3.4
Platinum	4.4	Thorium	3.4
Tantalum	4.3	Aluminium	3.0
Molybdenum	4.3	Magnesium	2.7
Carbon	4.1	Titanium	2.4
Silver	4.1	Lithium	2.35
Copper	4.0	Sodium	1.82
Bismuth	3.7	Mercury	4.4
Tin	3.8	Calcium	3.4
Iron	3.7		

The difference between any two of these values gives the contact potential difference between the corresponding substances.

We shall see that the contact potential difference plays an important part in thermionic amplifiers and detectors of electromagnetic waves that are so designed as to operate on small plate voltages.

The values of electron affinities given in the above table are of such order of magnitude that under normal conditions only very few of the electrons in the substance possess sufficient kinetic energy to enable them to escape by overcoming the force of attraction at the surface. In order to make use of electrons for the purpose of discharge through vacuum tubes they must first be dislodged from the parent substance.

We shall now proceed to a discussion of the agencies whereby

[1] Most of these are averaged values compiled by LANGMUIR (Trans. Am. Electro-chem. Soc., Vol. 29, p. 166, 1916) from measurements of RICHARDSON, MILLIKAN, HENNING, LANGMUIR, and others.

the dislodgment of the electrons can be effected. As was stated in Section (11) these agencies are: (1) heat; (2) electromagnetic radiation, and (3) impact of electrons. These agencies form the basis of the subjects of Thermionics, Photo-electricity and Secondary Electron Emission, respectively.

19. Elements of Thermionics. The first of these agencies which is the most important for our immediate purposes has been known for a considerable time. In fact, it has been known for over one hundred years that when a metal is brought into a state of incandescence the air in its neighborhood becomes a conductor of electricity. The phenomenon was studied in detail by Elster and Geitel during the years of 1882–1889. They found that when a metallic filament was placed near a plate the latter acquired a charge when the filament was heated to incandescence. At red heat the plate acquired a positive charge, but when the temperature of the filament was raised to white heat the plate charged up negatively. If the filament and plate were placed in an enclosure which could be evacuated, the tendency for the plate to charge up negatively was increased.

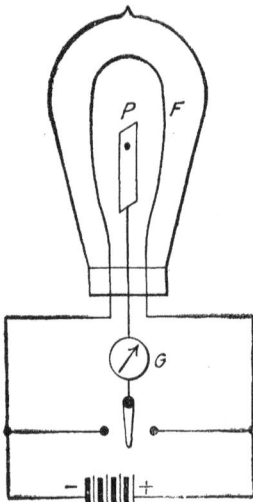

Fig. 7.

This effect also came to the notice of Edison in 1883. He noticed that if a metallic plate be inserted in the vacuous space of an incandescent lamp and this conductor be connected to the positive end of the filament, a current was established in the shunt circuit so formed, namely, the circuit *PFG* (Fig. 7). But if the plate was connected to the negative end of the filament, the galvanometer showed no deflection. A study of this effect, which is sometimes called the " Edison Effect " was made by J. A. Fleming in 1896,[1] but the true nature of the phenomenon was not understood until the work of J. J. Thomson and O. W. Richardson. In 1899, the former showed[2] that the phenomenon was the result of negative electricity given off from the hot filament in the form

[1] J. A. Fleming, Phila. Mag., Vol. 42, p. 52, 1896.
[2] J. J. Thomson, Phil. Mag., Vol. 48, p. 547, 1899.

of electrons. This explained, for example, why in the Edison effect a current was observed to flow through the galvanometer G. When the plate was connected to the positive end of the filament, the filament had a negative potential with respect to the plate, and the electrons given off by the filament were driven to the plate. Since the time of Thomson's experiment there has been no doubt in the minds of physicists that the carriers of electricity from the filament are electrons, but the mechanism of the emission of these electrons from the hot filament was not known until O. W. Richardson [1] showed, in 1901, that the electrons are emitted solely in virtue of their kinetic energy and need no chemical reaction at the surface of the filament. This result of Richardson's work was the first definite expression of what may be termed a pure electron emission.

Richardson's theory was based on an assumption that had previously been made and successfully applied, that the electrons in a metal, which are free to move under the influence of an electric field, behave like the molecules of a gas, that is, they have velocities distributed according to Maxwell's law. It was stated in Section 16 that these electrons are held in the substance by a force existing at the surface of the substance. There is still some speculation regarding the exact nature of this force which seems to be closely related to the structure of the atoms or molecules of the substance. At ordinary atmospheric temperatures very few electrons possess sufficient kinetic energy to overcome this force. The number escaping at such temperatures is therefore extremely small. According to Maxwell's law of velocity distribution some electrons will at one moment have zero velocity, others again will have extremely high velocities, while the majority will possess velocities ranging between these two extreme values. Only the few electrons with the very high velocities will be able to escape through the surface. The energy w which an electron must expend to overcome the force of attraction at the surface is related to the number of electrons per cubic centimeter inside and outside the surface by equation (2). From this equation it is seen that as the temperature is raised the number n of electrons outside the surface increases. Now, in vacuum tubes we are not so much concerned with the relative number of electrons inside and outside

[1] O. W. RICHARDSON, Proc. Camb. Phil. Soc., Vol. 11, p. 286, 1901. Phil. Trans. Roy. Soc., Vol. 11, p. 497, 1903.

of the surface in the state of equilibrium as with the rate at which they escape when they are carried away as fast as they are emitted. This can be ascertained by applying a potential difference between the body which emits the electrons and a conductor placed in its neighborhood. It will be understood in the following that this potential difference is always high enough to draw all the electrons away as fast as they are emitted. Applying the principles of the kinetic theory of gases it can be shown that, to an approximation which is sufficiently close for our purposes, the number n' of electrons that pass per unit time through unit area of the surface from the inside is given by:

$$n' = n\sqrt{\frac{kT}{2\pi m}}, \quad \cdots \cdots \cdots \quad (5)$$

where m is the mass of the electron, and n is the number of electrons per cubic centimeter outside the surface.

This number of electrons can be obtained in terms of the number N per cubic centimeter inside the surface by combining the relation (5) with equation (2). Thus:

$$n' = N\sqrt{\frac{kT}{2\pi m}}\,\epsilon^{-\frac{w}{kT}}.$$

n' is the number of electrons that would move per second to a conductor which is charged positively and placed in the neighborhood of the emitting substance. If e be the electronic charge, then $n'e$ is the saturation current per square centimeter surface of the emitting substance, or

$$I_s = n'e = Ne\sqrt{\frac{k}{2m}}\,T^{1/2}\epsilon^{-\frac{w}{kT}} \quad \cdots \cdots \quad (6)$$

or

$$I_s = A\,T^{1/2}\epsilon^{-\frac{b}{T}}. \quad \cdots \cdots \cdots \quad (7)$$

The constant b in this equation is a temperature and is expressed in absolute (Kelvin) degrees. It is, however, more convenient to use the equivalent constant ϕ expressed in volts. The relation between ϕ and b is as follows: From equations (6) and (7) $w = kb$, and from equation (1) $w = \phi e$, where ϕ is expressed in electrostatic units, or

$$\phi = \frac{300w}{e} \text{ volts.} \quad \cdots \cdots \cdots \quad (8)$$

Now k is the gas constant for one electron and is equal to 1.36×10^{-16}, while e is the electronic charge in electrostatic units. Hence

$$\phi = \frac{300kb}{e} \text{ volts,}$$

i.e.,

$$\phi = 8.6 \times 10^{-5} b \text{ volts.} \quad . \quad . \quad . \quad . \quad . \quad (9)$$

The constant ϕ is the electron affinity, values of which for a number of substances are given in the table on p. 29. It can be determined experimentally with a simple device consisting of a filament of the substance to be investigated and an anode placed in its neighborhood, the structure being enclosed in a vessel which can be evacuated to such an extent that the residual gas

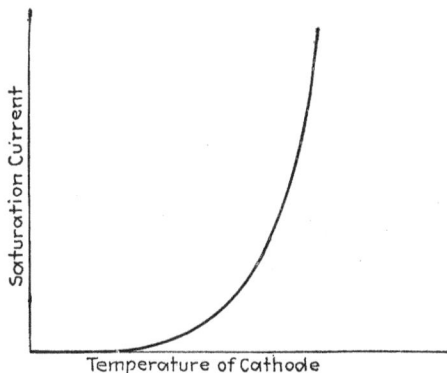

Fig. 8.

has no appreciable influence on the discharge. (The influence of gas will be considered in a later chapter.) Care must be taken that the voltage applied between the filament and the anode is so high that any further increase in the voltage does not appreciably increase the current. The current obtained under these conditions is then the saturation current given by equation (7). If the current is observed for different values of the filament temperature a curve is obtained such as that shown in Fig. 8. In order to evaluate the constant b or ϕ we can take logarithms of equation (7). Thus:

$$\log_{10} I_s - \tfrac{1}{2} \log_{10} T = \log_{10} A - \frac{.4343b}{T}. \quad . \quad . \quad (10)$$

By plotting the expression on the left-hand side against $\dfrac{.4343}{T}$ a straight line is obtained the slope of which gives b. We shall have occasion to return to this equation when we come to consider efficiency problems connected with the thermionic vacuum tube.

It may be remarked here that the constant b can also be determined by photo-electric means. The relation between photo-electric and thermionic phenomena will become apparent when we come to consider the photo-electric effect.

20. Influence of Surface Conditions on Electron Affinity. By applying the theory of images Debije [1] has shown that it is easier for an electron to escape from a sharp point than from a smooth flat surface. The theory of images involves a purely mathematical process that tells us little or nothing about the nature of the processes going on when a electron escapes from a surface and can be applied only to that part of the process when the electron is so far away from the surface that molecular irregularities in the surface can be neglected. We can nevertheless obtain an indication of the manner in which the configuration of the surface affects the electron emission, if we comply with the conditions that govern the application of the theory of images. This theory states that the force of attraction between a charged body and a conductor can be determined by assuming that the force is the same as if the conductor were replaced by another charge which is, in respect to size, shape and position, the optical image of the first charge reflected in the surface of the conductor, but is of opposite sign. Thus a charge $-e$ at a distance x from a plane surface would produce an image $+e$ at a distance x behind the surface. The force of attraction between the charge $-e$ and the plane surface is therefore $\dfrac{-e^2}{4x^2}$ and the work that must be done to remove the charge from a distance x_0 from the surface to infinity is:

$$w_1 = -\int_{x_0}^{\infty} \frac{e^2}{4x^2}dx = \frac{e^2}{4x_0}, \quad \cdots \cdots \quad (11)$$

which, when expressed in equivalent volts becomes:

$$\phi_1 = \frac{300e}{4x_0},$$

[1] P. Debije, Ann. d. Phys., Vol. 32, p. 465, 1910.

the condition being that the distance x_0 is large compared with molecular dimensions.

The quantity w_1 does not represent the total amount of work which an electron must do to escape from a plane surface. There is still to be added the work w_2 done in moving from the interior of the conductor through the interface and up to the point distant x_0 from it. Schottky [1] and Langmuir [2] have made certain assumptions regarding the force of attraction within this region which lead to the result that the work w_2 is equal to w_1, so that the total amount of work done is $\dfrac{e^2}{2x_0}$. Since the nature of the force very close to and inside the surface is not known and

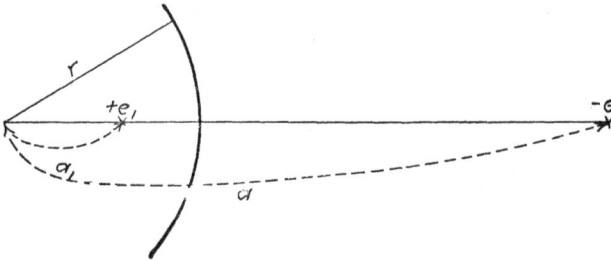

FIG. 9.

probably depends very materially on the molecular structure of the material of the conductor,[3] we shall confine our considerations to the force at distances which are large compared with the molecular diameters, and proceed to compute the corresponding part w_1, of the work which an electron must do to escape from a curved surface of radius r.

Let the surface be convex toward the electron (Fig. 9). Let the electron $-e$ be at a distance a and its image $+e_1$ at a distance a_1 from the center of curvature of the surface. Then

$$\left. \begin{array}{l} aa_1 = r^2, \\[2mm] \dfrac{r}{a} = \dfrac{e_1}{e}. \end{array} \right\} \quad \cdots \cdots \cdots \quad (12)$$

[1] W. Schottky, Phys. Zeitsch., Vol. 15, p. 872, 1914.

[2] I. Langmuir, Trans. Am. Electrochem. Soc., Vol. 29, p. 163, 1916,

[3] J. Frenckel, Phil. Mag., Vol. 33, p. 297, 1917,

Now the force of attraction between $-e$ and $+e_1$ is:

$$f = \frac{-ee_1}{(a-a_1)^2}. \quad \ldots \ldots \ldots \quad (13)$$

From equations (12) and the geometry of the system, we have, if x be the distance of the electron from the surface:

$$a = x + r$$

$$a_1 = \frac{r^2}{x+r}.$$

Substituting these values for e_1, a and a_1 in (13) the force of attraction becomes:

$$f = \frac{e^2\left(1 + \dfrac{x}{r}\right)}{4x^2\left(1 + \dfrac{x}{2r}\right)^2}. \quad \ldots \ldots \quad (14)$$

This equation holds for values of x greater than x_0, where x_0 is large compared with molecular dimensions. The work which an electron must do to move away from the point x_0 is obtained by integrating equation (14) between the limits $x = x_0$ and $x = \infty$. The integration gives:

$$w'_1 = \frac{e^2}{4x_0}\left(1 - \frac{x_0}{2r} + \frac{x^2}{4r^2} - \ldots\right) \quad \ldots \ldots \quad (15)$$

If we take the radius of curvature of the surface so large that x_0 is small compared with it we can write equation (15) in the form:

$$w'_1 = \frac{e^2}{4x_0} - \frac{e^2}{8r} \text{ erg.} \quad \ldots \ldots \ldots \quad (16)$$

The equivalent potential in volts is:

$$\phi'_1 = 300\left(\frac{e}{4x_0} - \frac{e}{8r}\right) \text{ volts.} \quad \ldots \ldots \quad (17)$$

This shows that the work of escape of an electron from a curved surface of radius r is less than that from a plane surface by an amount equal to $\dfrac{e^2}{8r}$. It also shows that if the surface is irregular

contact potential differences must exist between the protru-
sions and the hollows. These potential differences are very
small, but if irregularities are close to one another the resulting
electrostatic fields may be very large. Thus, if we consider a
protrusion and a hollow adjacent to it, each being regarded as
spherical surfaces having a radius of curvature of the order of 10^{-6}
cm., it follows that the electrostatic field tending to drive electrons
from the protrusion to the hollow may be of the order of several
thousand volts per centimeter. This would necessitate very
high plate potentials to overcome these fields and pull all the
emitted electrons over to the anode.

This result is obtained by the simple application of the theory
of images, which unfortunately does not tell us much about the
physical processes involved. A similar effect is to be expected
when the surface of the cathode contains impurities having
electron affinities which are different from that of the material
of the cathode itself. Langmuir [1] has ascribed the lack of saturation shown by tungsten filaments contaminated with thorium
to the local fields at the surface of the filament, due to the difference between the electron affinities of thorium and tungsten.
When the surface of the filament consisted either of pure tungsten
or pure thorium the saturation curve was substantially parallel
to the voltage axis. But when the surface was a mixture of tungsten and thorium the thermionic current continually increased
with the applied voltage.

The oxide-coated cathode is an example of a cathode which
generally has an irregular surface. It is obtained by coating a
platinum wire or ribbon [2] with oxides of the alkaline earths. The
coated filament has a much lower electron affinity and therefore
a higher thermionic efficiency than the metals used as sources of
thermionic current. Its surface, however, is rough and possibly
is not uniformly active thermionically. These filaments do not
give such well-defined saturation currents as metallic filaments
do. Lack of well-defined saturation currents is generally not a
disadvantage in thermionic tubes. It is, in fact, sometimes a
distinct advantage, as will become evident from the considerations
given in the following chapters.

[1] Paper read at Chicago meeting of Am. Phys. Soc., December, 1919.
[2] A. Wehnelt, Ann. d. Phys., Vol. 4, 425, 1904. Nicolson & Hull, U. S.
Pat. 1209324, Brit. Pat. 17580, 1915.

21. Photo-electric Effect. The process of the dislodgment of electrons from solid bodies by means of electromagnetic radiation brings into play the same forces that attend the emission of electrons from hot bodies, and furnishes valuable evidence to show the generality of that characteristic constant of solids which plays such an important part in the operation of thermionic vacuum tubes, namely, the electron affinity.

In 1887 Hertz observed that when a spark gap was illuminated with ultra-violet light the discharge passed more readily than when the electrodes were in the dark. Soon after this Hallwachs discovered that the incidence of ultra-violet light on a zinc plate caused it to become charged positively, or when the plate was first charged to a negative potential and then insulated it lost its negative charge when exposed to the light. This has since been found to be a general property of all conductors and could be explained in the light of the electron theory. The energy of the light wave striking the substance stimulates the electrons in the atoms of the substance. They thus acquire sufficient energy to overcome the force of attraction at the surface of the substance and escape with a velocity which depends upon the energy in the light wave and the amount of energy they must expend to overcome the surface force. Thus, if the amount of energy acquired by the electron in the substance from the light is W and the work which the electron must do to overcome the surface force is w, then it escapes from the substance with a kinetic energy.

$$\tfrac{1}{2}mv^2 = W - w \quad . \quad . \quad . \quad . \quad . \quad . \quad (18)$$

where v is the velocity of escape and m the mass of the electron. It will be understood that of the electrons in the substance those that happen to be near to the surface have to overcome only the surface force, while those that are further in the interior will have to do an extra amount of work in forcing their way out. We can therefore expect electrons to be emitted by the light with velocities ranging from zero to a definite maximum value. This maximum value expressed in volts is the electron affinity. We shall in the following consider only those electrons that have this maximum velocity, and equation (18) will be understood to refer to these maximum values. If a plate is placed in front of the electron emitting substance or cathode, the emitted electrons can be driven back to the cathode by the application of a potential difference between it and the anode. If the electron emerges from the surface with a velocity v it will be capable of moving against an electric field until it has spent its kinetic energy $\frac{1}{2}mv^2$, where m is the mass of the electron. If the maximum voltage against which the electron can move in virtue of its own kinetic energy is V, then $Ve = \frac{1}{2}mv^2$. The velocity of an electron is commonly expressed in terms of the voltage V, instead of centimeters per second. Equation (18) can then be written

$$Ve = W - w \quad \cdot \quad \cdot \quad \cdot \quad \cdot \quad \cdot \quad \cdot \quad \text{(19)}$$

where e is the electronic charge.

This voltage can be determined with the arrangement shown in Fig. 10. A is the photo-electric cathode which can be illuminated with ultra-violet light, and B is the anode. By means of

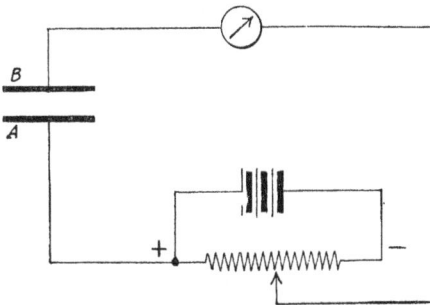

FIG. 10.

the potentiometer shown the voltage between A and B can be adjusted to any desired value tending to drive the electrons in the direction B to A. Unless this voltage exceeds a definite amount the electrons emitted from A under the influence of the light will travel all the way across to B in virtue of the velocity with which they are emitted, and the resulting current established in the circuit can be measured with a current-measuring device. If we now measure the current for increasing values of the voltage V, the current decreases until the voltage is large enough to return all the emitted electrons to the cathode before they can reach the anode. By plotting the photo-current against the voltage V a curve is obtained such as that shown in Fig. 11. The voltage is reckoned negative when the receiving plate B is negative with respect to the emitting plate A. The point at which the curve cuts the voltage axis gives the maximum velocity with which the electrons are emitted from the cathode.

Experiment has shown the remarkable result that the maximum velocity of emission is independent of the temperature of the cathode. The maximum velocity of photo-electric emission is furthermore independent of the intensity of the light with which the cathode is illuminated. If the intensity of the light is increased only the number of electrons emitted increases but their velocity stays the same, provided that by changing the intensity of the light we do not at the same time change its quality, that

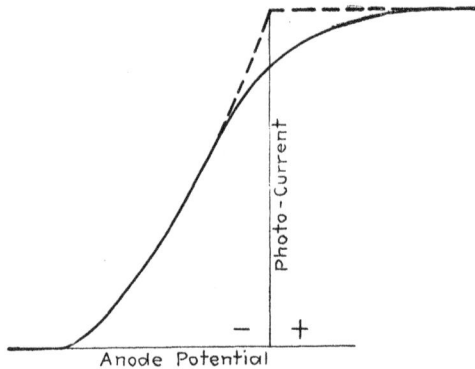

Fig. 11.

is, its wave length distribution. The frequency of the incident light is the only factor that influences the velocity of emission when dealing with one substance. For the same light frequency and different substances, the emission velocity depends upon the electron affinity of the substances. Millikan[1] has shown that if the maximum voltage necessary to keep the emitted electrons from reaching the anode be plotted against the frequency of the light the linear relation shown in Fig. 12 is obtained. Thus:

$$V = \frac{h\nu}{e} - C, \quad \cdot \quad \cdot \quad \cdot \quad \cdot \quad \cdot \quad \cdot \quad (20)$$

where h is a constant and ν the frequency of the light falling on the cathode. Referring to equations (18) and (19) it is seen

[1] R. A. MILLIKAN, Phys. Rev., Vol. 7, p. 355, 1916.

that W, the energy acquired from the light by the electron in the substance is equal to $h\nu$, and the constant C is equal to $\frac{w}{e}$. This extremely important experimental result shows that light energy can be expressed by the product of the frequency of the light and a constant. Indeed, Millikan found that this constant is the same as Planck's constant of action. Furthermore, C has actually been found to be equal to the electron affinity ϕ. We therefore have as the fundamental photo-electric equation:

$$Ve = h\nu - \phi e. \quad . \quad . \quad . \quad . \quad . \quad . \quad (21)$$

This equation was originally deduced theoretically by Einstein [1] on an assumption that he has since abandoned. But Millikan's ex-

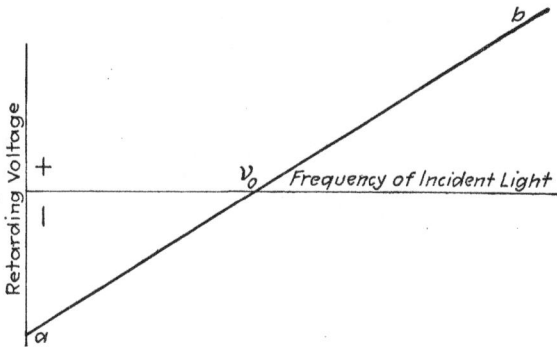

FIG. 12.

periments have shown that this equation holds with a high degree of accuracy and have placed beyond doubt the correctness of this very simple expression for the light energy necessary to dislodge an electron.

So far we have considered only the voltage applied between the plates A and B. This is, however, not the only voltage that affects the motion of the electrons. There remains to be considered the contact potential difference between the plates, which is equal to the difference between the electron affinities of the two plates. This potential difference, which can be measured by the method explained in Section 18, must be added to the applied voltage.

[1] A. EINSTEIN, Ann. d. Phys. (4) Vol. 20, p. 199, 1905.

Referring now to Fig. 12, it is seen that there is a definite frequency v_0 of the light at which the voltage necessary to drive the electrons back becomes zero. This means that for frequencies lower than this value no electrons escape at all. Putting $V=0$ in equation (21) we get:

$$v_0 = \frac{e}{h}\phi. \quad \cdots \quad \cdots \quad (22)$$

This limiting frequency, commonly referred to as the *photo-electric long wavelength limit* is a fundamental property of solids, and, is equal to a constant multiplied by the electron affinity—the same constant that plays such an important part in the emission of electrons from hot filaments in the thermionic vacuum tube.

The quantities h and e are universal constants, that is, their values are independent of the matter under investigation and the conditions of the experiments. Their values are $h = 6.55 \times 10^{-27}$ erg. sec., and $e = 4.77 \times 10^{-10}$ E.S. units.

22. Control of Space Current by Means of an Auxiliary or Third Electrode. A convenient and what has proved to be a very valuable means of varying a space current is obtained by placing a third electrode in the neighborhood of the cathode and applying potential variations to it. This scheme was used by de Forest to control the electron discharge in his audion detector in 1907.[1] He later gave the auxiliary electrode the form of a wire gauze or grid placed in the path of the discharge between cathode and anode.[2] About the same time von Baeyer [3] also used an auxiliary electrode to control a thermionic current from a hot filament. In von Baeyer's arrangement the anode was a cylinder and the cathode a wire placed along its axis. The third electrode was a wire gauze bent into the form of a cylinder and placed between cathode and anode. A similar scheme was also used by Lenard [4] in 1902 in connection with photo-electric experiments. It is hardly necessary to say that the insertion of the grid has made the audion a device of immense practical importance and enabled it to perform functions that would otherwise have been impossible.

[1] LEE DE FOREST, U. S. Patent No. 841387, 1907.
[2] LEE DE FOREST, U. S. Patent No. 879532, 1908.
[3] O. VON BAEYER, Verh. d. D. Phys. Ges., Vol. 7, p. 109, 1908.
[4] P. LENARD, Ann. d. Phys., Vol. 8, p. 149, 1902.

The quantitative effect of the third electrode was first given by the author.[1] The nature of this effect can be understood from the following: In Fig. 13 F is the cathode, P the anode and G the auxiliary electrode, which may be in the form of a wire grid or gauze. The battery E_p maintains the anode at a positive potential with respect to F, while G can be given any desired negative potential by means of the battery E_g. The positive potential on P has the effect of drawing the electrons through the grid to P, whereas the negative potential on the grid tends to drive them back to the cathode, and by increasing E_g a value can be reached for which all the emitted electrons are returned to the cathode. E_g therefore takes the place of V in the arrange-

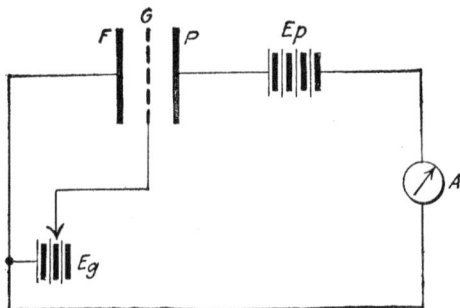

FIG. 13.

ment shown in Fig. 10. But there is this difference that while in Fig. 10 the electric field between the plates is due only to applied voltage V and the contact potential difference between the plates, in the present case there is a third factor which influences the field between the cathode and grid, namely, the potential difference due to the battery E_p. Thus, if E_g is zero and the contact potential difference between F and GP be supposed for the present to be also zero, then the electric field between F and G is not zero but has a definite value depending upon the structural parameters of the device and the potential difference between P on the one hand and F and G on the other. (F and G are now supposed to be metallically connected.) This is due to the fact that the potential of P causes a stray field to act through the openings of the grid. If the potential difference between P and FG be equal to

[1] H. J. VAN DER BIJL, Verh. d. D. Phys. Ges., Vol. 15, p. 338, 1913.

E_P the field at a point near F is equivalent to the field that would be sustained at that point if a potential difference equal to $\dfrac{E_P}{\mu}$ be applied directly between the cathode A and a plane coincident with that of the grid. For the usual connection in which the plate P is positive the direction of this field is such as to draw electrons away from the cathode. But it does not draw the electrons to the grid, as would be the case if a potential difference were applied directly between the grid and cathode; it tends to draw electrons to the anode through the openings of the grid.

Besides this stray field there is also the contact potential difference K between F and GP. Hence, if K be reckoned positive when it tends to draw electrons away from the cathode, and if the maximum velocity of emission, expressed in volts, of the electrons liberated from the cathode be V, then in order to drive all the emitted electrons back to the cathode we must apply between cathode and grid a potential difference equal to

$$E_s = \frac{E_p}{\mu} + \epsilon, \quad \cdot \ \cdot \ \cdot \ \cdot \ \cdot \ \cdot \quad (23)$$

where $\epsilon = K + V$. This expression can be regarded as the effective voltage when the potential difference between grid and cathode is zero. If this potential difference be made equal to E_g the effective voltage becomes

$$\frac{E_p}{\mu} + E_g + \epsilon. \quad \cdot \ \cdot \ \cdot \ \cdot \ \cdot \ \cdot \quad (24)$$

In this expression E_p and E_g are the potentials of the anode and grid with respect to that of the cathode, which can be regarded as the zero of potential. Hence, when E_g is varied the potential difference between grid and plate also changes. When I first established this linear stray field relation in 1913, I expressed the result by the equation

$$v = \frac{k}{d}V + \eta. \quad \cdot \ \cdot \ \cdot \ \cdot \ \cdot \ \cdot \quad (25)$$

where v is the potential difference between cathode and grid and V that between grid and anode. I also stated that k is a constant depending on the grid and d the distance between grid and anode. In testing this relation the grid remained grounded while the potential of the cathode was varied. This made it possible to keep the potential difference V between grid and anode nearly constant while varying the potential difference between cathode and grid. The accuracy with which equation (25) was found to hold is shown by Fig. 14.[1] In the case of the lower curve the

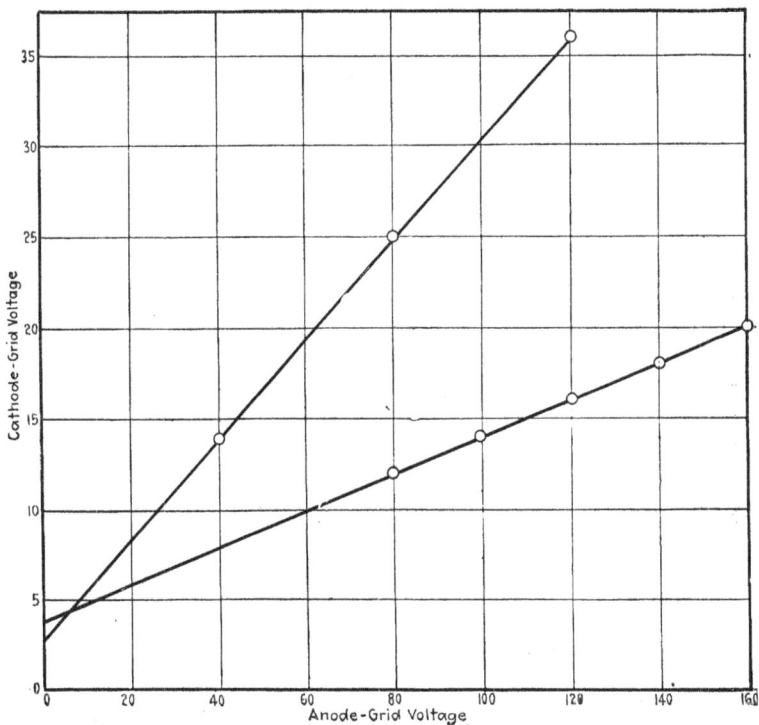

FIG. 14

distance d between grid and anode was 6.7 mm., while the upper curve was obtained with $d = 2.5$ mm.   If, instead of plotting V as abscissæ, we plot $\dfrac{V}{d_1}$, the two curves coincide.

It can readily be seen that if we substitute $-E_g$ for v and E_p, the potential of the anode *with respect to the cathode* for V, the

[1] Loc. cit., p. 339.

potential of the anode *with respect to the grid*, then (24) and (25) give the same result provided that

$$\eta = \frac{\mu\epsilon}{\mu - 1}$$

and

$$\frac{d}{k} = \mu - 1. \qquad \ldots \ldots \ldots \quad (25a)$$

I have since verified this relationship between μ and the structural parameters of the tube on the basis of an extensive series of experiments carried out in the research laboratories of the Western Electric Company, and have also expressed k in terms of the mesh of the grid and the diameter of the grid wires. (See Chapter VII, p. 231.)

The constant μ is a very important constant of the three-electrode tube and, as will be shown later, expresses the maximum voltage amplification obtainable.

Expression (24) or (25) can be regarded as the fundamental relationship of the three-electrode vacuum tube. The current in the circuit FPA (Fig. 13) is obviously a function of the expression (24). Hence, if the potential of the cathode be maintained constant, the fundamental expression for the current in a three-electrode tube is

$$I = f\left(\frac{E_p}{\mu} + E_g + \epsilon\right), \qquad \ldots \ldots \quad (26)$$

where E_p and E_g are the potentials of the anode and grid with respect to that of the cathode. We shall have occasion to make extensive use of this relationship in dealing with the three-electrode thermionic tube.

It will be shown later that a device like that shown in Fig. 13 and whose current-voltage characteristic can be expressed by the function (26) can be used as amplifier, radio detector, oscillation generator, etc.

A device which depends for its current on the emission of electrons by photo-electric means is not as suitable for these purposes as thermionic devices, because photo-electric currents are generally very small and the emission of electrons by heat is much more practical than emission under the influence of light.

23. Secondary Electron Emission. Delta Rays. We now come to a consideration of the third agency whereby electrons can be dislodged from substances, viz., by the impact of electrons.

When electrons are made to impinge on a metallic plate they give rise to the radiation of X-rays. These X-rays comprise a *general X-radiation* superimposed upon which there is under certain conditions a so-called *characteristic radiation*. The same effects are produced when the plate is exposed to X-rays. The frequency of the characteristic radiation is closely related to the atomic properties of the substance from which it originates, and is not produced unless the velocity of the impinging electrons, which we shall call the primary electrons, exceeds a certain definite value. When the characteristic radiation is produced there is also a copious emission of electrons from the plate. These electrons are, of course, returned to the plate by the potential difference between it and the cathode from which the primary electrons proceed, and in order to measure them a special circuit arrangement, such as will be described below must be used. Every metal possesses a number of characteristic frequencies and although in general X-rays have so far been investigated extensively only at high frequencies—at voltages corresponding to several thousand volts, they are also produced by low velocity electrons. Dember [1] has, for example, produced X-rays and measured their effect at a voltage as low as 17 volts. These rays are so soft that they cannot penetrate the walls of the containing vessel. Their effects must therefore be studied inside the vessel. Likewise in the case of the secondary electron emission it is not necessary for the primary electrons to strike the electrode with high velocities.

The impact voltage at which secondary electrons are emitted depends on the nature of the surface of the emitting electrode. The phenomenon of secondary electron emission, which is sometimes referred to as " Delta rays " shows this important property that as the velocity of the primary electrons is increased the number of secondary electrons emitted per impinging electron increases. In fact, if the applied voltage, i.e., the velocity with which the primary electrons strike the emitting electrode, be increased to a sufficiently high value one primary electron can expel as many as twenty secondary electrons.

[1] H. DEMBER, Verh. d. D. Phys. Ges., Vol. 15, p. 560, 1913.

The presence of secondary electrons can easily be demonstrated by means of the circuit arrangement shown in Fig. 15. The plate P is kept at a constant positive potential with respect to the filament F by the battery E. When no potential difference exists between filament and grid G, the current in the circuit FGA is very small, because practically all the electrons emitted from the filament are drawn through the openings of the grid and thrown on to the plate. If now the potential of the grid (positive with respect to the filament) be increased the current to the grid [1] at first increases, as shown by the part OA in the curve in Fig. 16. When the grid potential reaches a certain value the current, as indicated by the ammeter A (Fig. 15) begins to decrease, drops

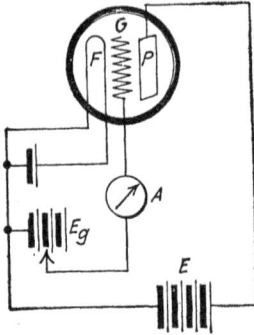

FIG. 15.

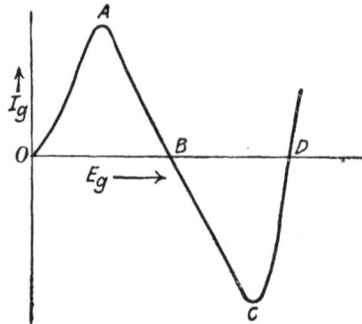

FIG. 16.

down to zero at B and then flows in the reversed direction. The explanation for this is that while the potential difference between filament and grid is small, the electrons that strike the grid enter it, but as the positive grid potential is increased the electrons on striking the grid emit so-called " secondary electrons " from it and these are drawn to the plate which is maintained by the battery E at a positive potential with respect to the grid. The net current as indicated by the ammeter A is the sum of electrons entering the grid and those leaving it. When the velocity with which the electrons strike the grid increases beyond a certain value, one primary electron can knock out more than one secondary electron from the grid, and the current in the circuit FGA reverses. When the positive grid potential is increased to such an

---

[1] The direction of current is here, as throughout the following, taken to mean the direction in which electrons move.

extent that the grid becomes positive with respect to the plate, the secondary electrons are no longer drawn away to the plate, but are driven back to the grid so that the reversed current in the grid circuit again decreases, as shown at C (Fig. 16), and finally assumes the original direction.[1]

Considering now the current as indicated by the ammeter A and the voltage between filament and grid, we see that over the region ABC (Fig. 16), the current decreases as the voltage increases. The part of the curve ABC therefore represents a negative resistance characteristic. It will be shown later that a device which possesses a negative resistance can function as an amplifier and a generator of continuous oscillations.

[1] A. W. Hull, Phys. Rev., Vol. 7, p. 1, 1916; Proc. I. R. E., Vol. 5, p. 5, 1918.

CHAPTER IV

PHYSICS OF THE THERMIONIC VALVE

24. Current-voltage Characteristic of the Thermionic Valve.
In discussing the elements of thermionics we considered only the
saturation current, that is, the current obtained from a cathode
at any desired temperature by applying a voltage which is so high
that all the electrons emitted from the cathode are drawn away
to the anode. For all values of the applied voltage greater than
the minimum value necessary to secure this, the current is independent of the voltage.[1] The saturation current is important
in telling us what the maximum current is that can be obtained
from a cathode of given area at a given temperature. In operating
thermionic devices as amplifiers, detectors, etc., we make use of
the variation in current with variation in applied voltage. The
conditions under which the saturation current is obtained are,
therefore, unsuitable for these purposes. In this chapter will be
discussed the phenomena encountered when the applied voltage
is not high enough to draw all the emitted electrons away to the
anode as fast as they are emitted from the cathode.

Let us consider a simple thermionic device consisting of a
cathode which can be heated to any desired temperature and an
anode placed at a convenient distance from it, both being placed in
a vessel which is evacuated to such an extent that the residual
gas plays no appreciable part in the current convection between
cathode and anode. We shall for the present suppose that the
cathode is a plane equipotential surface, and the anode a plane
parallel to the cathode. If the cathode be maintained at a definite
temperature T_1 and the current to the anode be observed for

---

[1] In practice it is found that this is seldom strictly true. The current
usually increases somewhat with the voltage, but not nearly as fast as for
the lower voltages. If the applied voltage is raised to excessive values gas
can be liberated from the electrodes and then the current may again increase
rapidly with increase in the plate voltage. (See Chapter V.)

various values of the voltage applied between the cathode and anode, a curve OA_1 of Fig. 17 is obtained. Any increase in the voltage beyond the value given by A_1 causes practically no further increase in the current and we get the part A_1B_1. **In practice**

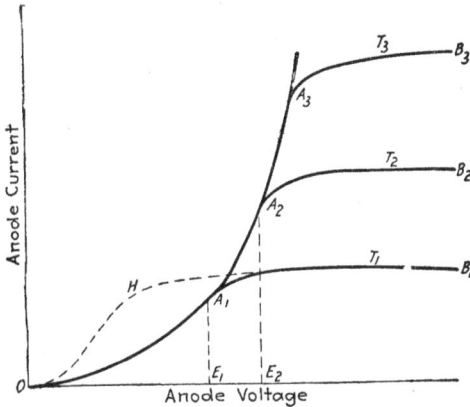

FIG. 17.

this line is seldom horizontal but usually slopes upward. This part of the curve corresponds to the condition when all the emitted electrons are drawn to the anode as fast as they are emitted from the cathode. The corresponding current is, therefore, the satura-

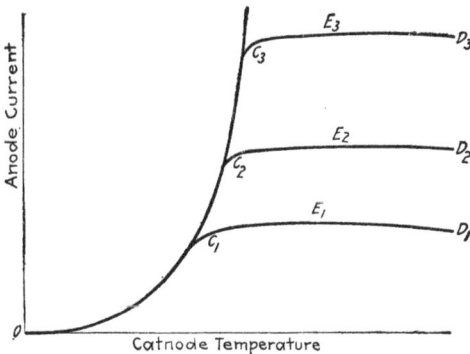

FIG. 18.

tion current. If the temperature of the cathode be increased to T_2 the curve OA_2B_2 will be obtained. If these values of the saturation current be plotted against the corresponding temperatures of the cathode, the curve shown in Fig. 18 will be obtained, the relation being given by Richardson's equation (equation (7), Chapter III).

The fact that a finite, and sometimes a considerable, voltage is needed to obtain the saturation current, shows that the current is limited by some means that is equivalent to a resistance. The current through a piece of wire is in a sense limited because, if the wire is of finite length, i.e., if it has a finite resistance, a finite potential difference must be applied to the ends of the wire to give a finite current. A device which obeys Ohm's law gives, of course, a linear relation between current and voltage, that is, when the voltages and corresponding currents are plotted on rectangular coordinates the result is a straight line passing through the origin, and the reason why the line does not coincide with the ordinate axis is because the wire has a finite resistance which limits the current. In the passage of current through a gas, the gas itself resists the flow of current, the resistance being partly due to the collisions of the carriers with the gas molecules. In a thermionic vacuum tube, however, this resistive medium is removed, because the vacuum is so high that the gas contributes practically nothing to the current convection through the tube.

If there were nothing to limit the current in a vacuum tube, the current would increase very rapidly with the applied voltage. This is, however, not the case; there are factors that have a very pronounced influence in limiting the current in a vacuum tube in such a way as to give a characteristic somewhat like that shown in Fig. 17.

One of these factors is the repelling effect of electrons in the space between cathode and anode on other electrons coming from the cathode. This is due to the volume density of electrification or space charge of the electrons in the space. Whenever current is carried by dislodged charges, their space charge must be taken into consideration. Thus, in the conduction of electricity through gases the fundamental equations always contain or involve the space charge equation, or Poisson's equation which was given on page 15, Chapter I. In conduction through gases we have to deal with the more general case where both positive and negative charges are present. The resultant space charge is the difference between that of the negative and that of the positive carriers, and since the positives or negatives generally do not move with the same speed, the problem is generally complicated. In the thermionic vacuum tube, on the other hand, we have to deal only with electrons, and therefore encounter in space charge a specific and comparatively simple manifestation of a general phenomenon always met with in convection of current by dislodged charges. The limitation of current by space charge when the current is carried only by electrons was noticed in the early experiments of Lenard, Stoletow and von Schweidler on the photo-electric effect. In these experiments the effect was very small on account of the smallness of the currents encountered in photo-electric phenomena.

In 1907 Soddy [1] made use of the property of metallic oxides discovered by Wehnelt [2] in 1904, namely, that they emit electrons copiously when heated. Soddy found that if the vacuum was made high by the vaporization of calcium, the thermionic current suddenly decreased to a small fraction of its value at the higher pressure. Soddy thought that this meant that Wehnelt cathodes became inactive in very high vacuum. An explanation of what Soddy had observed was given by O. W. Richardson and J. E. Lilienfeld.[3] The latter pointed out more specifically that the vacuum produced in Soddy's experiment by the vaporization of calcium resulted in the condition where the number of electrons carrying the current became large compared with the number of gas molecules in the path of the discharge. The number of positive ions formed by collision ionization of the electrons with the gas molecules became negligibly small, so that there was established a negative space charge due to the electrons and this reduced the current flow observed by Soddy.

We shall see later that the current in thermionic tubes is limited not only by space charge of the electrons but by other factors as well. For the present we shall consider only this factor, this being the simplest of the current limiting factors, and later explain how other effects contribute to give the current voltage characteristics of the thermionic tube.

The limitation of current by space charge can be demonstrated in a qualitative manner as follows: Let a definite voltage E_1 be applied between cathode and anode. We shall for the present assume that the cathode is an equipotential surface. If now the

[1] SODDY, Nature, Nov., 1907, p. 53.

[2] WEHNELT, Ann. d. Phys., Vol. 14, p. 425, 1904.

[3] O. W. RICHARDSON, Nature, Jan., 1908, p. 197; J. E. LILIENFELD, Phys. Zeitschr., Vol. 9, p. 193, 1908

current to the anode be observed as a function of the temperature of the cathode, the current will at first increase until it reaches a value indicated by C_1 (Fig. 18). Any further increase in cathode temperature causes no further increase in the current, and the part C_1D_1 of the curve is obtained. The current given by C_1D_1 is frequently referred to as the " temperature saturation current," and the condition characterized by this lack of increase of current with increase of cathode temperature as *temperature saturation*. The reason why under these conditions the current does not increase along $C_1C_2C_3$ as would be expected from Richardson's equation is because at cathode temperatures greater than that corresponding to C_1, so many electrons are emitted that the resulting volume density of their charge causes all other emitted electrons to be repelled, and these return to the cathode. The applied voltage E_1 is then not high enough to draw all the emitted electrons away to the anode. If now the voltage be increased to E_2 the current increases, since more electrons are now drawn away from the supply at the cathode, the full space charge effect being maintained by fewer electrons being compelled to return to the cathode. From Fig. 18 it is seen that with the voltage E_2 the cathode must be raised to a minimum temperature corresponding to C_2 before the full space charge effect can manifest itself. It is seen, then, that the higher the applied voltage, the higher must be the cathode temperature to obtain the full space charge effect. It is also seen that the part OC of Fig. 18 corresponds to the part AB of Fig. 17, and CD of Fig. 18 to OA of Fig. 17. The saturation current is obtained when the applied voltage is so high that a variation of voltage does not cause any appreciable variation in current, while the condition under which the thermionic tube operates, as a voltage operating device, is characterized by the condition that the cathode temperature is so high that the current does not vary appreciably with variation in cathode temperature.

25. Current-voltage Relation for Infinite Parallel Plates. To get an understanding of the quantitative effect on the current by the space charge of the electrons, it may be well first to consider the ideal and simple case that results when we neglect the complicating factors encountered in practice and then consider the modifications introduced by these factors. In deriving the equations for this simple case, we shall therefore assume that the electrodes are infinitely large parallel plates, capable of being maintained at any desired temperature. Both electrodes will be assumed to be equipotential surfaces. The cathode will be maintained at a high temperature, the anode remaining cold. It follows from Richardson's theory that the hot plate will emit electrons, the emission being the result of the kinetic energy of the electrons becoming sufficiently high to overcome the surface force that tends to hold the electrons within the cathode. It will be recognized that the energy and distribution of energy of the electrons play an important part in the mechanism of electron emission. A derivation of the relation between current and voltage, which takes into consideration the energy distribution between the electrons, is quite complicated. J. J. Thomson [1] has given the equations resulting from the assumption that the electrons all emerge with one initial velocity. In 1911, C. D. Child [2] gave the full solution, based on the assumption that the initial velocity of emission is zero.

Langmuir [3] of the General Electric Company and Schottky [4] also published derivations of the space charge equation and made a careful investigation of some of the phenomena observed in thermionic tubes.

We shall now derive Child's equation, making the same assumptions, and then consider the modifications introduced by a consideration of the factors neglected in the simple derivation, and more particularly how these factors contribute to produce the type of current-voltage characteristic generally obtained in practical thermionic tubes. We shall therefore assume that both cathode and anode are equipotential parallel surfaces of infinite extent, and that the electrons emerge from the hot cathode with zero velocity. The cathode C and anode P (Fig. 19) will be supposed to be in an enclosure in which a perfect vacuum is maintained. The degree of vacuum necessary to approximate this perfection will be discussed in the next chapter.

The cathode C can be raised to any desired temperature. Let the anode P be raised to a potential V_1, while the cathode remains grounded. As long as the temperature of C is so low that practically no electrons are emitted, the potential gradient between the

[1] J. J. THOMSON, Conduction of Electricity through Gases, 2d Ed., p. 223.

[2] C. D. CHILD, Phys. Rev. Vol. 32, p. 498, 1911.

[3] I. LANGMUIR, Phys. Rev., (2), Vol. 2, p. 450, 1913.

[4] SCHOTTKY, Jahrb. d. Radioaktivitat u. Elektronik. Vol. 12, p. 147, 1915.

plates is practically constant so that the potential distribution is a linear function of X, the distance from the plate C, and can be represented by a straight line OP. But this is no longer the case when C emits electrons. These electrons are pulled over to P by the applied electric field and their presence in the space between C and P modifies the potential distribution. In Section 9 it was shown that if the plates C and P are infinitely large so that the

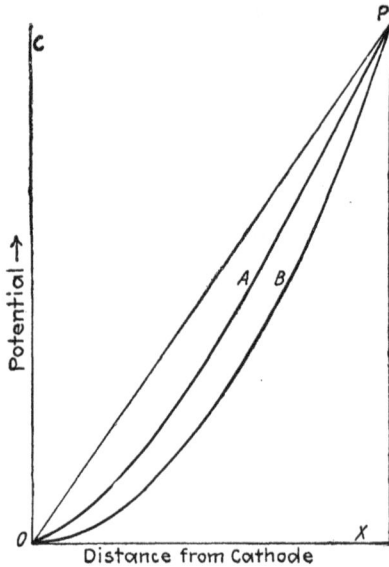

Fɪɢ. 19.

lines of force are straight the potential V at any point distant x from C is given by

$$\frac{d^2V}{dx^2} = 4\pi\rho, \quad \cdots \cdots \cdots \quad (1)$$

where ρ is the volume density of the charge. Now ρ is itself a function of x. Thus

$$\frac{d^2V}{dx^2} = 4\pi f(x). \quad \cdots \cdots \cdots \quad (2)$$

The relation between the potential V and the distance x cannot be obtained unless the form of the function x is known. We know that the density of electrons near C is greater than near P. It can therefore be seen in a general way that because of the presence of the electrons the potential distribution curve will be somewhat of the nature shown by OAP (Fig. 19). The potential gradient at the cathode is still positive, but if the temperature of the cathode be now raised until the potential difference applied between C and P is not high enough to draw all the emitted electrons away from C_1 and if it is assumed that the electrons are emitted from C with zero velocity, the potential distribution curve takes a shape somewhat like OBP, which has a horizontal tangent at O. This means that the potential gradient at O is zero. Any further increase in the number of emitted electrons would tend to depress the curve at O below the horizontal line OX. This would be equivalent to establishing a negative potential gradient at the cathode, which would tend to return the emitted electrons to the cathode.

The assumption that the electrons emerge from the cathode with zero velocity does not lead to a correct description of the true state of matters. The electrons are actually emitted with finite velocities which are distributed according to Maxwell's distribution law. This law forms the basis of Richardson's equation. In actual practice the finite velocities with which the electrons are emitted from C cause the potential distribution curve to be depressed slightly below the axis OX so that there actually exists at the cathode a slight negative potential gradient. This will be discussed more fully below. For the present we shall assume that the electrons start from the cathode with zero velocity. The velocity they acquire on their way to the anode P is then due entirely to the potential difference between C and P. The kinetic energy of an electron at a point distant x from C, i.e., where the potential is V, is given by

$$\tfrac{1}{2}mv^2 = Ve. \qquad \ldots \ldots \ldots \quad (3)$$

This is obtained from equation (4) Chapter I, by putting the initial velocity $v_0 = 0$.

It was shown in Section 4 that if ρ is the number of electrons per cubic centimeter multiplied by the electronic charge, that is, if ρ is the volume density of electricity or space charge, the motion of the electrons constitutes a current per square centimeter given by:

$$i = \rho v. \qquad \ldots \ldots \ldots \ldots \quad (4)$$

where v is the velocity.   Both ρ and v are functions of the distance from the cathode C, but the product ρv is constant, since the number of electrons passing per second through unit area perpendicular to the direction of the electric field, that is, the current i must be the same for all points between cathode and anode.

The quantities ρ and v in the above equations (1), (3) and (4), are unknown functions of x, and can be eliminated from these equations.   This gives:

$$\frac{d^2V}{dx^2} = 2\pi i \sqrt{\frac{2m}{eV}}. \quad \cdot \quad \cdot \quad \cdot \quad \cdot \quad \cdot \quad \cdot \quad (5)$$

The integration of this equation gives

$$\left(\frac{dV}{dx}\right)^2 - \left(\frac{dV}{dx}\right)_0^2 = 8\pi i \sqrt{\frac{2m}{e}}(V^{1/2} - V_0^{1/2}), \quad \cdot \quad \cdot \quad (6)$$

where V_0 is the potential of the cathode and $\left(\dfrac{dV}{dx}\right)_0$ the potential gradient at the cathode.   Now, the potential of the cathode is supposed to be zero, and since the initial velocity of emission of the electrons is assumed to be zero, the limiting condition for full space charge is that $\left(\dfrac{dV}{dx}\right)_0 = 0$.   Hence (6) becomes

$$\left(\frac{dV}{dx}\right)^2 = 8\pi i \sqrt{\frac{2mV}{e}}. \quad \cdot \quad \cdot \quad \cdot \quad \cdot \quad \cdot \quad (7)$$

Integrating this and putting as limits $V=0$ for $x=0$ and $V=E$ for $x=d$, the distance between the plates, we obtain the equation of the current as a function of E and d, namely:

$$i = \frac{1}{9\pi}\sqrt{\frac{2e}{m}} \cdot \frac{E^{3/2}}{d^2}. \quad \cdot \quad \cdot \quad \cdot \quad \cdot \quad \cdot \quad (8)$$

This is the so-called $\frac{3}{2}$-power equation first derived by Child, which gives the relation between the voltage and current carried by dislodged charges of one sign only.

By putting the value of $\dfrac{e}{m}$ in (8) and writing A for the area of the cathode the current can be expressed by

$$i = 2.33 \times 10^{-6} \frac{AE^{3/2}}{d^2}, \quad \cdot \quad \cdot \quad \cdot \quad \cdot \quad (9)$$

where i is the current in *amperes*, d the distance between the plates in *centimeters* and V the potential difference between the plates in *volts*.

It will be noticed that the space charge equation gives the current in terms of only the applied voltage and the geometry of the device. The space charge current does not depend, as in the case of the saturation current, upon the temperature or electron affinity of the cathode. On the other hand the saturation current is independent of the distance between cathode and anode, except that if this distance be made smaller the saturation current would be obtained at a lower minimum voltage. Referring to the curve OA_1B_1 (Fig. 17), it will be seen that the space charge equation gives the current for values of the voltage less than that corresponding to A_1. If the distance between cathode and anode be decreased the curve obtained would be OHB_1.

26. Quantitative Relation for Concentric Cylinders. Thermionic tubes are frequently made in the form in which the anode is a cylinder and the cathode a wire stretched along its axis. The quantitative relation for the characteristic of such a device was published by Langmuir [1] by making the assumptions that both cathode and anode are infinitely long, and both are equipotential surfaces. Here, also, the electrons are assumed to emerge from the cathode with zero velocity. In the next section it will be shown what modifications must be made in order to make the results conform more nearly with practical conditions.

The equation (3), for the velocity of the electrons, can be applied directly to this case. The differential equation for the potential distribution, however, now takes the form

$$\frac{d}{dr}\left(r\frac{dV}{dr}\right) = 4\pi\rho r, \quad \ldots \quad \ldots \quad (10)$$

where r is the distance from the cathode, and the equation for the current becomes

$$i = 2\pi r \rho v, \quad \ldots \quad \ldots \quad (11)$$

where i is the current per unit length.

These equations and equation (3) can be combined to give:

$$r\frac{d^2V}{dr^2} + \frac{dV}{dr} = i\sqrt{\frac{2m}{eV}}. \quad \ldots \quad \ldots \quad (12)$$

[1] I. Langmuir, loc. cit., p. 457.

Putting $V = E$, the potential difference between cathode and anode, when r is the radius of the cathode, the solution of this equation can be expressed in the form:

$$i = \frac{2}{9}\sqrt{\frac{2e}{m}} \cdot \frac{E^{3/2}}{r\beta^2}, \quad \cdot \quad \cdot \quad \cdot \quad \cdot \quad \cdot \quad \cdot \quad (13)$$

where β is a constant which is determined by the ratio $\frac{r}{a}$ of the radii of the anode and the filament. The relation between β and $\frac{r}{a}$ is given in the following table taken from Langmuir's paper.

TABLE I

$\frac{r}{a}$	β^2	$\frac{r}{a}$	β^2
1.00	0.000	5.0	0.775
1.25	0.045	6.0	0.818
1.50	0.116	7.0	0.867
1.75	0.200	8.0	0.902
2.00	0.275	9.0	0.925
2.50	0.405	10.0	0.940
3.00	0.512	15.0	0.978
4.00	0.665	∞	1.000

It is seen that for most practical cases β can be put equal to unity. If this is done and the length of the filament be put equal to l, equation (13) may be written:

$$i = \frac{2}{9}\sqrt{\frac{2e}{m}} \cdot \frac{l}{r} E^{3/2}. \quad \cdot \quad \cdot \quad \cdot \quad \cdot \quad \cdot \quad \cdot \quad (13a)$$

Substituting the value of $\frac{e}{m}$ and reducing i to amperes, V to volts and the radius r to centimeters, equation (13a) becomes

$$i = 14.65 \times 10^{-6} \cdot \frac{l}{r} E^{3/2} = CE^{3/2}, \quad \cdot \quad \cdot \quad \cdot \quad (14)$$

where l is the length of the filament in centimeters.

The only difference between equations (9) and (14), besides the numerical value of the constant, is that in the case of plane parallel electrodes the current varies inversely as the square of the distance d between the electrodes, whereas in the case of the cylindrical structure the current varies inversely as the first power of the radius r of the anode. This is an important property when considering the design of thermionic tubes with very small electrostatic capacity, such as is required when operating at extremely high frequencies.

27. Influence of Initial Velocities. The two main assumptions underlying the derivation of equations (8) and (13) are that the electrons emerge from the cathode with zero velocities and that

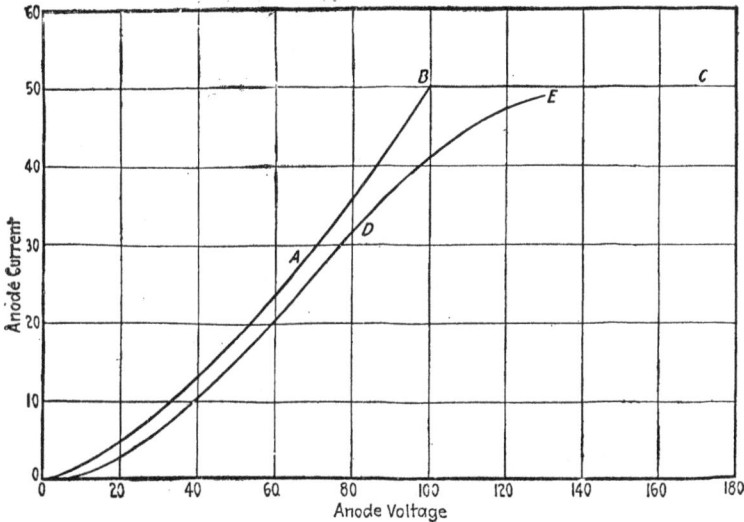

Fig. 20.

the cathode is an equipotential surface. Let us first see how the conditions are altered when we do not ignore the effect of the initial velocities.

By assuming zero initial velocities, the condition is obtained that the potential gradient at the cathode is zero. This means that the potential distribution curve OBP, shown in Fig. 19, has a horizontal tangent at O. The resulting equation (8) states that the current varies as the $\frac{3}{2}$-power of the applied voltage for all voltages up to that necessary to give the saturation current. This characteristic is shown in Fig. 20. The part OAB represents the current that is limited only by the space charge of the electrons in the space between cathode and anode. At B the applied voltage becomes large enough to pull all the electrons to the anode as fast as they are emitted from the cathode. The part BC represents the saturation current.

The effect of the initial velocities can be understood by referring to Fig. 21. Suppose the anode were insulated and connected to a

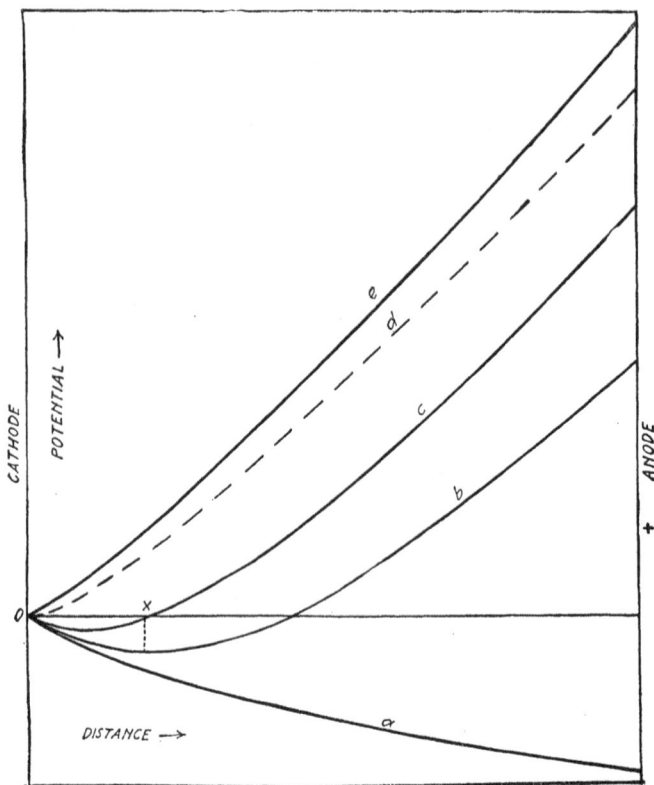

Fig. 21.

pair of quadrants of an electrometer. If the cathode be raised to a high temperature, the anode would acquire a negative charge which can be measured with the electrometer. Obviously in this case the potential at all points between cathode and anode must be negative and the potential distribution can be represented by a curve such as the curve a of Fig. 21. In this diagram, as in Fig.

19, the abscissæ represent distances from the cathode, and the ordinates the potentials with respect to that of the cathode which we shall take as zero. The anode will charge up negatively until the potential to which it rises is so high that the field at all points between anode and cathode is sufficiently strong to prevent any more electrons from reaching the anode under their own kinetic energy of emission. If the anode now be connected to the cathode through a battery which maintains the anode at a small positive potential, the potential distribution can be represented by the curve *b* of Fig. 21. This curve shows that the potential gradient is still negative at the cathode, but in passing from cathode to anode the gradient passes through zero at a distance *x* from the cathode and then becomes positive. The negative gradient at distances less than *x* is due to the space charge of the electrons and the fact that the energy of emission enables the electrons to move against a negative potential gradient. If the anode potential be increased the distance *x* of minimum potential shortens until the potential distribution finally takes the shape given by curve *d*. In this case *x* is practically zero, so that the potential gradient vanishes at the cathode. When this condition is reached we have the case corresponding to the point *B* of Fig. 20. If the plate potential be increased still further the potential distribution straightens out still more (curve *e*) and the current obtained is the saturation current.

Now, it is the condition represented by the potential distribution curve *d* that was assumed in the derivation of the $\frac{3}{2}$-power equation. But this is the condition obtaining when the space current just becomes equal to the saturation current. It is therefore to be expected that on account of the initial velocities, Child's $\frac{3}{2}$-power equation would hold, strictly speaking, only for currents approximating the saturation current.

The explanation given here for the influence of the initial velocities of the emitted electrons is based on the solution of the problem furnished me by T. C. Fry, who also computed the distance *x* of the minimum potential for cases approximating the conditions met with in practice. The values of *x* depend on the distance between the electrodes and the potentials applied to the anode.[1]

---

[1] The effect of the initial velocities has also been studied by W. Schottky (Phys. Zeitschr., Vol. 15, 1914), P. S. Epstein (Deutsch Phys. Gesell. Verh. 21, p. 85–99, 1919) and others.

The effect of the initial velocities on the shape of the current-voltage characteristic is demonstrated in Fig. 22, where the logarithms of the currents are plotted against the logarithms of the potential differences applied between anode and cathode. The characteristic shown in Fig. 20 gives on the logarithmic plot a straight line AB having a slope equal to $\frac{3}{2}$. On account of the initial velocities, the logarithmic plot takes the form given by the line CD, the slope of which approximates $\frac{3}{2}$ at the upper part.

28. Effect of Voltage Drop in the Filament. In practice the cathode is never an equipotential surface, but takes the form of a

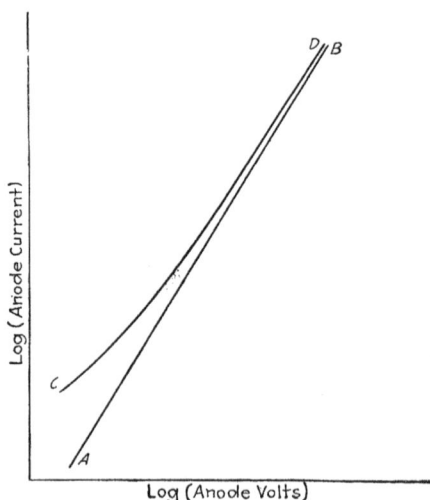

Fig. 22.

filament which is rendered incandescent by passing a current through it. There is consequently established a voltage drop in the filament which causes a marked deviation from the $\frac{3}{2}$-power equation. This deviation is in an opposite sense to that due to the initial velocities.

The equation for the characteristic resulting from a consideration of the effect of the voltage drop in the filament was given by W. Wilson.[1] This derivation is also based on the assumption that the initial velocities can be neglected.

[1] Paper read at the Philadelphia meeting of the American Physical Society, Dec., 1914.

Let us consider a structure in which the filament is a single wire stretched along the axis of a cylindrical anode and let the negative terminal of the filament be grounded so that we may regard its potential as zero. Let the potential drop in the filament, due to the heating current, be E_f and let the potential of an intermediate point at a distance x from the negative end be V. If l be the length of the filament we have for constant current in it:

$$\frac{V}{E_f} = \frac{x}{l} \quad \dots \dots \dots \quad (15)$$

If E be the potential difference between the anode and the negative end of the filament, the potential difference between the anode and a point of the filament which is at a potential V is

$$E - V = E - E_f \frac{x}{l},$$

and this is zero when

$$x = E\frac{l}{E_f}. \quad \dots \dots \dots \quad (16)$$

This means that current will only flow from the length x of the filament given by (16) since all points of the filament beyond x are positive with respect to the anode.

Since a very small length dx of the filament can be regarded as an equipotential surface, the current can be taken to vary as the $\frac{3}{2}$-power of the potential difference between the anode and the point x, if we neglect the factors such as the initial velocities, which cause a deviation from the simple $\frac{3}{2}$-power equation. The potential difference between the anode and the point x of the filament is $E - E_f\frac{x}{l}$. Hence if Δi be the current from the element dx of the filament, we get by equation (13a)

$$\Delta i = \frac{2}{9r}\sqrt{\frac{2e}{m}}\left(E - E_f\frac{x}{l}\right)^{3/2} dx.$$

In order to obtain the total current, we have to integrate over the length of the filament from which the electrons flow to the anode. We have to distinguish two cases: (a) When the voltage E between the anode and the negative end of the filament is less than the voltage drop E_f in the filament, due to the heating current, and (b) when E is greater than E_f.

Case (a) $E \leqq E_f$. Here the integration is to be performed over the part x of the filament given by equation (16). Thus

$$i = \frac{2}{9r}\sqrt{\frac{2e}{m}}\int_0^{E\frac{l}{E_f}}\left(E - E_f\frac{x}{l}\right)^{3/2}dx.$$

This gives

$$(i)_{E \leqq E_f} = \frac{2}{5}\cdot\frac{2}{9}\sqrt{\frac{2e}{m}}\frac{l}{rE_f}\cdot E^{5/2}. \quad . \quad . \quad . \quad . \quad . \quad (17)$$

When E is expressed in volts and i in amperes, we may write this equation:

$$\left.\begin{aligned}i &= 5.86\times10^{-6}\frac{l}{rE_f}E^{5/2} \\ &= KE^{5/2}\end{aligned}\right\} \quad . \quad . \quad . \quad . \quad . \quad (18)$$

Comparing this constant of proportionality K with C in equation (14) it will be seen that

$$C = \tfrac{5}{2}E_f K. \quad . \quad . \quad . \quad . \quad . \quad (18a)$$

Equation (18) shows that as long as the potential difference between the anode and the negative end of the filament is less than the voltage drop in the filament, the anode current varies as the $\frac{5}{2}$-power of the anode potential. Except for the fact that here the limitation of current by the voltage drop in the filament has been taken care of, this equation is subject to the same limitations as the $\frac{3}{2}$-power equations which were derived on the assumption that the cathode is an equipotential surface.

Case (b) $E \gtrless E_f$. In this case electrons flow from the whole surface of the filament to the anode. Hence the current is:

$$i = \frac{2}{9r}\sqrt{\frac{2e}{m}}\int_0^l\left(E - E_f\frac{x}{l}\right)^{3/2}dx,$$

which gives:

$$(i)_{E \gtrless E_f} = K[E^{5/2} - (E - E_f)^{5/2}], \quad . \quad . \quad . \quad . \quad (19)$$

where

$$K = \frac{4}{45}\sqrt{\frac{2e}{m}}\cdot\frac{l}{rE_f} = 5.86\times10^{-6}\frac{l}{rE_f}.$$

l = length of the filament;
r = radius of the cylindrical anode;
E_f = voltage drop in the filament.

Equation (19) may be expanded into the more convenient form:

$$i = CE^{3/2}\left[1 \mp \frac{3}{4}\frac{E_f}{E} \pm \frac{3}{24}\left(\frac{E_f}{E}\right)^2 \mp \cdots\right], \quad \cdots \quad (19a)$$

where $C = 14.65 \times 10^{-6}\frac{l}{r}$, the same constant as appears in the equation (14).

The lower signs in the series of (19a) pertain to the case in which the potential of the anode is reckoned with respect to the positive instead of the negative end of the filament.

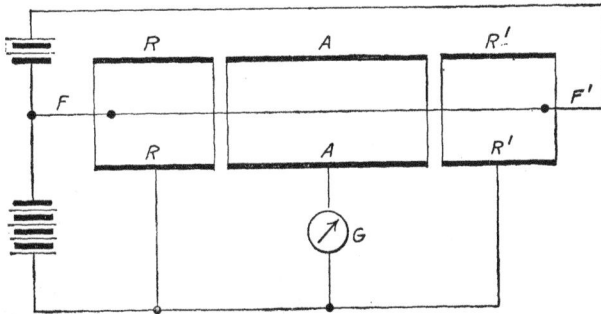

FIG. 23.

This series converges so rapidly that for all values of the anode potential greater than twice the voltage drop in the filament we can write for the current with close approximation.

$$i = CE^{3/2}\left[1 \mp \frac{3}{4}\frac{E_f}{E}\right]. \quad \cdots \quad (20)$$

In deriving these equations, the length of the filament (and that of the cylindrical anode) were put equal to a finite value l. Strictly speaking l should be infinitely long so that the distortion of the field at the ends of the anode can be neglected. This condition can be realized in practice with a device shown schematically in Fig. 23.

The anode AA is in the form of a cylinder and the filament is stretched along its axis. In order to insure straight lines of force the guard rings RR' are placed on either side of AA, the filament extending beyond the ends of the anode. The anode and guard rings are electrically connected but the galvanometer G

is inserted as indicated in the diagram, so that it registers only the electron current flowing to the anode. The effective length of the filament is then equal to the length of the anode.

The general effect of the voltage drop in the filament when the plate is connected to the negative end of the filament, is to make the space current smaller because the average potential difference between the filament and the plate is smaller than that between the negative end of the filament and the plate, this being the potential difference that is ordinarily measured with a voltmeter. The general effect is shown by the curve OD of Fig. 20. The curve OAB represents the theoretical curve in accordance with the simple $\frac{3}{2}$-power equation, and the part AB represents the ideal saturation current which is supposed to be independent of the applied voltage. The characteristic which is ordinarily observed is indicated by ODE. For the present we shall consider only the lower part OD of this characteristic. The deviation at voltages greater than the voltage corresponding to the point D is due to the limitation of the current by the electron emission from the filament and will be discussed in the next section. The line OAB is computed from the $\frac{3}{2}$-power equation (14), the constant C being put equal to 50×10^{-6} amperes. If we assume that the voltage drop in the filament is 10 volts, then referring to equation (18a), we find the constant K becomes equal to 2×10^{-6}. With the help of equations (18) and (19) we can then compute the current as a function of the potential differences E between filament and plate, by putting $E_f = 10$. The values so computed give the curve OD of Fig. 20. The percentage deviation of this curve from the theroretical curve OA is quite considerable at the lower voltages. It will be explained in the next chapter that it is desirable to so design thermionic tubes that the saturation current is obtained at the smallest possible voltage. In practice the voltage necessary for saturation seldom exceeds a few hundred volts. Fig. 24 represents two experimental curves plotted on the logarithmic scale and obtained in such a way that in the one case (curve 2) the voltage drop in the filament was effective and in the other (curve I) it was eliminated. To eliminate the effect of the voltage drop in the filament we can, as has been done in taking these curves, resort to a scheme used by von Baeyer [1] in 1909, which consists in connecting the ends of the filament and the

[1] O. von Baeyer, Phys. Zeits., Vol. 10, p. 168, 1909.

plate through a commutator which is so arranged that the filament
current and the plate voltage are applied alternatively for short
intervals of time, the plate voltage being applied only while the
filament voltage is cut off. If the alternations are frequent
enough, the filament does not get a chance to cool off markedly
during the time that the filament current is cut off. In this way
the plate current is measured only while there is no voltage drop
in the filament.

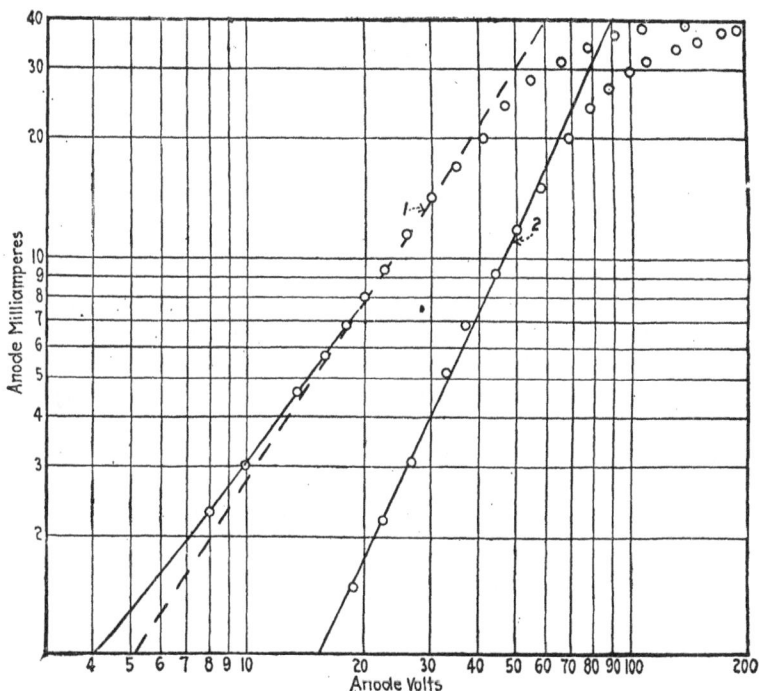

FIG. 24.

The dotted line in Fig. 24 is drawn to have a slope equal to $\frac{3}{2}$.
For the higher voltages curve 1 is probably close to the theoretical
line, but at the lower voltages it bends to the left on account of the
initial velocities of the emitted electrons. Curve 2 plainly shows
the effect of the voltage drop in the filament; it deviates everywhere to the right of the theoretical line. Hence, the effect of
the voltage drop in the filament is opposite to the effect caused by
the initial velocities. While the latter causes the logarithmic

plot of the characteristic to deviate to the left from the $\frac{3}{2}$-line, the voltage drop in the filament causes not only a deviation to the right but also a lateral shift of the whole curve to the right. These two effects sometimes contribute to give a better $\frac{3}{2}$ logarithmic line than is to be expected from theoretical considerations only, because their effects are opposite and tend to neutralize each other. This is especially the case when the filament is operated at a very high temperature, because the higher the temperature of the filament, the greater will, of course, be the effect of the initial velocities.

As far as the lower part of the characteristics of thermionic valves is concerned, it follows, therefore, that the current is limited not only by space charge but by the initial velocities of emission of electrons and by the voltage drop in the filament as well. If we consider the whole characteristic over which it is sometimes operated, we find that the current is also limited by another factor which we will now proceed to explain.

29. Influence of Limitation of Current by Thermionic Emission. If we consider the characteristics as actually obtained in practice, we find that on the upper parts they deviate even more from the theoretical line than the lower part. Referring, for example, to Fig. 20, the curve ODE represents more nearly what is actually obtained with thermionic tubes, which is quite different from the theoretical curve OAB. As we proceed up the characteristic from the lower voltage values, we find that the curve deviates from the theoretical curve on account of the voltage drop in the filament, but as we proceed higher up the characteristic the curve bends over gradually toward saturation. The two things to be noticed are that this transition region between the lower part and the saturation part of the characteristic generally starts at comparatively low voltages and, secondly, the saturation current itself usually increases gradually up to very high voltages instead of becoming flat, as indicated by AB. This deviation, which is shown by the part DE, Fig. 20, is due to the fact that in the neighborhood of D the voltage approaches values at which the current begins to be limited by electron emission. The gradual bending of the curve and the length of this transition region depend on the surface conditions of the cathode the voltage drop in the filament and on the shape of the anode. With the Wehnelt cathode the transition part of the curve is generally much longer than with Tungsten cathode.

Furthermore, if the anode is in the form of a wire or a small plate, the transition part is also longer than when the anode is, for example, a cylinder surrounding the cathode placed along its axis.

The transition part of the characteristic can best be seen from diagrams in which are plotted a number of characteristics obtained with the same tube but with different filament temperatures such as shown, for example, in Fig. 25. Taking, for example, the lowest one of these curves, we see that the transition part sets in at *A*. In the second curve, it sets in at *B*, and so on.

Generally speaking, practical thermionic devices operate over the part of the characteristic indicated by *ODE* of Fig. 20. This part we can refer to as the infra-saturation part or the operating part of the characteristic. From the above explanations it follows that over this operating part of the characteristic the current is limited by space charge, initial velocities of emission, the voltage drop in the filament and by thermionic emission. There is another factor which limits the current on the lower voltages, namely, the formation of heavy negative carriers when a small amount of gas is present in the tube. This will be discussed more fully in the next chapter. In the case of the three-electrode tube which contains a grid inserted between the filament and the plate, the current is further limited by the grid (see Chapter VII). All these current limiting factors contribute to the production of a characteristic which is curved and which does not obey a simple power law when taken over the whole infra-saturation or operating range of the characteristic. The logarithmic plot of the characteristic is steepest at the lower voltages where the slope may be as high as $2\frac{1}{2}$. As the voltage increases the slope of the logarithmic plot decreases until finally it becomes less than unity when saturation is approached.

When dealing with certain small parts of the characteristic, we can advantageously apply a simple power law. Thus, in the case of a tube containing a grid, we operate the tube as an amplifier generally over a range on the lower part of the characteristic where the relation between current and voltage can be expressed as a simple quadratic relation (see Chapter VII). When using the three-electrode tube as an oscillation-generator, on the other hand, we use the whole characteristic ranging all the way up to saturation voltages.

In order to obtain a rough indication of the maximum current

that can be passed through the tube, we can apply the simple
$\frac{3}{2}$-power equation in special cases.   Thus, if the anode is a cylinder
and the cathode stretched along its axis, the application of equation

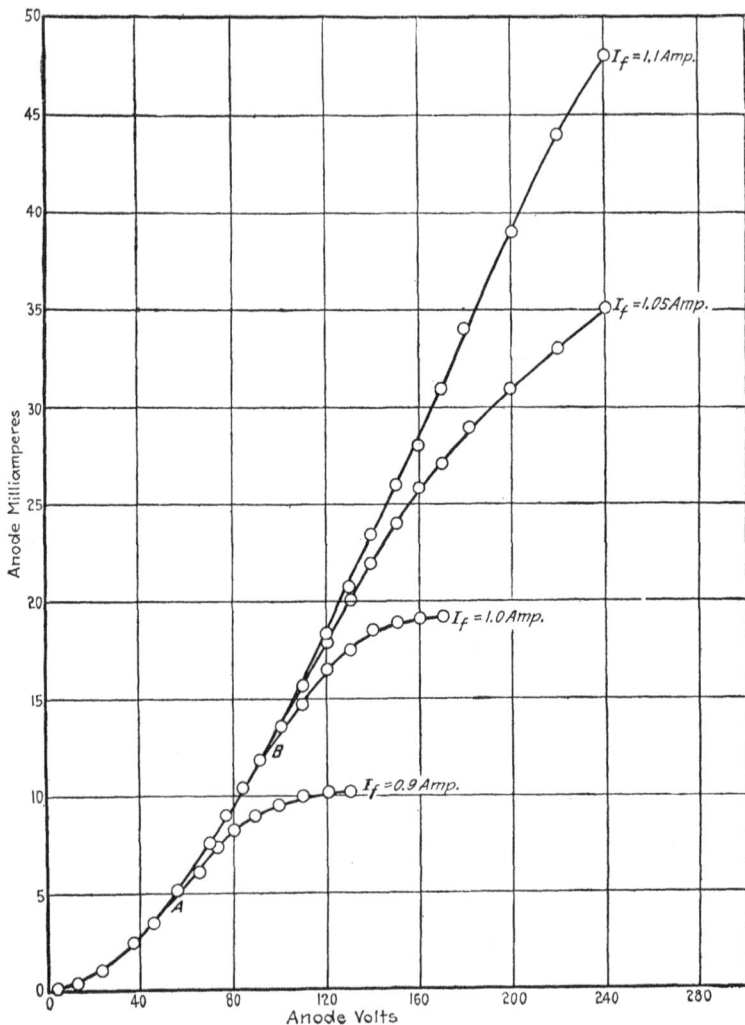

FIG. 25.

(14) will give an approximate indication of the maximum current
that can be passed through the tube at voltages approximating
to the saturation voltage.

The application of the $\frac{3}{2}$-power equation to the approximate determination of the current at the saturation voltage, is not based on the consideration in Section 27 where it was shown that theoretically the $\frac{3}{2}$-power equation holds for an equipotential cathode at voltages approximating the saturation voltage, because the limitation of current by electron emission which makes itself felt even on the lower part of the characteristic, causes quite a marked deviation from the theoretical curve, so that the $\frac{3}{2}$-power equation can be used only to give a very rough indication of the maximum current.

It is not to be concluded that since the initial velocities and the voltage drop in the filament both exert effects which become proportionately smaller as the voltage increases, an agreement with the $\frac{3}{2}$-power law can be demonstrated if the upper part of the characteristic gives a slope equal to $\frac{3}{2}$. The fact of the matter is that on account of the voltage drop in the filament the slope of the characteristic well below the saturation part is greater than $\frac{3}{2}$ and on the saturation part itself it is less than unity and sometimes almost zero. Since the slope changes actually from the high value to the low value, there must, of course, be a region where the characteristic has a slope equal to $\frac{3}{2}$. But, in this region the current is limited not only by space charge but by the voltage drop in the filament and by thermionic emission from the filament as well. This effect can readily be seen by plotting the curves shown in Fig. 25 on the logarithmic scale, as is shown in Fig. 26. It will be seen that each of these curves has a slope equal to $\frac{3}{2}$ over a more or less restricted region. The three different curves give three different lines each having a slope equal to $\frac{3}{2}$. This would mean that this one tube obeys three different $\frac{3}{2}$-power equations which is, of course, impossible. For the $\frac{3}{2}$-power equation to be obeyed, it is not only necessary that the logarithmic plot give a line, the slope of which is $\frac{3}{2}$, but that line must, of course, also have a definite intercept on the voltage axis which gives the constant of proportionality of the $\frac{3}{2}$-power equation.

30. Effect of the Curvature of the Characteristic. In the previous section it was explained how a number of factors contribute to limiting the current in such a way as to produce a characteristic which is curved. Of these factors the space charge of the electrons in the space between filament and anode is responsible for the fact that the characterictic curves upwards. The limitation of current by electron emission, on the other hand, causes the characteristic gradually to bend over to the right. The fact that the characteristic is curved is a very great disadvantage in a large number of uses to which the tube is applied. As will be explained later, it causes the tube to distort when it is used as an amplifier. When the tube is used as an oscillation-generator, the curvature of the character-

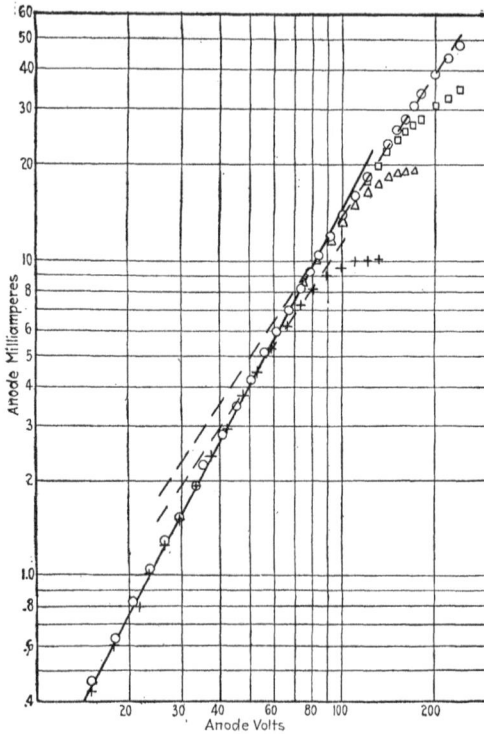

FIG. 26.

tic is also a disadvantage in that it makes the solution of the problem of the oscillator so much more difficult. In fact, the full solution for the curved characteristic has not yet been given. On the other hand, when the tube is used as a modulator or detector of high frequency oscillations the curvature of the characteristic is actually made use of. In considering the applications of the thermionic tube, we shall first deal with those for which it is desirable to have a straight characteristic and then consider the applications which make use of the curvature of the characteristic.

Whatever has been explained here with reference to the simple two-electrode valve applies in a general way and with certain qualifications also to the much more important type of thermionic device, the three-electrode tube which contains a grid between the filament and the plate.

31. Energy Dissipation at the Anode. When a thermionic tube is in operation its anode becomes heated to an extent depending on the size and nature of its surface and on the voltage and current at which the tube is operated. If v be the velocity with which an electron strikes the anode, the energy converted into heat at the anode is $\frac{1}{2}mv^2$. If the number of electrons striking the anode per second is n the power dissipated is

$$\tfrac{1}{2}nmv^2 = neE = IE$$

where I is the current through the tube and E the voltage between anode and cathode. The temperature of the anode will rise until the rate at which energy is transferred to it by the bombarding electrons becomes equal to the rate at which it is radiated from it. If A be the area of the anode and ϵ its thermal emissivity, the energy radiated per second can be given approximately by the Stefan-Boltzmann equation:

$$P = k\epsilon A(T^4 - T_0^4), \quad \ldots \ldots \quad (21)$$

where T is the temperature of the anode in Kelvin degrees and T_0 that of its surroundings. The radiation constant k is the power radiated by 1 cm.^{2} of a perfect black body at a temperature of $1°\ K$, its surroundings being at absolute zero of temperature.

For tungsten Langmuir [1] finds the following equation instead of the above equation (21):

$$P = 12.54A\left(\frac{T}{1703}\right)^{4.74}. \quad \ldots \ldots \quad (22)$$

When equilibrium is established the power radiated is equal to the kinetic energy of the bombarding electrons converted per second into heat. The term T_0^4 in equation (21) can generally be neglected, when the anode temperature is high and we obtain

[1] I. LANGMUIR, Phys. Rev., Vol. 34, p. 401, 1912.

the following relation between the power dissipation at the anode and the temperature to which it rises:

$$EI = k\epsilon A T^4. \quad . \quad . \quad . \quad . \quad . \quad . \quad . \quad (23)$$

The constant ϵ has its maximum value unity when the surface has the full radiating power of a black body. A shiny surface does not radiate as well as a dull black or dark surface, and is therefore not as suitable for use in high power tubes. Furthermore, the power radiated depends upon the area of the anode.

These factors have to be considered in the design of power tubes. The area of the anode must be so proportioned with respect to the product EI that its temperature does not rise beyond a certain value depending upon the nature of the anode and the extent to which it has been denuded of gas during evacuation of the tube.

If the anode temperature becomes too high three deleterious effects can come into play:

(a) It can cause the liberation of gas and so make the discharge depart from a pure electron discharge. This impairs the operation of the tube and often, when it does happen, causes the tube to blow out on account of the increase in current due to ionization and the consequent increase in power dissipation at the anode. Such impairment is often only transient since the gas is usually cleaned up by the hot filament. (See Chapter V.)

(b) It causes a volatilization of the anode resulting in the formation of a metallic deposit on the bulb ("blackening of the bulb"). Unless the parts where the lead-in wires are sealed in are shielded the metallic deposit impairs the insulation between the wires which must in most cases be very good.

(c) It can cause the emission of electrons from the anode. This is generally not so serious in the case of a three-electrode tube, consisting of filament, grid and anode because there usually exists a strong electric field between the grid and the anode which tends to return to the anode all electrons emitted from it. But in the case of a valve, that is, a tube containing only cathode and anode, care must be taken that emission of electrons from the anode does not take place, because if it did, the tube would not rectify completely.

32. Efficiency of the Cathode. The efficiency of a thermionic cathode is determined by two factors: its life, and the maximum thermionic current obtainable for a given amount of power expended in maintaining it at the desired temperature. The saturation current, i.e., the maximum obtainable current, depends upon the area of the cathode, its temperature and its electron affinity. (Equation 7, Chapter III.) The power necessary to maintain the cathode at a definite temperature depends upon its area, its temperature and its thermal emissivity (equation 23). By comparing these two equations it will be seen that the saturation current increases more rapidly with the temperature than does the power. Hence the saturation current per unit power increases with the temperature. The relation between these quantities is shown in the following table which gives the values for tungsten compiled from a paper by Dushman.[1] The second column gives the power in watts per cm.^{2} necessary to maintain the tungsten cathode at the temperature given in the first column, and the third column gives the saturation thermionic current in milliamperes per cm.^{2} of cathode surface. The numbers given under s in the last column are obtained by dividing I_s by p and can be taken as the thermionic efficiency of tungsten expressed in milliamperes per watt.

TABLE II

T Kelvin Degrees.	p Watts per Cm².	I_s Mils per Cm.².	s Mils per Watt.
1000	0.9	1.2×10^{-11}	1.25×10^{-11}
1500	6.9	6×10^{-4}	8.7×10^{-5}
1800	16.4	3×10^{-1}	1.8×10^{-2}
2000	26.9	4.2	1.6×10^{-1}
2100	34*	15.1	4.5×10^{-1}
2200	43*	48.3	1.12
2300	53*	137.7	2.6
2400	65*	364.8	5.6
2500	77.5	891.0	11 5
2600	90*	2044	22.7

* Obtained by interpolation.

For temperatures above 1800° K. the relation between the thermionic efficiency s and the power p per cm.^{2} expended in

[1] S. Dushman, G. E. Rev., Vol. 18, 156, 1915.

heating the filament can be expressed with a sufficient degree of accuracy by the simple equation

$$s = Cp^n \quad \ldots \quad \ldots \quad (24)$$

where C and n are constants.

From this it follows also that for the operating range of temperatures above 1800° K., we can instead of equation (7), Chapter III, use the simpler equation

$$I_s = Cp^{n+1}. \quad \ldots \quad \ldots \quad (25)$$

For tungsten the constants n and C have the values, $n = 4.13$ and $C = 1.812 \times 10^{-7}$ when s is expressed in milliamperes per watt, p in watts per cm.^{2} and i in milliamperes per cm.^{2}

The important quantity is s. In designing tubes it is not always necessary to know the temperature of the cathode. Instead of the temperature we can use the quantity p, because, after all, we are interested only in the power that must be expended in heating the filament to obtain the desired thermionic current. This power should not be made too high as this would decrease the life of the filament. If the desired life is known, we can, from a relation between the life and the power p dissipated per cm.^{2} of the filament, determine the value of p at which the filament must be operated. From the power that the tube is required to give in its plate circuit and the permissible voltage between filament and plate the saturation current can be determined. With the help of a table such as the above the required area of the filament can then be obtained. Once this area is known the length and diameter of the filament can be determined from its resistivity at the operating temperature. The length and diameter can, of course, be proportional to suit the voltage or current at which it is desired to operate the filament.

The most important of the factor that influences the thermionic efficiency of a filament is its electron affinity, i.e., the work which an electron must do to escape from the filament. A glance at the following table will show the enormous extent to which this constant can influence the saturation current. This table also gives an indication of the manner in which the saturation current is increased by increasing the temperature of the filament. The figures given in the table are only relative, and were computed from the equation

$$\log I_s - \frac{1}{2} \log T = \log A - \frac{.4343 \times 10^5}{8.6} \phi, \quad . \quad . \quad . \quad (26)$$

where A was put equal to 10^7. This is the logarithm equivalent of Richardson's equation of thermionic emission. (See equations (9) and (10), Chapter III.)

TABLE III

Temperature of Cathode.	Saturation Current in Amperes per Cm.² of Cathode Surface.			
	$\phi = 2$ Volts.	$\phi = 3$ Volts.	$\phi = 4$ Volts.	$\phi = 5$ volts.
° K.				
1000	25×10^{-3}	2×10^{-7}	2×10^{-12}	2×10^{-17}
1500	72	3×10^{-2}	13×10^{-6}	6×10^{-9}
2000	4×10^3	12	36×10^{-3}	1.1×10^{-4}
2500	4.6×10^4	43	4.2	4×10^{-2}

If we take two filaments having different values of electron affinity, but both filaments having the same area and thermal emissivity, place them in identical tubes and dissipate the same power in them, they would give the current voltage curves OA_1B_1 and OA_2B_2 (Fig. 27) where ϕ_2 is less than ϕ_1. If it is not necessary for the filament to give a greater saturation current than that given by A_1B_1 the filament with the smaller value of ϕ still offers a great advantage because in such case we could operate the filament at a lower temperature. The power saved in lowering the temperature could then be used in increasing the length of the filament. This would increase the total space current and the characteristic would take the form OHB_1. It will be shown later that the steepness of the characteristic is a very important factor in determining the efficiency of thermionic amplifiers, detectors, etc. The curve OH is therefore more suitable than OA_1.

It is to be seen, therefore, that it is very desirable to use a filament with as low an electron affinity as possible. This is obtained in the Wehnelt [1] cathode, which consists of a platinum filament coated with an oxide of the alkaline earths.

The type of filament used in Western Electric tubes is the result of efforts to reduce the electron affinity. A comparison of

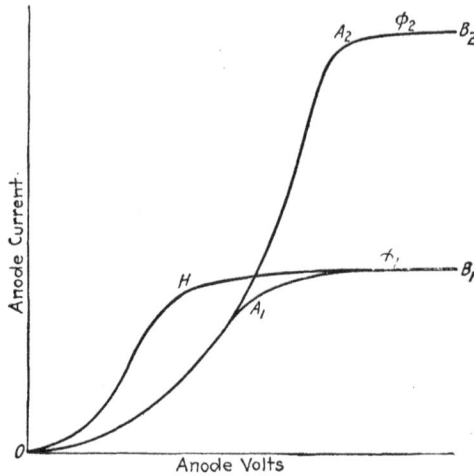

FIG. 27.

table II for tungsten filaments with the following, which gives the values for a type of Western Electric filament,[2] shows the relative thermionic efficiencies of the two types.

TABLE IV

p Watts per Cm².	I_s Mils per Cm².	$S = \dfrac{I_s}{p}$ Mils per Watt.
4	11	2.7
5	35	7
6	80	13
7	160	23
8	300	37.5
9	500	55.5
10	750	75

[1] A. WEHNELT, Ann. d. Phys., Vol. 14, p. 125, 1904.
[2] From measurements of C. J. DAVISSON.

The values for p, s and I_s given in this table also obey equations (24) and (25) but the constants n and C for this filament have the values: $n = 3.59$ and $C = 2.148 \times 10^{-2}$.

It is on account of the high thermionic efficiency that the oxide-coated platinum filament can be operated at such low temperatures. These filaments should never be heated above a reddish yellow (which corresponds approximately to $p = 8$ to 9 watts per cm.^{2}), whereas a tungsten filament can be heated to brilliancy.

The experimental verificaton of Richardson's equation for the Wehnelt type of filament presents greater difficulties than in the case of pure metallic filaments. For metallic filaments this equation was verified by Richardson in 1903,[1] and subsequently by several others. Accurate determinations of Richardson's constants for tungsten, molybdenum and other metals were made in the laboratories of the General Electric Company. The earliest experiments that were made to determine the electron affinity for oxide-coated filaments were those of Wehnelt.[2] Richardson's equation has been fully verified for oxide-coated filaments by investigations carried on in the research laboratories of the Western Electric Company. Some of this work is described by H. D. Arnold.[3]

The coated type of filament has now been used by the Western Electric Company since 1913, and has been in commercial use in the telephone repeater tubes of the Bell Telephone System since 1914. This filament is sufficiently constant in its behavior to meet the very rigid requirements called for by its use on the long distance telephone lines.

It consists of a core of platinum-iridium (6 per cent iridium with other impurities found in commercial platinum-iridium) covered with the oxides of barium and strontium. These oxides are applied alternately and after each application the filament is momentarily raised to a temperature of about 1000° C. The whole process consists of sixteen such applications. After that the filament is baked at about 1200° C. for two hours. If the

---

[1] O. W. Richardson, Trans. Roy. Soc., Vol. A-201, p. 497, 1903.

[2] A. Wehnelt, Ann. d. Phys., Vol. 14, p. 425, 1904. For a full discussion of these and similar experiments, see O. W. Richardson, " The Emission of Electricity from Hot Bodies " (Longmans, London).

[3] H. D. Arnold, paper read at the Chicago meeting of the American Physical Society, October, 1919.

filament is not exposed to moisture or carbon dioxide it does not deteriorate. If kept in vacuum containers they show no deterioration over a period of several years. The tubes containing these filaments are completely interchangeable even in repeater circuits where the requirements are held within very close limits.

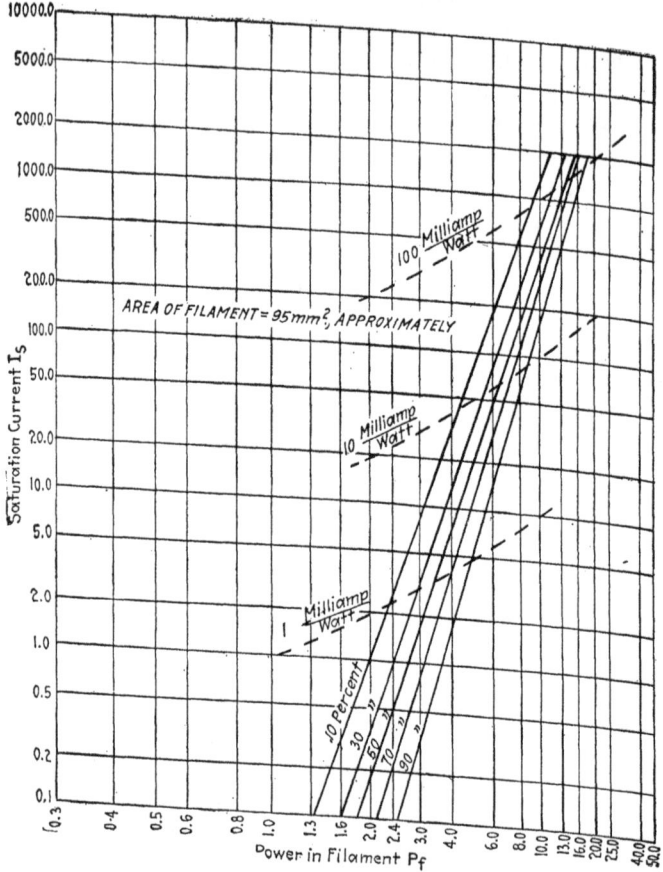

Fig. 28.

The investigation on the thermionic efficiency of the filaments was simplified by a coordinate system devised by Dr. C. J. Davisson, in which the abscissæ represent power supplied to the filament, and the ordinates the thermionic emission. The abscissæ of this system are curved, the coordinate lines being so proportioned that if the emission of the filament satisfies Richardson's equation, and the thermal radiation the Stefan-Boltzmann law, the relation between the thermionic emission and the power supplied to the filament when plotted on this chart is a straight line. Such a chart is shown in Fig. 28. The lines represent the average thermionic emission for a large number of different filaments. These filaments all have the same area, namely, 95 sq. mm. The further the line lies to the left, the greater is the thermionic efficiency. Each line shows the percentage of tubes that have a higher thermionic emission than that indicated by the line. For most purposes it is necessary only to insure that the thermionic emission is greater than a certain value. The thermionic efficiency is obtained by dividing the ordinates by the abscissæ $\left(s=\dfrac{I_s}{p}\right)$. The broken lines represent the lines of constant thermionic efficiency, the corresponding thermionic efficiencies indicated on this line being expressed in milliamperes per watt. The normal power dissipated in this standard coated filament is from 8 to 9 watts per square centimeter. From this it is seen that the efficiency of these filaments range from about 10 to 100 milliamperes per watt.

The constants of Richardson's equation $\left(I_s=A\,T^{1/2}e^{-\frac{b}{T}}\right)$ can be determined directly from these lines.

The following table gives the constants a and b of Richardson's equation for a number of different substances. The values for the Western Electric oxide-coated filament were obtained by C. J. Davisson from measurements covering about 4000 filaments.[1] The values for the other substances given in the table are taken from a paper by Langmuir.[2]

TABLE V

Substance.	A Amps/Cm².	b Kelvin Degrees.
Oxide coat (W. E. Standard).	$(8-24)\times10^4$	$(1.94-2.38)\times10^4$
Tungsten.	2.36×10^7	5.25×10^4
Thorium.	2.0×10^8	$3.9\ \times10^4$
Tantalum.	1.12×10^7	5.0×10^4
Molybdenum.	2.1×10^7	5.0×10^4

[1] H. D. ARNOLD, loc. cit.

[2] I. LANGMUIR, Trans. Am. Electrochem. Soc., Vol. 29, p. 138, 1916.

The thermionic efficiency is determined mainly by b; the smaller b the greater the efficiency.

To obtain the electron affinity ϕ the equation $\phi = 8.6 \times 10^{-5} \times b$ (Chapter III, equation (9)) can be used.

33. Life of a Vacuum Tube. The life of a tube is determined mainly by two factors:

(a) There is always a small amount of ionization by collision even in highly evacuated tubes. The positive ions so formed bombard the filament and this causes excessive local heating. In the three-electrode type of tube the grid acts as a partial screen to positive ion bombardment. The electric field in the region between grid and plate is usually much greater than between grid and filament. Most of the ionization, therefore, takes place between grid and plate and a large percentage of the resulting positive ions go to the grid instead of to the filament, since the grid is always negative with respect to the plate.

(b) The rate at which the filament volatilizes increases with its temperature. In the case of the metallic filaments, the volatilization causes the filament gradually to get thinner and so increases its resistance. If the filament is operated at constant voltage this will cause a reduction in the heating current, and the consequent lowering of the temperature lowers the thermionic emission as well as the thermionic efficiency. If the filament is operated at constant current the voltage increases, resulting in an increase of the temperature of the filament. This shortens the life of the filament. Whether the filament be operated at constant voltage or constant current, both effects are undesirable and must be taken into consideration in estimating the life of the filament.

The life of a metallic filament depends also on its diameter.[1] A 5-mil tungsten filament operated at a temperature of 2400° K. has a life of about 4000 hours, while the 10-mil filament operated at 2500° K. has a life of nearly 3000 hours. The thicker the filament the longer the life for the same operating temperature. Or, the same length of life can be obtained by operating the thicker filament at a higher temperature and so obtain a greater thermionic emission, as well as a higher thermionic efficiency, since the thermionic efficiency increases with the temperature.

The following table taken from Dushman's paper gives an idea of the effect of the diameter of the filament on its life:

[1] S. DUSHMAN, General Electric Review, Vol. 18, p. 156, 1915.

TABLE VI

Filament Diameter, Mils	Safe Temperature (Life > 2000 Hrs.).	I_s per Cm. length.	Watts per Cm. Length.
5	2475	30	3.1
7	2500	50	4.6
10	2550	100	7.2
15	2575	200	11.3

By " safe temperature " here is meant a temperature which is low enough to insure a life of at least 2000 hours. The quantities given in the third column give the thermionic emission per centimeter length of filament at the corresponding temperature, and the fourth column gives the power that must be expended in maintaining a centimeter length of the filament at that temperature. The thermionic efficiency can be obtained by dividing the values in the third column by those in the fourth. It is seen that the thermionic efficiency of the 15-mil filament is almost twice that of the 5-mil filament when both are operated at such temperatures as to give approximately the same life.

The coated type of filament retains a constant resistance throughout its life, because the heating current in this filament is carried mainly by the core, while what evaporates is mostly the coating. The nearing of the end of this filament is indicated by an increase in the temperature over sections of its length. These are commonly referred to as " bright spots." This warning is a desirable and important feature, especially where the tube is used as a telephone repeater, because it makes possible a timely replace‑ ment of the tube without interrupting the service.

Tubes containing the standard Western Electric filament have a life of several thousand hours, which depends, of course, upon the temperature at which the filament is operated. Such tubes have been operated in the laboratory for 20,000 hours continuously, during which period the thermionic current remained practically constant.

CHAPTER V

INFLUENCE OF GAS ON THE DISCHARGE

THE discussion given in the previous chapter and the current voltage relations that were obtained were based on the assumption that the residual gas in the device has a negligibly small influence on the discharge. It now remains to show under what conditions this assumption is justified and how these conditions can be realized in practice. It is important to know what are the sources of gas in thermionic tubes; how the gas influences the discharge and how the deleterious effects of gas can be eliminated.

There are two principal ways in which the presence of gas in a thermionic tube can affect the discharge. Firstly, gas in contact with the surface of a cathode can change the thermionic emission from the cathode and so change the saturation current, i.e., the total current obtainable from it at a definite temperature. This effect may be referred to as the *surface effect*. Secondly, the presence of gas in the space between cathode and anode will, if the velocity of the electrons coming from the cathode exceeds a certain small value, depending on the nature of the gas, give rise to the phenomenon of ionization by collision. This can be referred to as the *volume effect*.

34. Volume Effect of Gas. Ionization by Collision. In order to explain the effect of ionization by collision on the discharge, we shall assume that we have a characteristic corresponding to that obtained in a perfect vacuum and then see how this characteristic is changed when gas to a sufficiently high pressure is introduced into the tube. We shall also assume that the gas which is introduced is entirely neutral as regards the surface effect; that is, it is of such a nature that its coming in contact with the surface of the cathode does not change the electron emission from the cathode. In passing from cathode to anode, some of the electrons collide with the molecules of the gas and if they strike the molecules with a velocity exceeding a definite minimum amount ionization by collision sets in. The voltage through which an electron must drop to acquire this minimum velocity is called the *ionization voltage,* values of which are given on page 22.

If the voltage between cathode and anode is slightly greater than the ionization voltage, then, if an electron collides with a gas molecule just before reaching the anode, ionization will result, but a collision in the space nearer to the cathode will not result in ionization. In the latter case the electron may be reflected without any loss of energy from the molecule with which it collides, or it may lose part or all of its energy, this energy being transferred to the molecule, or it may combine with the molecule, thus forming a heavy negative carrier. It can readily be seen that if the voltage between cathode and anode be increased, collision of the electrons with molecules nearer to the cathode may result in ionization, and if the voltage just exceeds twice the ionization voltage, an electron which collides after having dropped through the ionization voltage in moving from the cathode, thus producing ionization, stands a chance of ionizing another molecule with which it may happen to collide just before reaching the anode. For low voltages, therefore, it is to be expected that the amount of ionization would increase with the applied voltage.

In practice we do not deal with a single electron moving from cathode to anode but with a stream of electrons, and under such conditions it is generally found that ionization sets in at applied voltages less than the ionization voltage. It is, for example, possible to maintain an arc in a gas or vapor by the application of a voltage which is not as great as the ionization voltage of the gas or vapor. This is because it takes a smaller amount of energy to disturb the equilibrium of an atom than it does to completely detach an electron from an atom. Once the equilibrium of an atom has been disturbed the potential energy of the atomic system is increased by an amount equal to the energy given up to the atom by the colliding electron. Such an atom is more easily ionized than the normal atom and therefore the potential difference through which any electron must drop in order to ionize this atom is less than the ionization voltage of the normal atom.[1]

The amount of ionization depends also on the pressure of the gas. The pressure of the gas may be so low that the electron does not strike a molecule at all in its flight from cathode to anode. On

[1] H. J. VAN DER BIJL, Phys. Rev., Vol. 10, p. 546, 1917.

the other hand, the pressure may be so high that the electron collides before it has acquired sufficient energy to ionize. The amount of ionization produced in this case depends on whether or not the gas is such that the collisions are elastic. If they are elastic the electrons will rebound from the molecules without losing their energy and may then strike the next molecules with a greater amount of energy than the first. If the collisions are inelastic the electrons lose some or all of their energy on colliding, but the energy which is transferred to the molecules is again radiated from them in the form of light, which causes photo-electric effects in the tube, resulting in a further dislodgment of electrons.

35. Mean Free Path of Electrons in Gases. The chance that an electron has of colliding with a gas molecule in its passage from cathode to anode depends on the mean free path of the electrons in the gas and upon the distance between cathode and anode. The mean free path is the average distance through which an electron can move freely without colliding with gas molecules. The following table gives an idea of the nature of this important quantity. The first column gives the number N of electrons, out of a total of 100 starting from the cathode, that can move freely through the distance d given in the second column of the table. The numbers in the second column are expressed in fractions of the mean free path L.

N	$\dfrac{d}{L}$
99	0.01
98	0.02
90	0.1
82	0.2
78	0.25
72	0.333
61	0.5
37	1.
14	2
5	3
2	4
1	4.6

This table snows that if the distance between cathode and anode is equal to the mean free path, only 37 per cent of the electrons starting from the cathode will strike the anode without having encountered molecules on their way, and as the ratio of the distance between cathode and anode to the mean free path is increased the number of collisions increases. On the other hand, if the pressure is so low that the mean free path is 100 times the distance between cathode and anode, only 1 per cent of the electrons will collide before reaching the anode.

The mean free path increases as the gas pressure is decreased; in fact, it is inversely proportional to the pressure of the gas. It also depends upon the size of the molecules. Thus it is greater for hydrogen than for oxygen. Consequently, since an electron is much smaller than a gas molecule, the mean free path of electrons in gases is greater than the mean free path of the gas molecules themselves. In order to obtain the mean free path of electrons in the gas from the mean free path of the gas molecules in the gas itself, we must multiply by the factor $4\sqrt{2}$. If L is the mean free path of the gas molecules at atmospheric pressure (760 mm. of Hg), then the mean free path of electrons in that gas, at a pressure p is:

$$l = 760 \times 4\sqrt{2}\frac{L}{p}, \quad \cdot \quad \cdot \quad \cdot \quad \cdot \quad \cdot \quad \cdot \quad (1)$$

where p is given in millimeters of Hg.

The mean free path for most of the common gases is given in tables of physical constants.[1]

The mean free path of a gas or vapor can be obtained if the coefficient of viscosity is known. The coefficient of viscosity is given by the equation

$$\eta = .31\rho\bar{c}L; \quad \cdot \quad \cdot \quad \cdot \quad \cdot \quad \cdot \quad \cdot \quad (2)$$

where

η = coefficient of viscosity;
ρ = density of gas;
$\bar{c}$ = mean molecular velocity;
L = mean free path;

the quantities being reduced to atmospheric pressure. Now, the pressure P is given by

$$\left.\begin{array}{l} P = \tfrac{1}{3}\rho\bar{c}^2 \\[2mm] \therefore \bar{c} = \sqrt{\dfrac{3P}{\rho}} \end{array}\right\} \quad \cdot \quad \cdot \quad \cdot \quad \cdot \quad \cdot \quad (3)$$

[1] See, for example, "Physical and Chemical Constants," by G. W. C. Kaye and T. H. Laby.

The mean free path for any gas for which the coefficient of viscosity is known can be obtained readily from the known mean free path and coefficient of viscosity of some other gas. For example, from equations (2) and (3) we obtain

$$L = \frac{\eta}{.31\sqrt{3}P\rho}, \qquad \cdots \cdots \quad (4)$$

and therefore if L_2 is the known mean free path of one gas at atmospheric pressure, the mean free path L_1 for the other gas at the same pressure is given by

$$\frac{L_1}{L_2} = \frac{\eta_1}{\eta_2}\sqrt{\frac{\rho_2}{\rho_1}} = \frac{\eta_1}{\eta_2}\sqrt{\frac{M_2}{M_1}}, \qquad \cdots \cdots \quad (5)$$

where M_1 and M_2 are the molecular weights of the two gases considered. The mean free path of the electrons in the gas at some other pressure can then be obtained from equation (1).

36. Ionization at Low Pressures. The application of the theory of ionization by collision when the pressure is of such order of magnitude that the mean free path is large compared to the distance between the electrodes is simpler than when the mean free path is of the same order as, or less than, the electrode distance. The relation between ionization current and the pressure is also simpler. Let us consider the case in which the pressure of the gas in the tube is so low that the mean free path is large compared with the distance between cathode and anode. If p is the pressure in millimeters of Hg and N the number of gas molecules per cubic centimeter at atmospheric pressure, the number of molecules per cubic centimeter at the pressure p is $\frac{pN}{760}$.

Let us suppose that cathode and anode are both in the form of infinitely large parallel plates, and let the number of electrons moving away from 1 square centimeter of cathode surface per second be n_1. In moving from cathode to anode some of these electrons will collide with the gas molecules. If the voltage between cathode and anode be so high that every collision results in ionization, the number of positive ions formed will be equal to the number of collisions, and since the mean free path is large compared with the electrode distance, the chance of an electron colliding more than once on its way to the anode will be extremely small. We can therefore imagine the molecules in the space projected on the plane of the anode and compute the ratio of the area covered by the molecules to area of the anode. This will then be proportional to the ratio of the positive ions formed by collision to the number of electrons moving from cathode to anode. Now, the cross-sectional area formed by the $\frac{pN}{760}$ molecules is $\frac{\pi r^2 pN}{760}$ where r is the molecular radius. Hence, if n_2 be the number of positive ions resulting from collisions in a column 1 square centimeter in cross-section, and n_1 the number of electrons moving to the anode per second from 1 square centimeter of cathode surface, we have

$$\frac{n_2}{n_1} = \frac{1}{k} \frac{\pi r^2 pN}{760}$$

or

$$p = \frac{760k}{\pi r^2 N} \cdot \frac{n_2}{n_1}, \quad \cdots \cdots \quad (6)$$

where k is a constant which becomes unity if all the molecules in the path of the electron stream are ionized.

The pressure p of the gas is therefore directly proportional to the ratio of positive ions to electrons. This linear relation has been observed experimentally by O. E. Buckley [1] and on the basis of this simple relationship he designed a thermionic gauge for the measurement of pressures below about 10^{-3} mm. of Hg. This gauge is described in Chapter X, page 375.

37. Effects of Ionization by Collision. When ionization takes place, the characteristic can be influenced in the following ways:

(a) The splitting of the gas molecules by bombardment of the electrons, in their passage from cathode to anode, results in the production of more dislodged charges so that the current is increased. This increase in current is small under the conditions prevailing in most thermionic tubes. Thus, if the pressure in the tube is 0.1 micron,[2] the increase in current due to this cause alone is less than 1 per cent.

(b) The positive ions resulting from the collisions move toward the cathode and since the total space charge is the difference between that due to the electrons and that due to the positive ions, the presence of the positive ions naturally reduces the total

[1] O. E. Buckley, Proc. National Academy of Science, Vol. 2, p. 683, 1916.
[2] 1 micron $= 10^{-3}$ mm. of Hg.

space charge, and this causes an increase in the current. The extent to which the space charge of the positive ions can reduce the negative space charge of the electrons depends on the number of positive ions compared with the number of electrons in the space at any particular time. It depends, therefore, on the speed with which the positive ions move toward the cathode; the lower the speed the greater will be the density of the positive ions. Thus when the tube contains oxygen the reduction in the negative space charge is greater than would be the case if the tube contained hydrogen at such a pressure that the number of positive ions formed is the same, because the oxygen ions are heavier and move more slowly than the hydrogen ions.

(c) The positive ions can, under certain conditions, combine with the electrons at the surface of the cathode and so form a layer of gas on it. This results in a surface effect which will be explained in Section 39.

(d) There is still another way in which ionization by collision can affect the operation of the tube. If the voltage between cathode and anode is sufficiently high, the bombardment of the cathode by the positive ions causes an abnormal heating of the cathode. This increases the saturation current because of the increase in the temperature of the cathode. There may also be a direct emission of electrons from the cathode under the bombardment of its surface by the positive ions. The undue heating caused by the bombardment wears away the cathode and has a very deleterious effect on the life of the tube.

(e) When the velocity of the electrons is less than the value necessary to cause ionization by collision, the electrons attract the neutral gas molecules and so form heavy negative carriers. The ease with which this formation of negative carriers takes place depends on the nature of the gas. Such gases as argon and mercury vapor do not readily form negative carriers, while hydrogen and oxygen combine with electrons more easily. The effect of this negative carrier formation is to counteract the reduction in the negative space charge occasioned by the heavy positive ions formed by collision ionization. The positive ions are atoms of the gas from which one or more electrons have been removed. The ions therefore have very nearly the same weight as the gas atoms. The negative carriers, on the other hand, may consist of an atom or molecule to which has been attached an electron. It is also possible that the attraction between an electron and the neutral gas molecules can result in the formation of clusters consisting of more than one molecule held together by the electron. These negative carriers, therefore, move as slowly as, and sometimes more slowly than, the positive ions and consequently have a relatively great effect in counteracting the tendency of the positive ions to reduce the negative space charge of the electrons.

38. Influence of Ionization on the Infra-Saturation Part of the Characteristic. To confine our attention to this part of the characteristic, let us assume that the tube contains gas which has no effect on the electron affinity of the cathode; in other words, it has no effect in either reducing or increasing the saturation current obtainable from the filament at any definite temperature. This condition can be realized, for example, with a tube containing a tungsten cathode and mercury vapor because mercury vapor is neutral as regards the electron emission from tungsten.[1] Let us suppose that we measure the current voltage relation of a tube containing a tungsten cathode and a tungsten anode, and to which is attached an appendix containing mercury which can be maintained at any desired temperature thus maintaining the pressure of the mercury vapor at any desired value. It is, of course, to be understood that all other gases and vapors have been driven out of the electrodes and the walls of the vessel. It will be appreciated that by doing this we do not simulate the conditions obtaining in practical thermionic tubes, because the gases remaining in the practical tubes seldom, if ever, contain mercury vapor, but do comprise those gases that we are excluding from the experiment at present under consideration. For such an experiment it is desirable to use a vapor which can be maintained at any pressure by keeping the parent substance at the corresponding temperature.

Let us first suppose the appendix containing the mercury is immersed in liquid air. Under such conditions the characteristic will be that obtained in a high vacuum. Then, to study the effect of the mercury vapor in increasing the current on the infra-saturation part of the characteristic we can, instead of using liquid air, maintain the mercury at other temperatures by dipping the mercury tube into freezing mixtures or water baths. A set of such characteristics is shown in Fig. 29. The characteristic marked

[1] IRVING LANGMUIR, Physik. Zeitsch., Vol. 15, p. 519, 1914.

No. 1 was obtained with the mercury tube immersed in liquid air. The other characteristics were obtained with the mercury at temperatures ranging from $-7.5°$ C. to $10°$ C. The temperatures below $0°$ C. were obtained by keeping the tube containing the

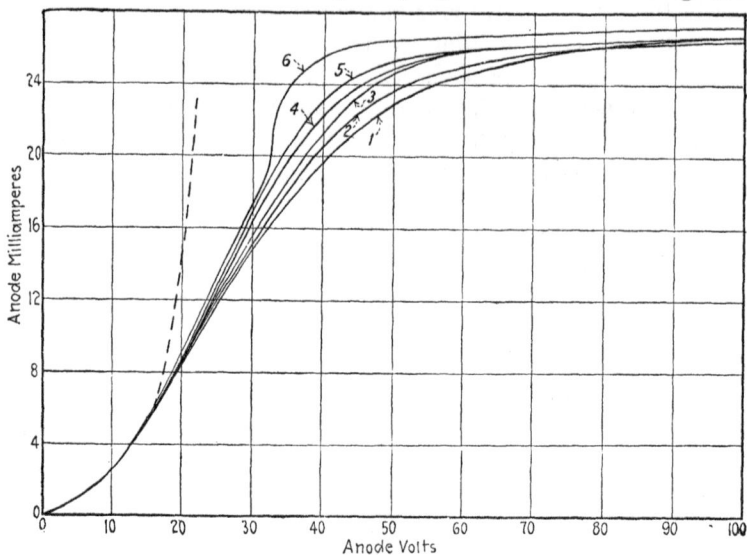

Fig. 29.

mercury immersed in ice and salt freezing mixtures. The following table shows the temperatures of the mercury and the pressures of the mercury vapor corresponding to the set of characteristics shown in Fig. 29.

TABLE VII

Curve No.	Temperature of Hg, ° C.	Pressure of Hg Vapor, Micron.
1	-185	0.009
2	-7.5	0.085
3	0.0	0.18
4	5.0	0.3
5	7.5	0.4
6	10.0	0.5

The pressures of the mercury vapor at these temperatures are obtained from a table given by Knudsen.[1] Referring to this table

[1] M. Knudsen, Ann. d. Phys., Vol. 29, p. 179, 1909.

and to Fig. 29, it will be seen that as the pressure of the mercury vapor is increased the current voltage curve becomes steeper and steeper. When the pressure rises to a value of 0.5 of a micron, the current shows a rather sudden increase at about 33 volts (curve 6). Below this pressure there is no sudden increase in current on the infra-saturation part. For a still higher pressure of the mercury vapor, about 2 microns or more, the current increases sharply at about 15 to 20 volts, as indicated by the dotted curve in Fig. 29.

At these pressures of the mercury vapor, the formation of the positive ions by collision of the electrons with the mercury molecules neutralizes the space charge of the negative electrons so that the current that can flow through the tube undergoes a considerable increase. For this neutralization of the negative space charge it is necessary, firstly, that the potential difference between cathode and anode be high enough to cause a considerable amount of positive ionization. In the case of mercury vapor, this voltage is of the order of 5 volts or less. Secondly, it is necessary that the number of positive ions formed by collision be great compared with the number of negative carriers formed by combination of the electrons with the neutral molecules. The formation of heavy negative carriers accounts at least partly for the fact that the negative space charge does not become neutralized until the voltage becomes considerably greater than the voltage through which the electrons must drop to produce ionization by collision. Thus, the current does not increase suddenly until the filament-plate voltage reaches about 15 volts or more, although actually a considerable amount of ionization by collision takes place at much lower voltages. The reason why the current does not increase when ionization takes place at these low voltages is because a relatively large number of negative carriers are formed by the combination of electrons with the neutral gas molecules, and these negative carriers tend to neutralize the space charge of the positive ions. Another factor which tends to counteract the reduction of the space charge of the electrons is the recombination of the positive and negative charges.

The extent to which the current is increased by ionization by collision depends of course not only on the pressure of the gas vapor, but also on the distance between cathode and anode. The important quantity that determines the amount of collision ionization is the ratio of the mean free path of the electrons in the gas or vapor to the electrode distance. Since the pressures corresponding to the curves shown in Fig. 29 are known, we can from these curves find the ratio of the mean free path to the electrode distance for the maximum pressure at which there is no sudden change in the characteristic due to the presence of the gas. These curves show that this maximum pressure of mercury vapor is about 0.4μ for the tube with which they were obtained.

Now, the effects as shown by these curves are much greater for mercury vapor than for the gases that commonly remain as residual gases in thermionic tubes. In order to get an indication of the extent to which the other gases increase the current on the infra-saturation part of the characteristic, we have to distinguish between mercury vapor and such gases in three respects: Firstly, mercury molecules are very heavy compared to the molecules of the ordinary gases, such as hydrogen, oxygen and nitrogen, and, therefore, have a greater effect in reducing the negative space charge. Secondly, the mean free path of electrons in mercury vapor seems to be considerably less than the mean free path of electrons in the common gases. Thirdly, such gases as oxygen and hydrogen show a greater tendency to form negative carriers by combining with electrons than mercury vapor. It is also possible that the rate of recombination of the positive ions with electrons is different for different gases.

In attempting to obtain an indication of the minimum value of the ratio of mean free path to the electrode distance necessary to give a characteristic which on the infra-saturation part is not influenced to a disturbing extent by the presence of gases other than mercury vapor, the differences mentioned above must be considered. Little is known with regard to the difference in the coefficient of recombination or the rate of formation of negative carriers in different gases and vapors. We can, however, obtain some indication of the maximum allowable pressure of such gases as hydrogen and oxygen by considering only the velocity of the ions and the mean free path of the electrons in the gas.

Suppose that the minimum value of the ratio of mean free path to electrode distance for mercury has been determined from a set of curves like that shown in Fig. 29. The extent to which positive ions reduce the negative space charge depends upon the velocity of the ions in the electric field, due to the potential difference between anode and cathode and is inversely proportional to it. Thus, if l_1 is the mean free path of the electrons in mercury vapor at the maximum permissible pressure, and d the distance between the electrodes, then the ratio of $\dfrac{l}{d}$ for any other gas is

$$\frac{l}{d} = \left(\frac{l}{d}\right)_1 \frac{v_1}{v} = \left(\frac{l}{d}\right)_1 \sqrt{\frac{M}{M_1}}, \quad \ldots \ldots \quad (7)$$

where M_1 and M are the molecular weights of mercury and the gas considered, and v_1 and v are the corresponding velocities of their ions under the same electro-static field. Take, for example, the case of hydrogen and mercury vapor, since the molecular weight of hydrogen is 2 and that of mercury 200, the permissible ratio of mean free path to electrode distance is one-tenth as great for hydrogen as it is for mercury. This does not, however, mean that when the tube contains hydrogen the pressure can be ten times as high as when it contains mercury vapor; it can, in fact, be still higher because the mean free path of electrons in hydrogen is greater than that of electrons in mercury vapor at the same pressure. The relation between the pressure and mean free path is shown by equation (1). If L and L_1 be the mean free paths of the gas and of mercury vapor at atmospheric pressure, and l and l_1 the corresponding mean free paths at the pressures p and p_1, then we have

$$\frac{p}{p_1} = \frac{Ll_1}{L_1 l}.$$

Substituting the value of $\dfrac{l}{l_1}$ from equation (7), we obtain:

$$\frac{p}{p_1} = \frac{L}{L_1} \sqrt{\frac{M_1}{M}}. \quad \ldots \ldots \quad (8)$$

From this equation the maximum allowable pressure p for any gas can be obtained from the maximum allowable pressure for mercury vapor. In the case of hydrogen, for example, we have $L = 18.5 \times 10^{-6}$ cm. at atmospheric pressure. L_1, the mean free path of mercury vapor at atmospheric pressure, can be taken to be about 6×10^{-6} cm.[1] If these values be inserted in equa-

---

[1] If the mean free path of mercury vapor is computed from equation (5) by putting its coefficient of viscosity equal to 162×10^{-6} (figure given by

tion (8) we find that if the tube contains hydrogen the maximum allowable pressure is about thirty times as high as when the tube contains mercury. When the tube contains oxygen or nitrogen the pressure can be about four times as high as in the case of mercury vapor.

For the practical operation of thermionic devices it is necessary that the current over the operating range should not show erratic changes. It will be apparent from the previous discussions that the pressure necessary to secure a discharge that is not appreciably influenced by gas is of such an order of magnitude that it can readily be obtained. But it is important also to maintain the pressure constant enough to prevent any appreciable changes in the effects of ionization on the characteristic. To secure this the electrodes and walls of the vessel must be freed of gas to such an extent that the heating of these parts during the operation of the tubes does not cause the evolution of enough gas to bring about such pressure changes. The part of the characteristic on which the great majority of thermionic devices operate is the infra-saturation part that we have discussed in the previous pages. The effect of gas on the saturation part of the characteristic will be discussed in the following section.

39. Effect of Gas on the Electron Emission. Surface Effect. It was shown in section 19 that the relation between the saturation thermionic current and the temperature of the cathode can be expressed by Richardson's equation:

$$I_s = A T^{1/2} e^{-\frac{b}{T}},$$

where A is a constant depending on the number of electrons per cubic centimeter of the cathode, b a measure of the work which an electron must do to escape from the cathode, and T the temperature of the cathode in absolute Kelvin degrees. If the vacuum in the tube is supposed to be perfect and the electrodes entirely void of gas, the constants A and b of the above equation have definite fixed values that are determined only by the nature of the cathode. Richardson's equation holds for any hot cathode and is

KAYE and LABY) it is found to be 3.5×10^{-6} cm. at atmospheric pressure. This value of the viscosity coefficient is obtained by extrapolation and possibly involves a considerable error. It is likely that the value 6×10^{-6} cm. for the mean free path of mercury vapor is more nearly correct.

not dependent on the structural dimensions of the device, it being understood that the thermionic current I_s is the current obtained from unit area of the cathode surface when the potential difference between cathode and anode is high enough to draw all the emitted electrons to the anode as fast as they are emitted from the cathode. It follows then that if the cathode contains impurities, Richardson's equation must still hold but the constants A and b will have different values. From the nature of the equation it is seen that while the current changes in proportion to a change in A, a small change in b causes a very considerable change in the current. It has been known for a long time that small traces of gas occluded in the cathode can cause large changes in the thermionic emission. H. A. Wilson [1] has found, for example, that when a platinum wire is freed of the hydrogen occluded in it, the thermionic current drops to a very small fraction of the value obtained from a platinum wire not so treated. J. J. Thomson [2] and O. W. Richardson have pointed out that the effect of the gas occluded in the surface of the cathode is to change the work necessary to detach an electron from the cathode, that is, to change the constant b in Richardson's equation and so produce relatively very large changes in the thermionic current. Thus, if $b = 5 \times 10^4$ ($\phi = 4.3$, see equation (9), Chapter III) and the temperature of the cathode is 2000° K., an increase in b of 25 per cent can decrease the current to approximately $\frac{1}{500}$ of its original value. Such changes can readily be produced by very small quantities of gas coming in contact with the cathode. The amount of gas that is necessary to produce great changes in the saturation part of the characteristic is often so small that its presence does not noticeably affect the infra-saturation part of the characteristic. This is shown, for example, in Fig. 30. The curves shown in this figure were obtained with a bulb containing two tungsten filaments, one of which was used as cathode and the other as anode. The bulb was not baked during the process of evacuation, so that a small trace of gas and water vapor remained in the tube. It is seen from curve 1 that the pressure of the residual gas was not sufficient to cause any appreciable increase in the current on the lower or operating part of the character-

[1] H. A. Wilson, Phil. Trans., Vol. 202, p. 243, 1903.
[2] J. J. Thomson, " Conduction of Electricity through Gases," 2d Ed., p. 202.

istic. As the voltage and current were increased, however, the heating of the bulb by the energy dissipated in the tube caused the liberation of a sufficient amount of gas to give the irregular curve, as evidenced at voltages higher than about 200 volts. The readings were obtained in the order indicated by the arrow. Curve 2 was obtained while the whole tube remained immersed in liquid air. It therefore represents the high vacuum characteristic.

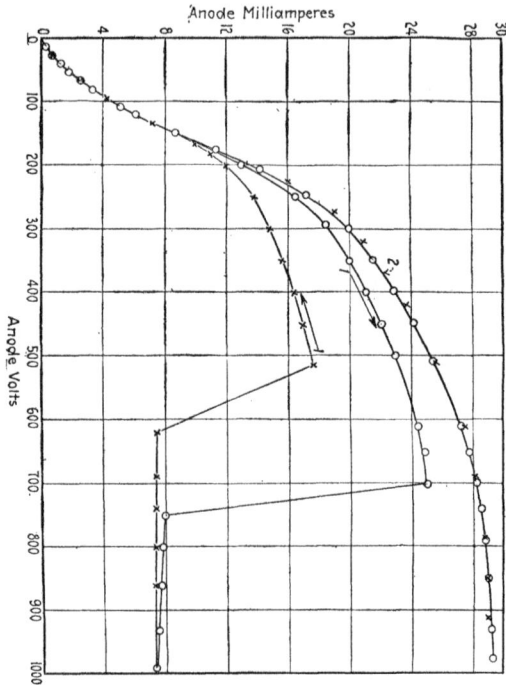

Fig. 30

The reduction in the electron emission caused by the presence of gas generally becomes more pronounced when ionization takes place, because this has the effect of directing the flow of gas towards the filament. When the gas is not ionized the electric field has no effect on the motion of its molecules, and the chance of their striking the surface of the filament is determined by he laws of the kinetic theory of gases applicable at low pressures. When ionization by collision takes place, however, the molecules in the space between cathode and anode become positive ions which are directed by the electric field towards the cathode where they recombine with the electrons into neutral gas molecules. When they recombine before reaching the cathode the resulting neutral molecules or atoms have an increased momentum in the direction of the cathode, due to the momentum acquired in the electric field while they were ions. Ionization by collision therefore causes more gas molecules to come in contact with the surface of the cathode. It has been found, for example, that the receiver type of three-electrode tubes, containing oxide-coated filaments and that are evacuated sufficiently well to operate very satisfactorily as amplifiers, may still contain a sufficient amount of gas to paralyze the tubes when operated as oscillation generators. (See Chapters VIII and IX.) When the tube is used as an amplifier most of the ionization of the gas takes place between the grid and the anode, and since the grid is negative with respect to the anode, the positive ions formed by collision ionization in this region are attracted to the grid. On the other hand, when the tube operates as an oscillation generator, the grid is subject to large potential variations, so that in this case positive ions are also formed in the space between filament and grid. These positive ions move to the filament and there combine with the electrons to form neutral molecules. If the gas remaining in the tube is of such a nature as to decrease the electron emission when coming in contact with the surface of the filament, this effect can easily become so large that the space current is reduced to practically zero. This accounts for the phenomenon that has been observed that a tube will start to oscillate and after a time, ranging from a fraction of a second to several seconds, the space current will drop to zero and the tube become inoperative. The normal condition of the tube can be restored readily by heating the filament to a higher temperature so as to drive off the gas. Generally it will recover automatically after a period of time depending upon the temperature of the filament. This period may range from a fraction of a second to several seconds or even minutes. The best way to prevent this paralysis of the tube is to evacuate it more thoroughly.

Langmuir [1] has made extensive investigations on the effects of gas on electron emission from tungsten filaments. Langmuir

---

[1] I. LANGMUIR, Phys. Rev., Vol. 2, p. 450, 1913; Phys. Zeitschr., Vol. 15, p. 516, 1914

finds that argon, mercury vapor and dry hydrogen have no direct effect on the emission of electrons from tungsten, but water vapor has a very large effect.

Pure dry nitrogen has been found by Langmuir to have no appreciable direct effect in reducing the electron emission when the amount of nitrogen left in the tube is so small that there is no appreciable ionization of the nitrogen molecules. But when the voltage is raised so high that ionization becomes appreciable the nitrogen ions can bombard the filament with sufficient velocity to combine with the tungsten. This causes a decrease in the electron emission from the tungsten. The higher the velocity with which the nitrogen ions strike the tungsten filament the greater seems to be the effect on the electron emission, so that in the presence of nitrogen the current at first increases in the manner shown in the characteristics of most thermionic devices, and then suddenly starts to decrease when the voltage is still further increased. Hence, instead of getting a curve which, for voltages higher than the saturation voltage, becomes substantially parallel to the voltage axis, the curve obtained at these voltages has a negative slope.

40. Influence of Occluded Gases. From the explanations given in Section 38 it follows that the influence of ionization by collision of the residual gases in a tube on the infra-saturation part of the characteristic is generally not disturbing for pressures lower than of the order of one-tenth to one micron. Such a pressure is easily obtained. Hence, as far as removing the gas in the space of the tube is concerned, there would be no difficulty in obtaining a sufficiently high vacuum to realize what may be called a " pure electron discharge." What is necessary, however, is to maintain the vacuum in the tube while it is in operation, and it is therefore necessary not merely to remove the gas from the volume of the tube, but also to free the electrodes and walls of the vessel of occluded gases to a sufficient extent. If the electrodes of a device remain cold during operation the occluded gases are not liberated very readily, but when the electrodes become hot during the operation the occluded gases are liberated. In all vacuum devices using hot electrodes it is therefore necessary previously to free the electrodes of gases. An incandescent lamp is such a device and therefore it has always been common practice, in evacuating incandescent lamps, to heat the bulbs and raise the filaments to abnormally high incandescence during the evacuating process. In thermionic devices usually only the filament is such that it can be heated during evacuation by passing a current through it. The other electrodes are heated by applying a positive potential to them which is sufficiently high to enable the electrodes to rise to high temperatures by the bombardment of the electrons coming from the hot filament.

The extent to which the electrodes and walls of the vessel must be freed of gas depends on the temperature to which these parts of the tube rise during operation. If, for example, a tube is designed to operate on voltages ranging, say, from 15 to 50 volts, such as is the case with the type of tube commonly used as detector and amplifier in radio-receiving stations, it is not necessary to evacuate the tube so well that it can also operate satisfactorily at much higher voltages. Such a tube while operating satisfactorily as a pure electron discharge device over the operating voltages stated above, may undergo a sufficient liberation of gas from electrodes to spoil the tube when the voltage is raised to, say, 100 volts or more. Tubes that are to operate on higher voltages and currents must have their electrodes more thoroughly freed of gas during the process of evacuation.

The way in which the characteristic is influenced by the liberation of gas when a tube is subjected to voltages higher than those for which it is designed is shown in Fig. 31. These curves were obtained with a standard Western Electric $VT1$ tube. It contains an oxide-coated filament and is designed to operate on voltages not higher than 100 volts. If the voltage is raised much beyond this value the electrodes become hot enough to liberate some of the gas occluded in them, and the amount of gas liberated increases as the potential difference between filament and plate is raised. This increases the amount of ionization by collision and causes the filament to be bombarded by the positive ions. The bombardment of the filament raises its temperature and increases the space current over the value that it would have if there were no positive ion bombardment. When the voltage becomes high enough the current increases rapidly. Such a rapid increase in the current is shown in the case of curve 1 (Fig. 31) to take place at about 400 volts, and in the case of curve 2 at about 300 volts. Tungsten filaments do not seem to be so sensitive to positive ion bombardment. It sometimes happens that the gas liberated from the electrodes has a very pronounced effect in reducing the electron emission from the cathode and then the current instead of increasing may decrease at the higher voltages.

If the voltage is raised to an excessive amount so much gas may be liberated as to spoil the tube permanently. On the other hand, the gas liberated can be cleaned up by the hot filament so that the tube automatically restores itself. This is especially the case with tungsten filaments.

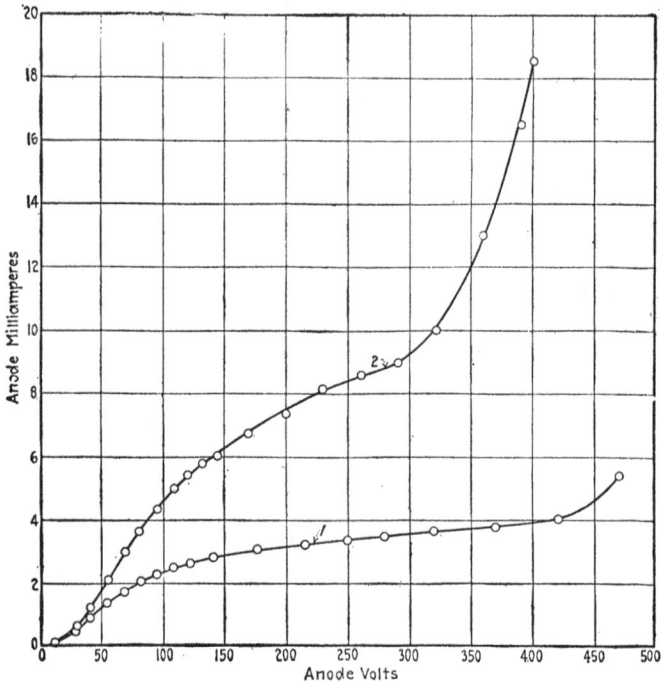

Fig. 31.

If the amount of gas liberated by applying an over-voltage is not excessive the tube will behave like a high vacuum tube on the lower or operating part of the characteristic, even after the gas has been liberated at the voltages corresponding to the saturation part. This is shown, for example, by Fig. 32, which represents a curve also obtained with a *VT*1 tube. When the voltage was raised to about 250 or 300 volts, the current began to increase rapidly, as shown by the continuous line. The broken line shows the currents obtained for decreasing voltage after the voltage had been raised to about 340 volts. It will be seen that on the upper part of the characteristic the currents obtained for decreasing voltages differ considerably from those obtained for increasing voltages. On the infra-saturation part of the characteristic, however, the two curves coincide very well, showing that the amount of gas in the tube is not sufficient to cause any appreciable deviation for voltages lower than about 100 volts.

Thermionic tubes as they are used to-day perform a large number of different functions, most of which require that the characteristic be steady and reproducible. This is, for example, the

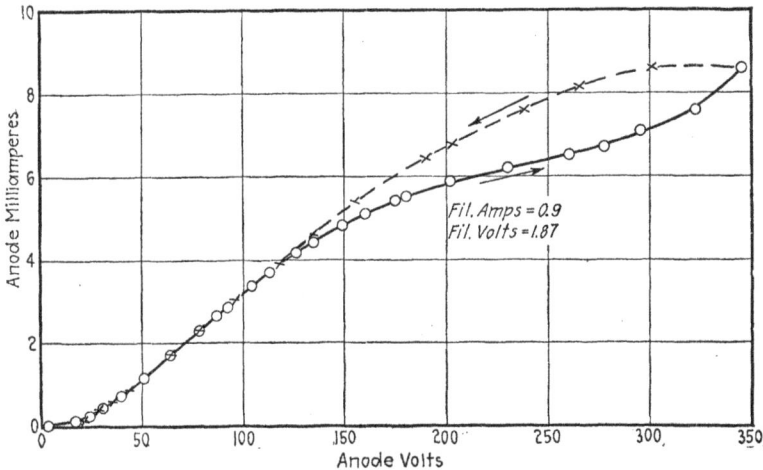

Fıg. 32.

case where the tube is used as a telephone repeater. In order to insure this the procedure commonly adopted in practice is to apply a potential for about a minute or so, to the plate, which is higher than the normal operating voltages, and then test the tube to see if it performs properly the function for which it is designed. This test is commonly referred to as the "over-voltage test." If the tube is not sufficiently well evacuated the gas is liberated from the electrodes during the time that the over-voltage is applied. If the tube still functions properly after the application of the over-voltage, it means that the amount of gas liberated was not sufficient to have any deleterious effect on the operating part of the characteristic. The difference between the over-voltage to be used in this test and the highest normal operating voltage depends on the margin of safety that it is desired to secure.

41. Ionization at High Pressures. The ionization phenomena encountered in thermionic tubes belong to a class where the mean free path of the electrons in the gas is generally great compared with the distance between the electrodes. The phenomena resulting from the discharge at such pressures that the mean free path is smaller than the distance between the electrodes are more complicated and show the effect of cumulative ionization.

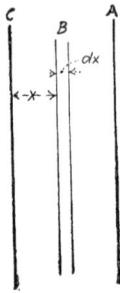

In order to show this in an elementary way let us consider what happens when the mean free path is so small that in passing from cathode to anode an electron has a chance of colliding several times with gas molecules. Let the number of electrons starting from the cathode be n_0, and let the number of electrons formed by collision ionization in traversing a distance x from the cathode be n. (See Fig. 33.) The total number of electrons arriving at a plane distant x from the cathode is then $n+n_0$. If each electron in moving through unit distance can dislodge α other electrons by collision, the number dislodged in a region of thickness dx at a distance x from the cathode will be·

$$dn = (n_0+n)\alpha dx.$$

To find the total number N arriving at the anode, we have to integrate this equation between the limits of $n=0$ when $x=0$, and $n_0+n=N$ when $x=d$, the distance between cathode and anode. This gives:

$$N = n_0 e^{\alpha d}.$$

This equation shows that the number of electrons reaching the anode, and therefore also the current through the tube, increases with increasing distance d between cathode and anode. This is in marked contrast to the discharge in high vacuum thermionic tubes in which, as was shown in the previous chapter, the saturation current is independent of the distance between cathode and anode, while the infra-saturation current decreases as the distance between cathode and anode is increased.

FIG. 33.

42. Difference between Gas-free Discharge and Arc Discharge.
There are other important differences between these two different types of discharge. The mercury arc is an example of a practical device which depends for its operation on ionization by collision, the gaseous medium being mercury vapor in equilibrium with the liquid mercury used as cathode. In a device containing a considerable amount of gas and cold electrodes the discharge will not pass unless the applied voltage be made so high that the few electrons in the space can cause a small initial ionization. The positive ions so formed bombard the cathode and give up sufficient energy to the cathode to enable the electrons to overcome the force of attraction at the surface of the cathode and so escape from it. The discharge may also be started, as is done in the mercury arc, by bringing the cathode in contact with an auxiliary electrode and then striking the arc by separating them. This furnishes the initial ionization necessary to start the discharge. Thus, while electrons are liberated from the cathode in the pure electron device simply by external heating of the cathode such as passing a heating current through it, in the gas-filled tube the electrons are liberated by bombardment of positive ions and also by photoelectric effects in the tube.

The positive ions formed by collision ionization move toward the cathode and the electrons toward the anode. There is, thus, a predominance of positive space charge in the neighborhood of the cathode and a predominance of negative space charge near the anode. When the conditions are such that an arc discharge passes, the total space charge is small compared with that in the gas-free tube, where the space charge is negative only and has a maximum value near the cathode. The resistance of the arc is therefore lower than that of a gas-free tube. In order to maintain an arc steady it is necessary to connect it in series with an external resistance. The gas-free tube, on the other hand, does not need an external resistance to stabilize the discharge. On account of the frequent collisions in an arc there is also a great deal of recombination and this causes a pronounced blue glow in the tube. The gas-free tube, on the other hand, shows no blue glow. If a blue glow does accidentally appear, it is because the tube has been over-taxed and it may cause the tube to become inoperative.

Another important difference between a pure electron discharge and an arc discharge is that the latter has a " falling characteristic "; that is, its relation between current and voltage is given by a curve such as AB (Fig. 34). The gas-free device, on the other hand, has a characteristic similar to OC. The difference between tubes containing these characteristics becomes apparent when we consider the corresponding a-c. resistances. The a-c. resistance for small current or voltage variations at any definite voltage is given by the reciprocal of the slope of the characteristic at a point corresponding to that voltage. Since the slope of the curve AB is negative, the arc has a negative resistance, while the resistance of a gas-free tube is positive.

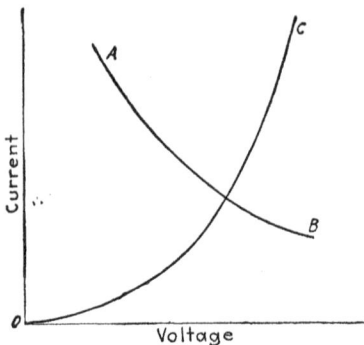

Fig. 34.

It is the negative resistance of the arc which enables it to produce sustained oscillations. It will be shown in Chapter VIII that a device containing only two electrodes can only produce sustained oscillations if it has a negative resistance or a falling characteristic. The principle involved in the production of sustained oscillations by the audion or three-electrode thermionic tube is entirely different and depends on the controlling action of the grid on the electron flow from filament to anode.

CHAPTER VI

RECTIFICATION OF CURRENTS BY THE THERMIONIC VALVE

43. Conditions for Rectification. Let us consider a device on which can be impressed a simple harmonic voltage and let the current through the device be any function $f(e)$ of the voltage, This function can always be expressed in a Fourier series, thus:

$$f(e) = I_0 + \sum_1^n a_n \sin nx + \sum_1^n b_n \cos nx, \quad \ldots \quad (1)$$

where I_0, a_n and b_n are constants.

The summation terms of this series are simple harmonic functions, and will therefore vanish when integrated over a complete period. On the other hand I_0, being a constant, will be independent of such integration and can be measured with a d. c. measuring instrument. Hence, provided that I_0 be not zero, the function $f(e)$ will be such that the device will rectify. The fundamental condition for rectification by any device is therefore·

$$\int_0^T f(e) dt \neq 0. \quad \ldots \quad \ldots \quad (2)$$

This will be the case either when

$$\int_0^{\frac{T}{2}} f(e) dt = 0 \quad \text{or} \quad \int_{\frac{T}{2}}^T f(e) dt = 0 \quad \ldots \quad (3)$$

or when

$$\int_0^{\frac{T}{2}} f(e) dt \gtrless \int_{\frac{T}{2}}^T f(e) dt. \quad \ldots \quad \ldots \quad (4)$$

Devices which comply with condition (3) are: (1) those which conduct current only in one direction and for which $f(e)$ may be any function of e in the transmission half period as, for example, the thermionic rectifier (Fig. 35); (2) those which conduct current only in one direction and for which $f(e)$ is any finite function of e for all values of e greater than a minimum value e_1. The electrolytic rectifier practically complies with this condition;

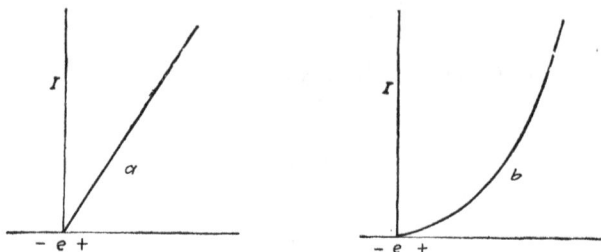

FIG. 35.

during the transmission half period it does not conduct unless the applied voltage exceeds its back E.M.F. (Fig. 36.)

Devices which comply with condition (4) are: (a) those which conduct current in both directions but for which $f(e)$ is unsymmetrical with respect to the axis of current (Fig. 37); (b) those for which $f(e)$ is a linear function of e, provided the input voltage exceeds a minimum value e_1 (Fig. 38).

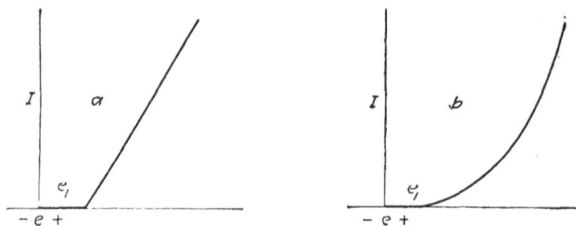

FIG. 36.

The three-electrode thermionic detector, or audion, cannot be called a rectifier, as far as the plate current is concerned, because it does not rectify the incoming current. This current only releases energy in the plate circuit which is supplied by the local plate battery, and the characteristic of the device is such that more energy is released during the one-half period than during the other. It will be seen that devices represented by the conditions (3) can be made to comply with condition (4) by inserting a local battery in the rectifier circuit so as to shift the axis of current. But even with this modification they differ from the audion detector because they actually resolve the *incoming currents* into d-c.

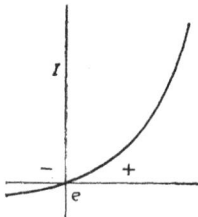

Fig. 37.

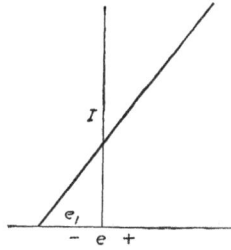

Fig. 38.

and a-c. components in the measuring circuit. The three-electrode device or audion therefore differs radically from these other types of radio detectors.

A full discussion of the operation of the various types of rectifiers is beyond the scope of these pages. We shall therefore confine our attention to the thermionic rectifier or valve.

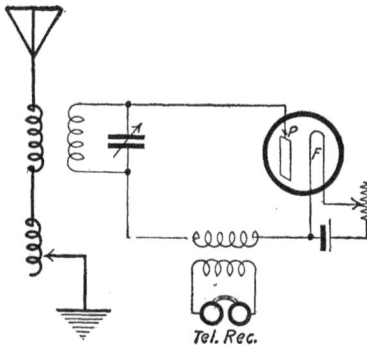

Fig. 39.

44. The Fleming Valve. This device satisfies condition (3) and has a characteristic such as that shown in Fig. 35 (*b*). It consists of a filament which can be heated to incandescence and a plate, both placed in a highly evacuated bulb. In 1905 Fleming[1] recognized the use of the rectifying properties of this device for the indication of high frequency oscillations, and used it as a

[1] J. A. FLEMING, Proc. Roy. Soc., Jan., 1905, p. 476; U S. Pat., 803, 684.

radio detector. The circuit in which Marconi used this device as a radio detector is shown in Fig. 39.

45. Valve Detector with Auxiliary Anode Battery. By our present standard of measurement the two-electrode tube is a very inefficient detector. It can be, and has been used more efficiently by operating on a chosen point of the current-voltage characteristic, thus making it fall in the class of rectifiers given by condition (4) instead of that represented by condition (3). This can be done by inserting a local battery in the circuit as shown in Fig. 40.[1]

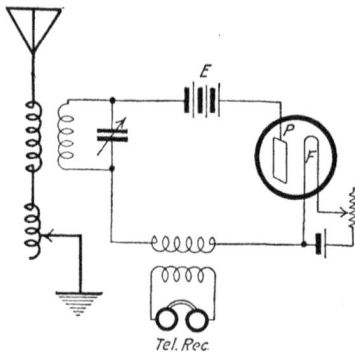

Fig. 40.

The operation of the device when used this way can be understood from the following: By the insertion of the battery E there is established in the circuit FPE a constant direct current which has a finite value even when no oscillations are impressed from the antenna. The current through the device can therefore be represented by a function of the form

$$I+i=f(E+e \sin pt), \quad \ldots \quad \ldots \quad (5)$$

where E is the local source of direct voltage, I the direct current due to E and i the superposed a-c. due to e. This can be expanded into a power series:

$$f(E+e \sin pt) = f(E)+f'(E)e \sin pt$$
$$+f''(E)\cdot\frac{e^2 \sin^2 pt}{2}+ \ldots +f^{(n)}(E)\frac{e^n \sin^n pt}{\underline{|n}}.$$

[1] See Lee de Forest, Proc. A.I.E.E., Vol. 25, p. 719, 1906, and J. A. Fleming, Proc. Roy. Inst., Great Britain, June, 1909, p. 677.

that is

$$I+i=f(E)+f''(E)\frac{e^2}{4}+ \ . \ . \ . \ +f'(E)e \sin pt$$

$$-f''(E)\frac{e^2 \cos 2pt}{4}+ \ . \ . \ . \quad (6)$$

If now this be integrated over a complete period the sine and cosine terms vanish. Of the remainder the term $f(E)$ represents simply the direct current established in the circuit FPE by the battery E, and the series $f''(E)\frac{e^2}{4}+ \ . \ . \ .$ represents direct current component established by the incoming oscillations. This series represents second and higher derivatives of the characteristic and hence if f is a linear function, the series vanishes and the device does not rectify. But the characteristic of the thermionic valve is not linear; the derivative series is therefore finite. This series generally converges so rapidly that all the terms except the first, $f''(E)\frac{e^2}{4}$, can be neglected, so that the rectified current can be given by the second derivative of the characteristic. It also follows from this that the rectified current is proportional to the square of the input voltage e.

As an example let us take an arbitrary case in which the current in the device is proportional to the nth power of the voltage, thus: $I=aE^n$. The rectified current is then given by

$$\frac{e^2}{4} \cdot n(n-1)E^{n-2}.$$

If, for example, the current varies as the $\frac{3}{2}$-power of the applied voltage $(n=\frac{3}{2})$ the rectified current is inversely proportional to the square root of the locally applied voltage. If $n=2$ the rectified current is independent of the local voltage while for higher values of n, it increases with the local voltage.

It is well known that none of these cases applies to the valve when used as a radio detector, but that the rectified current shows a maximum for a definite value of the local voltage. This is due to the fact that the cathode is not an equipotential surface but a filament in which is established a voltage drop due to the heating current.

In Chapter IV it was shown that if the voltage drop in

the filament is taken into account the characteristic of the valve
can be represented by the two following equations:

$$(I)_{E \leqq E_f} = \frac{k}{E_f} E^{5/2} \quad \cdots \cdots \cdots (17')$$

for all values of E less than the voltage drop in the filament, E_f,
and

$$(I)_{E \gtreqless E_f} = \frac{k}{E_f}[E^{5/2} - (E - E_f)^{5/2}] \quad \cdots \cdots (18')$$

for values of E greater than E_f.

If I be computed from these equations for arbitrary values of
E within the respective limits, the two curves obtained will fit
to form a continuous curve such as the curve OA in Fig. 17 (p. 51).
But this is not the case with the relation between the voltage
and the second derivative of the current. The second derivatives
of the above equations are

$$\left(\frac{d^2I}{dE^2}\right)_{E \leqq E_f} = \frac{k}{E_f} E^{1/2} \quad \cdots \cdots \cdots (7)$$

$$\left(\frac{d^2I}{dE^2}\right)_{E \gtreqless E_f} = \frac{k}{E_f}[E^{1/2} - (E - E_f)^{1/2}]. \quad \cdots \cdots (8)$$

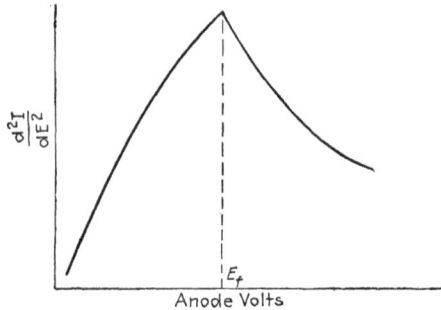

Fɪɢ. 41.

If these expressions be plotted for arbitrary values of E, the
result is a curve which shows a distinct maximum at a value
of the local voltage E equal to the voltage drop [1] in the filament,
(Fig. 41). This therefore accounts for the observed maximum

---

[1] It is to be understood that this voltage E is the sum of the voltage applied
and the contact potential difference between filament and anode.

in rectified current obtained at a suitably adjusted voltage of the battery in the valve circuit.

But even with this adjustment the valve falls far short of the three-electrode device which has now completely superseded it as a radio detector. The discussion of the valve as a radio detector will therefore be limited to the elementary theoretical considerations given above, which are generally applicable and can with some modification be adapted to the treatment of the three-electrode device.

46. Thermionic Valve as High Power Rectifier. The thermionic valve has considerable practical value as a rectifier of

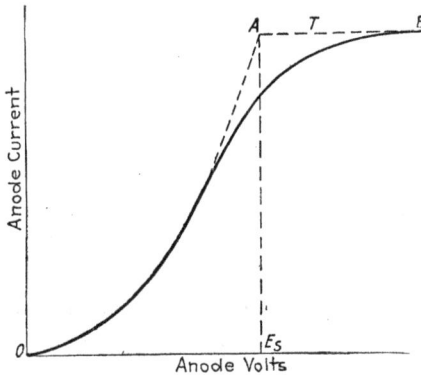

FIG. 42.

currents for high-power purposes, and has been used successfully for the production of unidirectional currents at high-voltages. By a proper arrangement of circuits it can also be used to convert alternating current into steady direct current. When this device is used for rectifying currents at high voltages it does not mean that it must be so constructed as to transmit current with these high voltages at its terminals. As a matter of fact, even in cases where the voltage to be rectified is as high as 100,000 volts and more, the voltage drop in the valve when it transmits current is only a few hundred volts and sometimes much less and the power dissipated in the valve is, comparatively speaking, very small.

In order to explain the operation of the thermionic valve in a rectifier circuit, let us consider its d-c. characteristic curve, shown in Fig. 42. When operated under the right conditions the device

conducts practically no current during the half cycle when the
filament is positive with respect to the anode. During the other
half cycle a current wave shape is obtained depending on the
value of the applied voltage and upon the shape of the d-c. char-

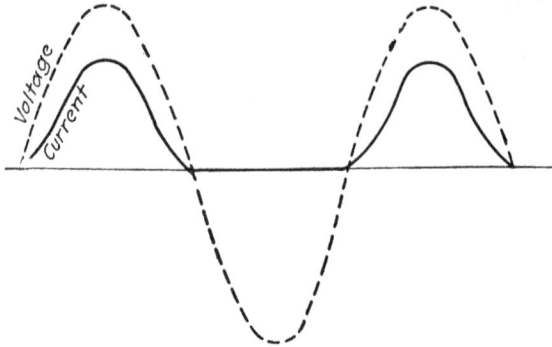

FIG. 43.

acteristic of the valve. As long as the peak value of the applied
simple harmonic voltage is less than E_s (Fig. 42) when the filament temperature is T the current wave shape takes the form
shown by the continuous curves in Fig. 43, the applied voltage
being given by the broken curve in arbitrary scale. The departure

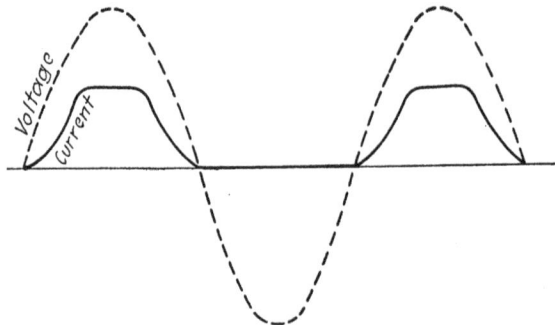

FIG. 44.

of the current curve from the simple harmonic shape is due to the
non-linear current-voltage characteristic of the valve.

If the voltage applied between the terminals of the valve
exceeds the value given by E_s (Fig. 42) the current curve will be
flattened as shown in Fig. 44,

47. Optimum Voltage for Rectification. An increase of the voltage beyond E_s causes no further increase in current and the valve then obviously operates inefficiently. Furthermore, if the applied voltage is less than E_s the valve also operates inefficiently; because in this case more power is expended in heating the filament than is necessary. This can be understood by remembering that the saturation current increases with the temperature. It is seen, therefore, that the valve operates most efficiently when the maximum value of the voltage applied between its terminals is equal to the d.c. voltage that is just sufficient to produce the saturation current. How this voltage depends on the constants of the tube can be seen from the following consideration. The saturation is given by Richardson's equation, $I_s = A_c A T^{1/2} e^{-\frac{b}{T}}$, where A_c is the area of the cathode, T the temperature of the cathode and A and b constants the meaning of which was explained in Section 19. This equation holds for the flat portion AB of the characteristic (see Fig. 42).

As was explained in Chapter IV the current on the lower part of the characteristic is limited by the space charge, the voltage drop in the filament, etc., so that in general the current cannot be taken to vary as the $\frac{3}{2}$-power of the voltage between filament and anode. The limitation of the current by the voltage drop in the filament, for example, causes the current to be smaller than in the theoretical case of an equipotential cathode, but to increase at a greater rate than the $\frac{3}{2}$-power of the voltage. We shall therefore assume that the current on the infra-saturation part of the characteristic is proportional to the nth power of the applied voltage.

Considering that we are interested in determining the optimum voltage which occurs at the knee of the characteristic, it must be noted that the limitation of current by electron emission from the filament, which usually extends into the lower part of the characteristic and is not confined only to the saturation region, causes the characteristic to bend over to the right, so that it is not possible to express the whole infra-saturation part of the characteristic by a simple power relation between current and voltage. We can, however, in the present consideration neglect this effect and determine the point A at the intersection of the saturation current and the current given by $I = CE^n$. This would correspond sufficiently closely to the optimum voltage.

The current below A will furthermore depend on the areas of the anode and cathode and the distance between them.   Hence:

$$I = f(A_c, A_a, d)E^n. \qquad \ldots \ldots \quad (9)$$

where A_c = area of cathode;
   A_a = area of anode;
   d = distance between cathode and anode.

At the point A the characteristic is obeyed by both equations $(I = I_s)$.   Hence:

$$E_s^n = \frac{A_c A T^{1/2} e^{-\frac{b}{T}}}{f(A_c, A_a, d)}. \qquad \ldots \ldots \quad (10)$$

This gives the voltage that should be applied to the terminals of the rectifier to make it operate most efficiently.

Instead of using Richardson's equation for the saturation current we can make use of the simpler equation given on p. 78 which holds with sufficient accuracy over the whole range of temperatures that one might want to use in practice.   This equation is

$$I = A_c C p^{n+1}, \qquad \ldots \ldots \quad (11)$$

where I is the current from a cathode of area A_c, p the power dissipated in heating the cathode and n an empirically determined exponent.

The voltage E_s can then be obtained by eliminating I from this equation and equation (9).   Before doing so let us obtain an expression for the function f in equation (9).   This equation was written in the functional form to apply it to the case in which both filament and anode are of finite size.   For such a tube in which the anode is in the form of a plate or plates the current is practically proportional to the area of both anode and cathode and inversely proportional to the square of the distance between them. Equation (9) can therefore be written approximately:

$$I = K \cdot \frac{A_c \cdot A_a}{d^2} \cdot E^n. \qquad \ldots \ldots \quad (12)$$

Hence from (11) and (12) we obtain

$$E_s^n = \frac{C}{k} \cdot \frac{d^2}{A_a} \cdot p^{n+1}, \qquad \ldots \ldots \quad (13)$$

where E_s is the value of E .corresponding to the point where the characteristic bends over.   The constant C depends upon the thermionic efficiency of the filament (see equation (24), Chapter IV). For example, for tungsten C has the value 1.8×10^{-7} and for a type of Western Electric filament the value 2.15×10^{-2}.

It is therefore seen that the optimum voltage drop in the rectifier depends upon the thermionic efficiency of the cathode, the power per cm.² used for heating the cathode, the area of the anode and the distance between cathode and anode.

The minimum voltage E_s necessary to obtain the saturation current is an important quantity, not only in dealing with the

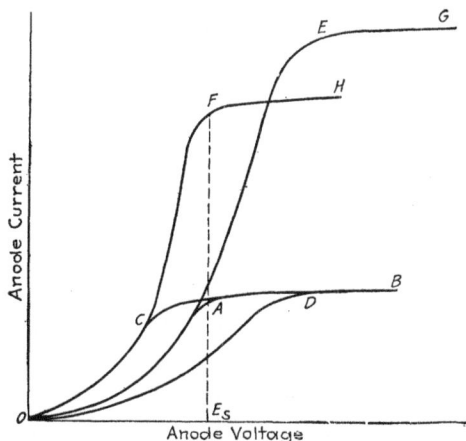

FIG. 45.

rectifier but also in designing the three-electrode type of ther‹ mionic tube. We shall therefore make the above relationship somewhat clearer by considering the current-voltage characteristic. Let us start with a tube having certain values for the parameters appearing in equation (13), and let the current-voltage characteristic of this tube be represented in arbitrary scale by OAB (Fig. 45). The voltage E_s of equation (13) is given by the projection of the point A on the voltage axis. Now let the size of the anode be increased to have double the area. The infra-saturation current (given by the curved part of the characteristic) will be doubled, but the saturation current (given by the horizontal portion) is independent of the area of the anode. The characteristic will therefore take the form OCB and the optimum voltage E_s will be less, although the maximum current is the same as before. The result is therefore a better rectifier. Now, suppose the distance d between cathode and anode be increased. This will not change the saturation but will reduce the infra-saturation current, and the characteristic may be represented by a curve such as ODB. This increases the voltage E_s. If the anode and cathode be kept as initially but the power dissipated per cm.^{2} at the cathode, i.e., its temperature, be increased the characteristic takes the form OEG. This increases the voltage E_s but at the same time the total current is increased. Furthermore, an increase in the thermionic efficiency of the cathode produces a similar effect to that resulting from an increase in the temperature, so that an appropriate increase in the thermionic efficiency would also change the characteristic to the curve OEG.

It will be noticed that the area of the cathode does not enter into equation (13), which means that the optimum voltage is independent of it. This can easily be understood if we consider that a change in the cathode area, say by changing the length of the filament, produces an equal change in both the space charge and saturation currents. Thus, if the filament length be doubled the characteristic obtained will be OFH instead of OAB, and it is seen that the voltage E_s has the same value as before, namely that corresponding to the point A or F, but the total current will be increased.

48. Types of Thermionic Valves. In designing a valve to operate on very high voltages certain important factors must be taken into consideration. In the first place the high potential difference existing between the electrodes during part of the blocking half period causes a mechanical strain which tends to pull the filament over to the anode, and this force will be the greater the smaller the distance between anode and filament. On the other hand, it was shown above that a decrease in this distance decreases the voltage drop in the valve. Hence to keep this voltage drop small the valve has to be designed so as to prevent the filament and anode from being short-circuited by the strain.

Arcing across the glass is another factor which has to be reckoned with in designing high voltage valves. Fig. 46 shows a General Electric valve designed by Dushman to rectify 100,000 volts.[1] Arcing across the glass is prevented by placing anode and filament terminals at opposite ends of the elongated parts of the tube. The elongation necessary for the use of such high voltages necessitates special means of rigidly supporting the electrodes.

For lower voltages the distance between the terminals can be smaller, which greatly facilitates the construction. A simple type

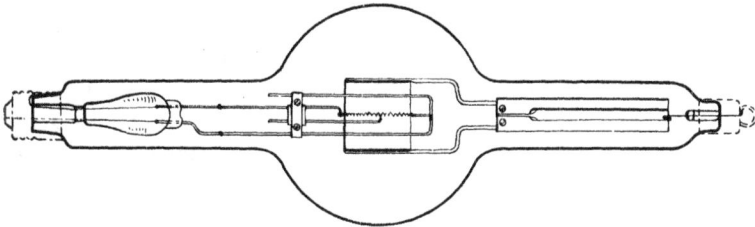

FIG. 46.

of valve is shown in Fig. 47. The re-entrant tubes are made long enough to give rigid support to the electrodes. The anode can be in the form of two parallel plates placed on either side of the filament or in the form of a cylinder or preferably a flattened cylinder completely surrounding the filament.

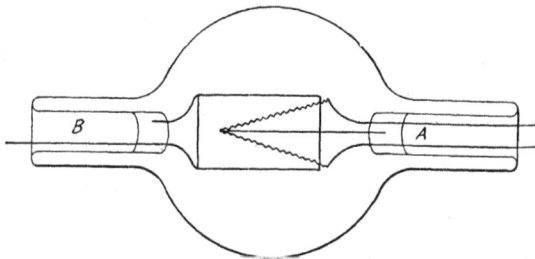

FIG. 47.

For voltages not exceeding a few hundred volts both electrodes can be supported from the same press, as shown in Fig. 48. This is desirable because the whole structure can then be assembled and the spacing accurately adjusted before sealing it into the bulb.

If the voltage exceeds a few hundred volts sparking may take place between the wires in the press. This deleterious effect is

---

[1] General Electric Rev., p. 156, March, 1915.

mainly due to the anode becoming so hot that it volatilizes and

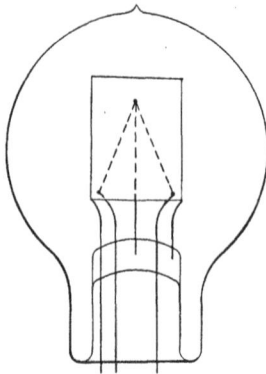

FIG. 48.

forms a metallic deposit on the press. It can to some extent be overcome in the tube shown in Fig. 49. In this tube both electrodes are mounted on the same re-entrant tube but the leakage path along the glass between the electrodes is considerably lengthened by taking the anode lead through the side tube *PA* which is closed by a small press at *P*. *B* is simply a glass rod to give added support to the anode. The possibility of sparking in this tube depends almost entirely on the air space between the wires in the re-entrant tube and on the insulation of the base plate *CD*.

The glass press can also be protected against metallic deposit by means of a metallic shield placed near the press, as was for example done by Dushman.[1]

Although the voltage drop in the valve when current passes is usually not high and should never exceed the optimum voltage, which generally does not amount to more than a few hundred volts, the full voltage of the generator is impressed on the valve during the blocking half period. The tube must therefore be evacuated to such an extent that the high voltage does not start a glow discharge. This necessitates clearing the electrodes and walls of the bulb of gases during evacuation to an extent depending on the temperature to which these parts are

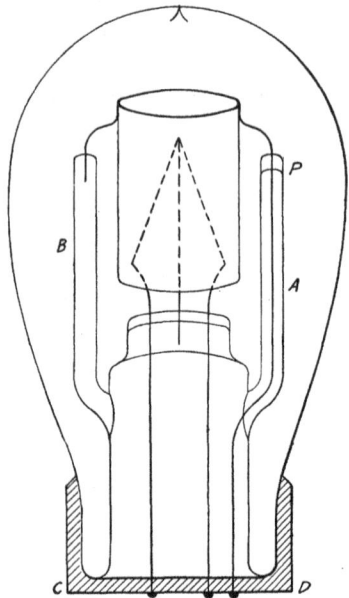

FIG. 49.

heated during operation of the tube. If the anode is of tungsten

[1] U. S. Pat. 1,287,265.

the power that could safely be dissipated in it is about 10 watts per cm^2.[1] This corresponds to a temperature of about 1600° K. If the anode temperature rises above this value the anode emits electrons at a sufficiently high rate to make the rectification imperfect. If the anode is of nickel the safe power dissipation in it is about 5 watts per cm^2. The size of the anode must therefore be chosen in proportion to the power loss in the rectifier. This is also the case with the size of the glass container. Of the power radiated from the internal structure, part is transmitted through the glass, and part is absorbed and radiated. This part causes a heating of the glass. The size of the glass tube or bulb must be so chosen that its area in square inches is not less than the total power in watts dissipated by the internal structure. This includes power radiated from the filament as well as from the anode.

49. Rectification Efficiency. Let us consider the operation of the thermionic valve in a circuit such as that shown in Fig. 50 in which the valve is supposed to be connected to a source of constant a-c. voltage e. During the half cycle that the filament is positive with respect to the anode no current is transmitted by the valve and the potential difference established between its terminals is equal to the full voltage supplied by the generator. During the other half cycle electrons emitted from the hot filament

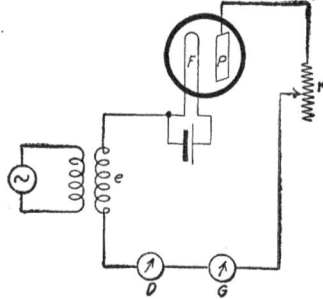

FIG. 50.

are driven to the anode and current flows in the circuit *FPGD*. The voltage drop in the valve now depends upon its resistance and the load resistance r. Suppose that r is initially so large that the voltage drop E in the valve is less than the optimum voltage E_s. If r be now decreased the current increases and the voltage drop in the rectifier also increases. How this change takes place can readily be seen by considering the d-c. values. Suppose a direct voltage equal to E_0 be applied instead of the alternating voltage, the filament being negative with respect to the anode. If I be the

---

[1] S. DUSHMAN, General Electric Review, loc. cit.

current and E the voltage drop in the valve we have $E = E_0 - Ir$.
Putting $I = CE^n$ we get:

$$r = \frac{1}{C}\left(\frac{E_0}{E^n} - \frac{1}{E^{n-1}}\right). \quad \cdots \quad \cdots \quad (14)$$

This shows that as r is decreased the voltage drop E in the valve
increases slowly at first and then more rapidly. But this equation only holds until E becomes equal to the optimum voltage drop
in the valve and the current obtained is equal to the saturation
current. If the resistance r be still further decreased there is no
further increase in current and the voltage drop in the valve increases rapidly and may cause a blow out of the tube. By suitably choosing the external resistance r the valve can be made to
rectify extremely high voltages.

It is seen that the thermionic valve acts not only as a rectifier
but also as a current limiting device. When the filament is positive
the emitted electrons are returned to it and there is then consequently no current. When the filament is negative the current
is limited by the saturation value which it cannot exceed and which
is determined by the total number of electrons emitted per second
from the filament at the temperature used.

To obtain an expression for the rectification efficiency let us
consider the general case by supposing that the valve does not
completely block current during
the blocking half period. The
current wave will then have a
shape somewhat like the curve
I_1I_2 (Fig. 51). When the valve
is short-circuited the current
through the external resistance r
is given by the curve $I'I'$. The
introduction of the valve not only
limits the current in both half
periods due to the addition of its
resistance in the circuit, but it also
changes the shape of the current wave due to its non-linear
current-voltage characteristic. Let D (Fig. 50) be an a-c. measuring instrument such as a dynamometer and G a galvanometer to
read the d-c. component of the current. Let i_0 be the reading
of the a-c. instrument when the valve is in the circuit and i the

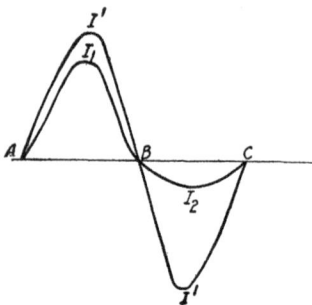

FIG. 51.

reading of the galvanometer. The efficiency is sometimes defined simply as the ratio $\dfrac{i}{i_0}$. To see what this means, let I_1 be the maximum instantaneous value of the current in the transmission half period and I_2 the corresponding maximum in the blocking half period. Then the dynamometer reading i_0 will be given by

$$i_0 = g\left(\frac{I_1 + I_2}{2}\right), \quad . \quad . \quad . \quad . \quad . \quad . \quad (15)$$

and the galvanometer reading by

$$i = \frac{g}{f}\left(\frac{I_1 - I_2}{2}\right), \quad . \quad . \quad . \quad . \quad . \quad (16)$$

where g is the "amplitude factor," given by the ratio of the R.M.S. to the maximum values of the current, and f is the "form factor," given by the ratio of the R.M.S. to the true mean current.[1] Both g and f depend upon the shape of the wave. For a pure sine wave $g = .707$ and $f = 1.111$. The above expression for the efficiency therefore takes the form

$$\frac{i}{i_0} = \frac{1}{f} \cdot \frac{I_1 - I_2}{I_1 + I_2}. \quad . \quad . \quad . \quad . \quad . \quad (17)$$

From this it is seen that if the valve conducts absolutely no current during one-half period, i.e., $I_2 = 0$ and if the current wave shape during the other half period is that of a pure sine wave, i.e., $f = 1.111$, the efficiency as expressed above has a maximum value of 90 per cent.

Instead of this method Fleming [2] makes use of the expression $\dfrac{I_1 - I_2}{I_1}$ and calls it the "rectifying power." By a simple transformation of equation (17) this can be expressed as:

$$\frac{I_1 - I_2}{I_1} = \frac{2f}{\dfrac{i_0}{i} + f}. \quad . \quad . \quad . \quad . \quad . \quad (18)$$

[1] J. A. FLEMING, " Alternating Current Transformer," Vol. I ᴅ. 585.

[2] J. A. FLEMING, Proc. Roy. Soc., Jan., 1905, p. 484.

This equation expresses the ratio of unidirectional current observed to the unidirectional current that would flow if the rectification produced by the valve were complete. It is seen that if $\dfrac{i_0}{i}$ is equal to the form factor f, that is if $I_2 = 0$ (equation 17), this ratio becomes unity, or 100 per cent, irrespective of the shape of the wave.

It is, however, not sufficient to know how well the device blocks current in the one direction. It is equally important to know how much current it transmits in the other direction, and this depends upon the wave shape of the transmitted current. Furthermore, the insertion of the valve in the circuit is equivalent to the introduction of an extra resistance. Let J be the dynamometer reading when the valve is short-circuited, i.e., the R.M.S. value of the current $AI'BI'C$ (Fig. 51). Then the ratio of useful rectified current to the available alternating current is obtained by dividing equation (16) by $J = gI'$. This gives:

$$\frac{i}{J} = \frac{I_1 - I_2}{2fI'}. \qquad \cdots \cdots \cdots \quad (19)$$

If E_1 be the maximum value of the applied generator voltage (see Fig. 50) and r the load resistance, we have $I' = \dfrac{E_1}{r}$ and $I_1 = \dfrac{E_1}{r + r_1}$, where r_1 is the resistance of the valve at the maximum voltage across its terminals. The resistance of the valve for any given voltage between filament and anode is given by the ratio of that voltage to the current and is a function of the voltage. Referring to the characteristic curve (Fig. 42) it will be evident that as the voltage increases from zero to the optimum value E_s, the resistance decreases from infinity to a definite value given by the reciprocal of the slope of the straight line joining A and O. What we are concerned with here is the resistance which obtains when the voltage has its maximum value. We shall refer to this as the *minimum resistance* and denote it by r_1. Combining the equations for I' and I_1 we obtain

$$I' = \left(1 + \frac{r_1}{r}\right)I_1,$$

and hence with the help of equations (18) and (19):

$$\frac{i}{J} = \frac{1}{\left(\frac{i_0}{i}+j\right)\left(1+\frac{r_1}{r}\right)}. \qquad \cdots \cdots \quad (20)$$

If the tube conducts no current in the one direction the form factor f is equal to $\frac{i_0}{i}$, the ratio of the dynamometer to the galvanometer reading when the valve is in the circuit. This follows from equation(17) for $I_2 = 0$. If furthermore r_1 is negligibly small compared with the load resistance r, equation (20) reduces to

$$\frac{i}{J} = \frac{1}{2f}.$$

Hence the rectification efficiency as given by equation (20) would reach a maximum of 50 per cent (1) if the form factor f were unity; (2) if the minimum resistance of the valve is negligibly small compared with the load resistance and (3) if the valve completely blocks current in one direction. Let us see how nearly these conditions can be satisfied.

(1) As regards the first condition, the form factor is always greater than unity for thermionic valves. If the relation between the applied voltage and the current were linear and a true sinusoidal voltage be impressed on the valve the current loop during the transmission half period would also be sinusoidal and then the form factor would be that of a sinusoid, namely 1.111. If this is not the case but if the current is, let us say, proportional to the nth power of the applied voltage the current wave will have a shape somewhat like that shown in Fig. 43 (p. 116), and the form factor f must be determined from the root mean square and true mean values of $\sin^n pt$. For such cases f and the amplitude factor g can be obtained from a table of gamma functions by expressing them in the forms:

$$f = \sqrt[4]{\pi}\, \frac{\Gamma\left(\frac{n}{2}+1\right)}{\Gamma\left(\frac{n+1}{2}\right)} \sqrt{\frac{\Gamma\left(\frac{2n+1}{2}\right)}{\Gamma(n+1)}}. \qquad \cdots \cdots \quad (21)$$

$$g = \sqrt{\frac{1}{\sqrt{\pi}}\, \frac{\Gamma\left(\frac{2n+1}{2}\right)}{\Gamma(n+1)}}. \qquad \cdots \cdots \cdots \quad (22)$$

The following table gives the values of f and g for a range of values of the exponent n:

Exponent n.	Form Factor, f.	Amplitude Factor g.
1.00	1.111	.707
1.25	1.141	.677
1.50	1.170	.652
1.75	1.199	.632
2.00	1.225	.612
2.50	1.275	.584
3.00	1.320	.560

To use the above expression for the rectification efficiency it is necessary first to determine the relation between the total applied voltage and the current through the valve and external resistance. To a first approximation this can be expressed by a simple power relation in which case the exponent n of the voltage can be determined by plotting the logarithms of the voltages against the logarithms of the currents. The corresponding values of f and g can then be obtained from the above table.

In this connection it must be pointed out that there is a difference between the current voltage characteristic of the valve itself and that of the circuit comprising the valve and external resistance. The effect of this resistance is to straighten out the characteristic, because it means the addition of an ohmic resistance which partially neutralizes the curvature of the characteristic due to the non-ohmic resistance of the valve. The effect is obviously the more marked the larger the external resistance compared with the valve resistance. However, since the applied voltage alternates the valve resistance continually changes. During the half period when the plate is negative with respect to the filament the resistance of the valve is infinite, and during the other half period the resistance decreases from infinity to a definite minimum value and then increases again to infinity. Hence, for low instantaneous values of the applied alternating voltage the external resistance does not exert a marked effect in straightening out the characteristic, so that even if the external resistance is large the current still tends to assume a shape somewhat like that shown in Fig. 43 although on the whole it approximates more closely to the shape of the sinusoid.

(2) Considering now the effect on the efficiency as given by equation (20) of the relative values of the external or load resistance to the minimum resistance r_1 of the valve itself, it is to be noted that when the valve is used to rectify high voltages its minimum resistance is generally negligibly small compared with the load resistance. It was shown in Section 47 that the valve is operated most efficiently when the peak value of the voltage applied to its terminals is just equal to the minimum voltage necessary to give the saturation current. This we called the optimum voltage and it is given by E_s in Fig. 42. If, for example, the saturation current of the valve at a certain filament temperature is 300 milliamperes and if the valve is so designed that the minimum voltage across its terminals necessary to give this current is 150 volts, the minimum valve resistance is 500 ohms. Now, if the peak value of the voltage to be rectified is 30,000 volts (about 21,000 volts effective) the total resistance of the circuit must be 100,000 ohms, which is very large compared with that of the valve. For higher voltages the device operates even more efficiently, so that the second factor in the denominator of equation (20) reduces to unity.

(3) Coming to the third condition mentioned we can for most practical purposes regard the thermionic valve as a perfect unilateral device; it completely blocks current in one direction, provided that the plate does not become so hot that it emits an appreciable number of electrons and provided that the valve be so constructed that leakage between its terminals across the glass does not take place. At high frequencies, however, the rectification becomes imperfect due to the capacity between the electrodes (see p. 134).

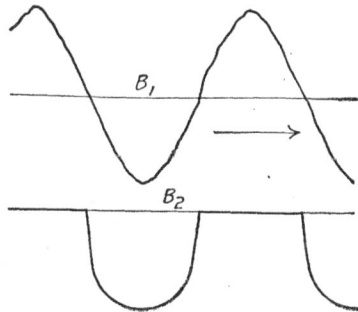

FIG. 52.

The completeness of the rectification produced by thermionic valves is shown in the oscillogram given in Fig. 52 and which was obtained by Dushman.[1] The upper curve gives the voltage across the tube and the lower curve the rectified current.

[1] S. Dushman, General Electric Review, loc. cit.

Assuming, therefore, that the load resistance is so large that we can regard the characteristic of the circuit as linear, we have $f = 1.111$ and $\frac{r_1}{r} = 0$. If the valve completely blocks current in one direction, $\frac{i_0}{i} = f$. Putting these values into equation (20) we find that the highest efficiency obtainable is 45 per cent.

If the resistance of a valve during one half period is infinite, and during the other half period zero, the valve is perfect, and the rectification efficiency, as expressed by equation (20), becomes independent of the resistance r, used in the external circuit, because under these conditions r_1 is zero. Actually, however, the resistance of valves during the transmission half period is not zero, so that according to equation (20) the efficiency increases the larger the external resistance r becomes in comparison with the resistance r_1 of the valve. Now, it was explained in Section 47 that for most efficient operation the voltage drop in the valve during the transmission half period should never exceed the value necessary to give the saturation current. This means that the external resistance should be so adjusted that the maximum voltage drop across the valve is equal to the optimum voltage. In other words, the ratio of r_1 to r, occurring in the second expression in the denominator of (20) becomes smaller the larger the voltage that is to be rectified. The rectification efficiency of the valve, therefore, increases with the voltage that is to be rectified and approaches the maximum efficiency that could be obtained with a perfect valve.

The efficiency can, of course, be doubled by making use of both half waves so that the rectified current is given by the continuous lines of Fig. 53 instead of Fig. 43. This can be done by using two tubes in the circuit shown in Fig. 54. It is seen that for both half periods the electron current in the load is in the direction of the arrow. This scheme necessitates dividing the input voltage on the secondary side of the transformer T. In order to make use of the full voltage of the transformer the circuit shown in Fig. 55 can be used.[1] Here again current flows in the load resistance during both half periods in the direction of the arrow. Such an arrangement requires four times the total power necessary to heat the filaments. The filament heating power is, however, small

---

[1] GRAETZ, Die Elektrizität und ihre Anwendungen, 15th Edition, p. 444.

compared with the power that becomes available in the form of
unidirectional current. This is the more marked the higher
the voltage that is to be rectified. Let us, for example, take the
case considered above in which a tube that could give 300 milliamperes was made to rectify 21,000 volts. Referring to the

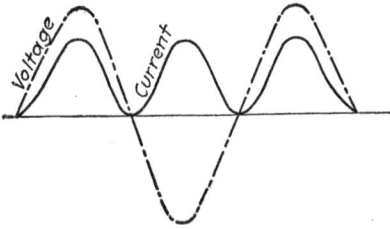

FIG. 53.

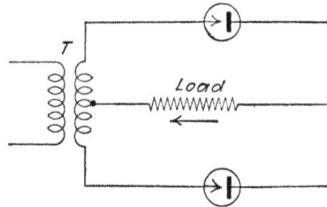

FIG. 54.

table on p. 77 it is seen that if the filament is of tungsten and is
operated at a temperature a little under 2500° K. the thermionic
efficiency is 10 m.a. per watt. Since the necessary saturation current I_1 is 300 m.a. the power necessary for heating the filament of
the valve under consideration is 30 watts. The power available
in the form of unidirectional current in the load can be obtained

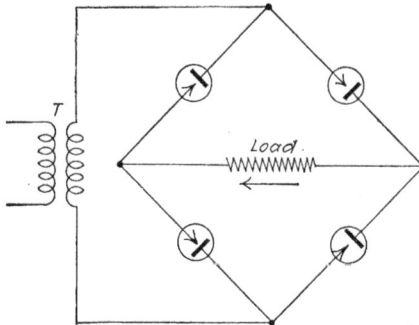

FIG. 55.

from equation (15). Putting $I_2 = 0$, the heating equivalent of the
current in the load resistance is $g \dfrac{I_1}{2}$ and the available power in
the form of unidirectional current is $g^2 \dfrac{I_1^2 r}{4}$. Assuming that the
current-voltage characteristic of the circuit is, in view of the
high load resistance of about 100,000 ohms, practically linear,

we can put $g = .707$. This makes the available power about $1KW$, which is quite large compared with the power necessary to heat the filament. It is, in fact, so much larger that it is advantageous to quadruple the filament heating power in order to double the output power with the arrangement shown in Fig. 55.

Now, what is ordinarily observed in practice is not the heating current i_0 which must be measured with an a-c. meter, but the true mean of the unidirectional current i measured with an ordinary d-c. ammeter. The output power can then be readily obtained from $i^2 f^2 r$, where f is the form factor (equation 17), assuming that the valve does not conduct current at all in one direction, an assumption which is justified in most practical cases.

50. Production of Constant Source of High Voltage with the Thermionic Valve. As a rectifier the thermionic valve offers three distinct advantages: It can be used to rectify voltages ranging up to the highest met with in practice; it rectifies currents of comparatively high frequency as well as low frequency currents; it completely blocks current in one direction, provided the frequency is kept below a certain limit depending on the voltage that is to be rectified. (This effect will be discussed below.) The value of these three advantages will become apparent in the following discussion:

When used as a rectifier under the conditions described in the previous paragraphs the valve produces a unidirectional pulsating current. In some practical applications, such as direct current high voltage transmission, testing of dielectric strength of insulators at high voltages, use of a d-c. source of high voltage for laboratory purposes, etc., it is necessary that the pulsating current be smoothed out into a constant direct current. We shall therefore, proceed to a discussion of the means whereby this smoothing out can be accomplished.

When the required direct current is small and it is not essential that the wave be completely smoothed out, we can resort to the simple and well-known expedient of shunting the load with a sufficiently large condenser, C_1 (Fig. 56). This condenser acts as a reservoir from which a practically constant current can be drawn continuously, it being charged up in alternate half periods and always in the same direction. The effect of this condenser (the inductance L being omitted) can be seen from Fig. 57.[1]

---

[1] A. W. HULL, General Electric Rev., Vol. 19, p. 177, 1916.

The sine wave represents the transformer voltage and the heavy line the output voltage across the condenser, which initially becomes charged up to the full peak value of the input voltage. During the rest of the period it receives no charge until the input voltage becomes greater than the value to which the condenser voltage has dropped in virtue of the current drain from it. It is

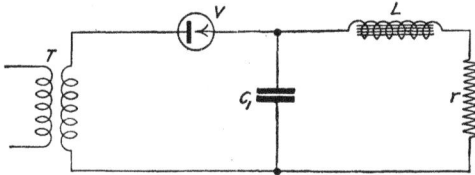

Fig. 56.

seen that the advantage offered by this type of valve that it almost completely blocks current in one direction is an important one; the condenser never discharges itself through the input circuit C_1TV. The rate at which the condenser discharges through the load is given by

$$-C_1\frac{dE_1}{dt}=i$$

where E_1 is the voltage across the condenser plates and i the average current, which can be regarded as practically constant.

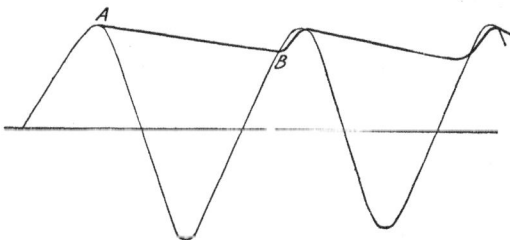

Fig. 57.

Hence, integrating between E_A and E_B and putting $E_A-E_B=\delta E_1$, the voltage variation across the condenser is:

$$-\delta E_1=\frac{it}{C_1}, \quad \cdots \cdots \cdots \quad (23)$$

where t is almost a complete period and can in the following calculations be regarded approximately as such. We can therefore write (23) in the form

$$\frac{\delta E_1}{E_r} = \frac{2\pi}{r\omega C_1} \quad \cdot \cdot \cdot \cdot \cdot \cdot \cdot \quad (24)$$

where r is the load resistance and E_r the direct voltage in it. This equation shows that the condenser C_1 alone will appreciably reduce the voltage fluctuation provided the load resistance is sufficiently large. The same result can, of course, also be secured by increasing the capacity and the frequency of the impressed generator voltage. If, however, the load resistance is small C_1 would have to be made so large as to make its use impracticable nor can the frequency be made very high, because then the capacity of the tube itself would become effective with the result that the tube would not rectify completely. That the limiting frequency is lower than is sometimes assumed will be seen from the following simple consideration.

To operate the valve most efficiently, the voltage across it during the transmission half period must not exceed the optimum value, which is generally of the order of a few hundred volts, so that in rectifying very high voltages, say 100,000 volts, the load resistance must be very high. But it must always be small compared with the resistance of the valve during the blocking half period. For low frequencies or d-c. this resistance of the valve is infinite, but at high frequencies the valve may on account of its electrostatic capacity have an impedance which is comparable with the load resistance and then currents of comparable magnitude will obviously flow in both directions in the load resistance. Thus, if the capacity of valve be $C = 10$ micro-microfarads, which is well within the range of the capacities of the valves used in practice, its impedance at a frequency of 16,000 cycles per second is 1 megohm. If now the voltage to be rectified is E_0 (peak value), the optimum voltage of the valve E_s and the maximum current obtainable from it I_s then the load resistance r is

$$r = \frac{E_0 - E_s}{I_s}$$

E_s can usually be neglected in comparison with E_0. If the voltage to be rectified is, say, 100,000 volts, $E_0 = 140,000$ volts, and if $I_s = 100$ milliamperes, r will be 1.5 megohms (approx.). Hence the impedance of the tube due to its electrostatic capacity at 16,000 cycles is of the same order of magnitude as the load resistance which is necessary when the voltage to be rectified is of the order of 100,000. The condenser will therefore partly discharge itself in alternate half periods through the input circuit $C_1 TV$ (Fig. 56). At lower voltages this effect is not so marked, so that higher frequencies could be used to advantage.

In order to smooth out the voltage fluctuations somewhat more effectively an inductance L is sometimes inserted in series with the load, as shown in Fig. 56. When the load resistance is low the inductance helps appreciably, but for high load resistances it serves little purpose. Considering this inductance, it is evident that the voltage fluctuation δE_r across r is to the voltage fluctuation δE_1 across the condenser C_1 as the ratio of r to the impedance of L and r in series. Thus:

$$\frac{\delta E_r}{\delta E_1} = \frac{1}{1+j\dfrac{L\omega}{r}} \quad \cdots \cdots \cdots \quad (25)$$

It is seen that for large values of the load resistance r the inductance contributes little to smoothing out the fluctuations, but it helps appreciably at the lower load resistances. Assuming that the coil L is a pure reactance, the d-c. voltage in r is the same as that across the condenser C_1. Hence the percentage voltage fluctuation can be obtained by combining equations (24) and (25). This gives

$$\frac{\delta E_r}{E_r} = \frac{2\pi}{\omega C_1 \sqrt{r^2 + L^2 \omega^2}}. \quad \cdots \cdots \quad (26)$$

The relation between $\dfrac{\delta E_r}{E_r}$ and $\log r$ is shown by curve II of Fig. 58. The curve I gives the relation when the inductance L is omitted. These curves were computed with the following values: $L = 100$ henrys, $C = 10^{-9}$ farad. The curves show that if the load resistance is greater than about a megohm, that is for resistances of the order of magnitude used when rectifying very high voltages, practically the same result can be secured by using only the condenser instead of adding the inductance. On the other hand the condenser alone is useless at load resistances less than a megohm.

Better results can, of course, be obtained by adding more sections to the wave filter LC_1. Such an arrangement is shown in Fig. 59, which at the same time shows a circuit that makes possible

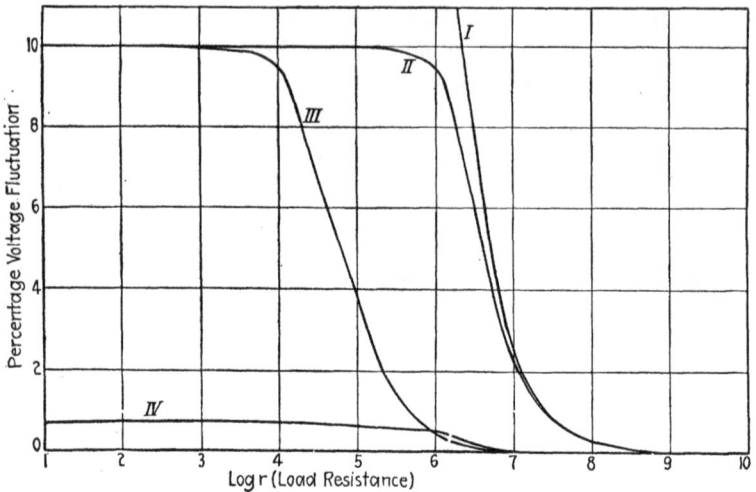

Fig. 58.

the use of both half periods by employing two tubes. It will be seen that current flows in the direction of the arrow (say) when A is positive and B negative as well as when A is negative and B

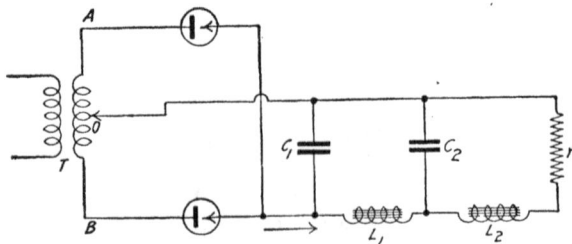

Fig. 59.

positive, the current being transmitted through the tubes alternately.

Let us now see to what extent the added filter section contributes in reducing the voltage fluctuations in r and how they depend on the value of r. Let δE_r, δE_2 and δE_1 be the voltage fluctuations at the terminals of r, C_2 and C_1 respectively. Then

$$\frac{\delta E_r}{\delta E_2} = \frac{r}{r+jL_2\omega}. \quad \ldots \ldots \ldots \quad (27)$$

If Z_2 be the impedance of the circuit C_2L_2r, as measured between the terminals of C_2, and Z_1 the impedance of L_1 and Z_2 in series, then

$$\frac{\delta E_2}{\delta E_1} = \frac{Z_2}{Z_1}.$$

Now since

$$\left.\begin{aligned} Z_2 &= \frac{r+jL_2\omega}{1-C_2L_2\omega^2+jrC_2\omega} \\ Z_1 &= \frac{r(1-L_1C_2\omega^2)+j\omega(L_1+L_2-L_1L_2C_2\omega^2)}{1-C_2L_2\omega^2+jrC_2\omega} \end{aligned}\right\} \quad \cdot \cdot \quad (28)$$

we get

$$\frac{\delta E_2}{\delta E_1} = \frac{r+jL_2\omega}{r(1-L_1C_2\omega^2)+j\omega(L_1+L_2-L_1L_2C_2\omega^2)}. \quad \cdot \cdot \quad (29)$$

The voltage fluctuation across the terminals of condenser C_1 is given by equation (24). Hence, multiplying together (24), (27) and (29) and expressing the impedances numerically instead of symbolically the ratio of the voltage fluctuation in r to the d-c. voltage in r is:

$$\frac{\delta E_r}{E_r} = \frac{2\pi}{\omega C_1[r^2(1-L_1C_2\omega^2)^2+(L_1\omega+L_2\omega-L_1L_2C_2\omega^3)^2]^{1/2}}. \quad (30)$$

A. W. Hull[1] described a high voltage rectifying set in which he used two condensers but only one inductance, L_1. Putting $L_2=0$ in equation (30) the voltage fluctuation for Hull's set becomes:

$$\frac{\delta E_r}{E_r} = \frac{2\pi}{\omega C_1[r^2(1-L_1C_2\omega^2)^2+L_1{}^2\omega^2]^{1/2}} \quad (L_2=\text{zero}) . \quad . \quad (31)$$

and if $C_2=0$ this equation reduces to (26) which is the equation for the circuit shown in Fig. 56. On the other hand, if the inductance, frequency and second condenser have such values as to put L_1 and C_2 in resonance, i.e., when $L_1C_2\omega^2=1$, then the

---

[1] A. W. Hull, loc. cit. Hull's equations are not the same as these, since he did not add the imaginary terms in quadrature.

arrangement corresponding to equation (31) is worse than that of Fig. 56 in which the second condenser is omitted altogether. It follows that for this circuit to be better than that of Fig. 56 we must make $L_1C_2\omega^2>2$. This condition is easily satisfied in practice. In Hull's set, for example, the values of L_1, C_2 and ω happen to be such that their product is about 60. However, although this circuit, containing one inductance and two capacities is a decided improvement over the simpler one shown in Fig. 56, it is better to split the inductance and use the circuit of Fig. 59. This circuit has a decided advantage at lower load resistances, even when the inductances L_1 and L_2 are each one-half of the value of L_1 when $L_2=0$. The percentage ratio of $\dfrac{\delta E_r}{E_r}$ as a function of $\log r$ for these two cases is shown by curves III and IV of Fig. 58. Curve III was computed for the following values: $C_1=C_2=10^{-9}$ farad, $L_1=100$ henrys, $\omega=2\pi\times4000$. In curve IV the values were the same except that $L_1=L_2=50$ henrys. It can readily be seen that a frequency of 4000 is obtained in the filter circuit when the frequency of the voltage impressed at T is 2000 cycles, since by using two tubes as shown in Fig. 59, the condensers are charged up every half period of the voltage in T.

Curve IV shows the value of a circuit like that shown in Fig. 59, when it is desired to have a rectifying set which is to operate with large variations in the load resistance.

It will be evident that these circuits simply represent a type of wave filter which is supposed to filter out all frequencies except zero, that is, the direct current. The waves obtained in the output of these circuits comprise not only the fundamental frequency that we considered in the above computations, but also a number of harmonics which are generally weak compared with the fundamental. It will be evident that harmonics must necessarily be present, considering that the wave, which has the form shown in Fig. 53 is not a pure sinusoid. Such a wave can always be expressed in a Fourier series (equation 1). It will also be seen from the nature of the above equations for $\dfrac{\delta E_r}{E_r}$ that the harmonics will be damped out more effectively than the fundamental. They were therefore left out of consideration in the above calculations.

Another type of circuit that could be used for smoothing out the voltage fluctuations was suggested to me by Mr. T. C. Fry, and is a special case of Campbell filter (Fig. 60). It has the advantage that the capacities and inductances necessary are relatively small, which is always a good thing when rectifying very high voltages in view of the difficulty of constructing condensers of high capacity for high voltage work.

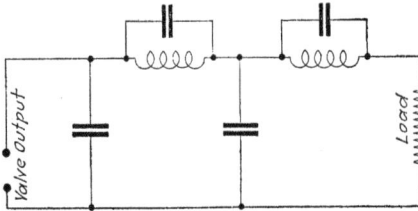

Fɪɢ. 60.

The characteristic of this filter is seen from Fig. 61, where the current attenuation produced by the filter is plotted against the frequency. The capacities and inductances can be so chosen that the fundamental frequency, $\frac{\omega}{2\pi}$, is that which gives infinite attenuation. This frequency will therefore not be present in the

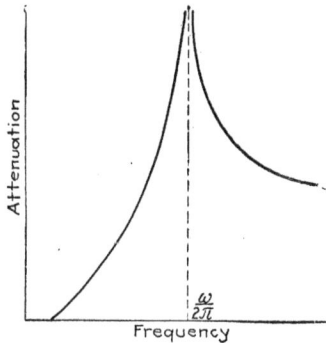

Fɪɢ. 61.

load resistance. The filter would transmit lower frequencies than $\frac{\omega}{2\pi}$, but such frequencies, except zero, are not present when $\frac{\omega}{2\pi}$ represents the fundamental. Hence, for all frequencies below

$\frac{\omega}{2\pi}$ only direct current is transmitted. The higher frequencies will be transmitted and they will be present in the form of harmonics. These are, however, so weak that when attenuated to the extent shown by the curve to the right of $\frac{\omega}{2\pi}$, their effect in the load is practically *nil*. It will be evident from the nature of the attenuation curve that when using such a filter the frequency of the input must be adjusted rather accurately to the value determined by the constants of the filter.

Figs. 59 and 60 show only two filter sections. If desired, better results can be obtained by adding more sections.

Before leaving this subject let us discuss briefly the relative value of a few types of circuits, considering mainly the arrangements of the valves irrespective of the type of filter used in the output circuit.

The circuit shown in Fig. 59 is arranged to make use of both half waves. The voltage fluctuation at the condenser C_1 will therefore be of double the frequency of the wave supplied through the transformer T. The potential of points A and B (Fig. 59) will always be 180° out of phase, but only when they are positive with respect to O will the voltage be effective in charging up the condensers. If the potentials of A and B be represented by the broken lines A' and B' (Fig. 62) the potential fluctuation at the condenser

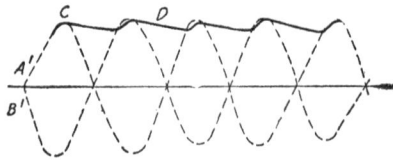

Fig. 62.

will be represented by the curve CD, which possesses a fundamental whose frequency is twice that of the waves A' and B'. This is an advantage because it follows from equation (32) that the higher the frequency, the more effectively will the fluctuation be smoothed out by the filter. On the other hand this circuit has the disadvantage that the voltage impressed on the valves is only half that supplied by the transformer. In order to use the full transformer voltage we could resort to the arrangement shown in Fig. 55, replacing r by the filter and load resistance shown in Fig. 59.

Fig. 63 shows a circuit whereby the transformer voltage [1] can be doubled. When the transformer voltage is such that D is at a positive potential with respect to O, an electron current will flow in the direction of the arrow through the valve AD, thus charging the condenser C such that A is positive with respect to O. But during this half period no current will flow through DB. During

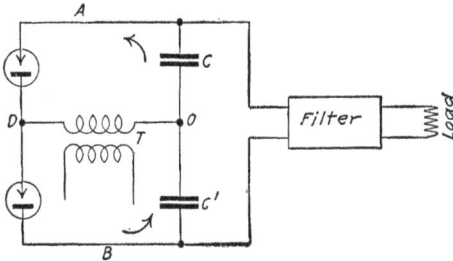

FIG. 63.

the next half cycle current flows only through DB, charging B negatively with respect to O. The potential difference between A and B (if the condensers did not discharge themselves) would therefore be twice the transformer voltage. What actually happens is that the one condenser discharges through the load while the other is being charged. Hence if the broken line (Fig. 64)

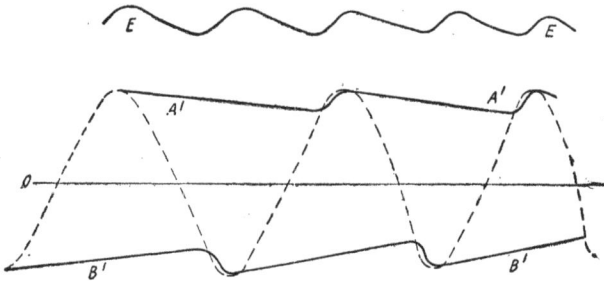

FIG. 64.

represents the potential of the point D with respect to O, the curves $A'A'$ and $B'B'$ will represent the potentials of A and B respectively with regard to O. The potential difference between A and B is therefore obtained by adding the curves A' and B' and is given by EE. Thus, although the condensers are charged only in alternate

[1] H. GREINACHER, Verh. d. D. Phys. Gesell., Vol. 16, p. 320, 1914.

half periods, the voltage fluctuation in the circuit leading to the filter is double the frequency of the impressed voltage, while the mean voltage on the filter is approximately twice the impressed voltage.

It will be observed from the above discussion that there are various ways in which thermionic valves can be used for increasing frequency.

For laboratory work it is often necessary to have a source of fairly high constant voltage supplying very small currents, such as would be needed, for example, in the study of photo-electric phenomena, ionization of gases by radium or X-rays, measurement of the intensity of X-rays with the ionization chamber, etc. For such purposes the thermionic valve could be used to replace the rather troublesome high voltage batteries frequently used in university laboratories, which consist of a large number of miniature storage or dry cells. In fact, the high voltage desired could be obtained from any standard storage or dry cell battery of a few volts, which forms part of the equipment of any physical laboratory, by connecting the primary of the transformer to the low voltage battery through an interrupter. This could, for example, be done by using a small Ruhmkorff coil with an ordinary hammer break. Since the desired current is small the valves could be designed to operate with very small power expenditure in the filament. It must, however, be remembered that when using the device for the purposes mentioned, the load resistance is usually very high, and hence, in order to prevent the condensers from discharging through the valves during the blocking half periods, the valves should be designed to have the lowest possible electrostatic capacity and the frequency of interruption of the primary current should not be very high.

51. The Thermionic Valve as a Voltage Regulator. The rapid increase in the saturation thermionic current with increase in the filament temperature, or filament heating current, as is shown by Richardson's equation, can be utilized to control the voltage of the generator of varying speed. A scheme whereby this can be done, and which was devised by H. M. Stoller, is shown in Fig. 65. Here the tube is used to regulate the voltage supplied by a wind-driven generator such as has been used on airplanes. The generator is designed to supply a high voltage for the plate circuit of thermionic tubes and a low voltage for heating the filaments. D and M are the differential and main field windings of the generator. The thermionic valve is inserted as indicated

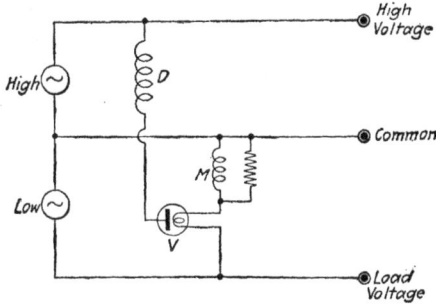

FIG. 65

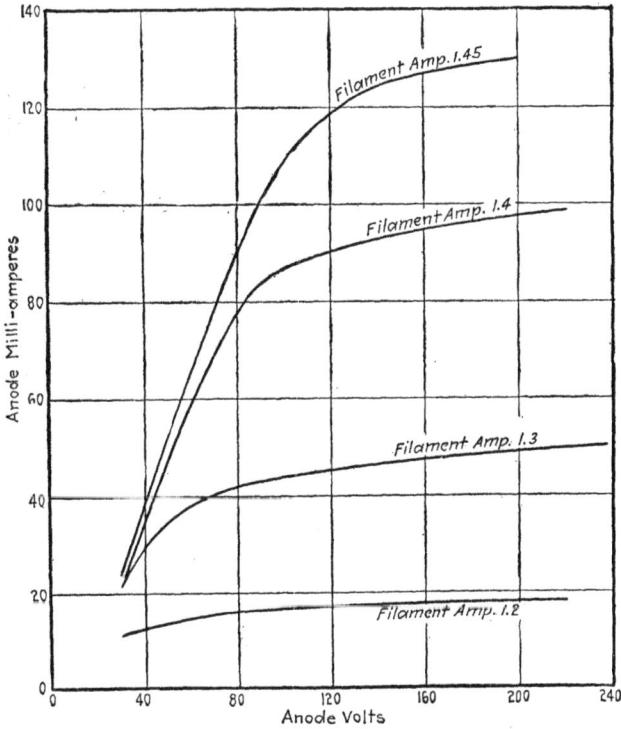

FIG. 66.

at V. The characteristics of such a valve are shown in Fig. 66. Suppose, now, that the speed of the generator is so low that the current flowing through the filament of the valve is 1.2 amperes. With this filament current the thermionic current through the valve and the differential winding is small and practically the full voltage is obtained. If, now, the speed of the generator increases, the filament current of the valve increases, but this causes a proportionately much greater increase in the thermionic current which flows through D. Thus Fig. 66 shows that a small change in the filament current of from 1.2 to 1.4 amperes causes a five-

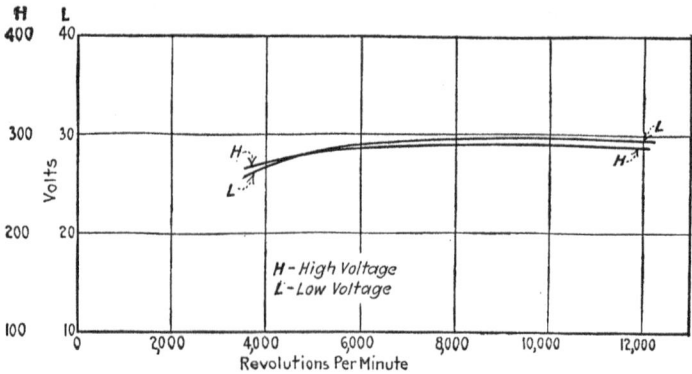

FIG. 67

fold increase in the thermionic current. This causes a decrease in the field flux of the generator, thus restricting the increase in the output voltage.

The regulation obtained with such a device is shown in Fig. 67. HH and LL represent the high and low output voltages as a function of the speed of the generator, and it will be seen that although the speed changes from about 4000 to over 12,000 R.P.M., the voltage output remains practically constant.

CHAPTER VII

THE THERMIONIC AMPLIFIER

EXCEPT for the derivation in Chapter III, of a few fundamental relationships that govern the discharge in three-electrode devices, we have so far considered only the simple type of device containing two electrodes. The physical principles underlying the thermionic tubes discussed in the previous chapters are applicable also to the three-electrode type of thermionic tube which it is our purpose to treat in this and the following chapters. This device consists essentially of a highly evacuated vessel, containing a thermionic cathode, usually in the form of a filament which can be heated by passing a current through it, an anode and a discharge-controlling electrode which generally takes the form of a wire mesh or grid, and placed between the cathode and anode. This third electrode can, however, be of any form, since a controlling effect on the discharge can be obtained by so positioning a conductor with respect to the path of the discharge that potential variations applied to it will cause variations in the current flowing between cathode and anode. The controlling electrode may, for example, be in the form of a plate placed on the side of the cathode opposite to that of the anode or in the form of a wire or a plurality of wires galvanically connected and placed in the plane of the cathode parallel to that of the anode. The theory of operation of the device to be given in the following applies to these various structures, but will be explained with particular reference to the case in which the auxiliary or discharge-controlling electrode takes the form most commonly used in practice, namely, a grid placed between cathode and anode. This was suggested by Lee de Forest.[1] Originally he used the device, which he called the " audion " as a radio detector. It has since developed, however, that its use is not by any means limited to this field, it being now

[1] U. S. Patents No. 841387, 1907; No. 879532, 1908.

used extensively also as amplifier, oscillation generator, and in a large number of widely varying applications. Fig. 68 shows a commercial type of thermionic amplifier.

52. Action of the Auxiliary Electrode. It was shown in Chapter III, page 42, that the relation between the electron current to the anode or plate, and the potentials applied to the grid and plate with respect to the filament, can be expressed in a simple way by making use of the writer's linear stray field relation:

$$E_s = \frac{E_p}{\mu} + \epsilon \quad . \quad . \quad . \quad . \quad . \quad . \quad (1)$$

which means that if the grid and filament be at the same potential, a potential difference E_p between filament (or grid) and plate, causes a stray field to act through the openings of the grid which is equivalent to the field that would be produced if a potential difference equal to $\frac{E_p}{\mu}$ were applied directly between the filament and a plane coincident with that of the grid. The small quantity ϵ represents an intrinsic potential difference between the filament and the system constituted by the grid and plate. The constant μ depends on the structure of the device (see p. 226). If we now apply a potential difference E_g directly between filament and grid, the effective voltage in the tube is obtained simply by adding E_s and E_g, and the current can be expressed as a function of this sum, thus

FIG. 68.

$$I = f\left(\frac{E_p}{\mu} + E_g + \epsilon\right). \quad . \quad . \quad . \quad . \quad (2)$$

Before discussing this relationship, let us look more fully into the functions of the two quantities E_s and E_g. To simplify matters somewhat we shall neglect the effect of the small quantity ϵ. In Fig. 69, the distribution of the field intensity in the region

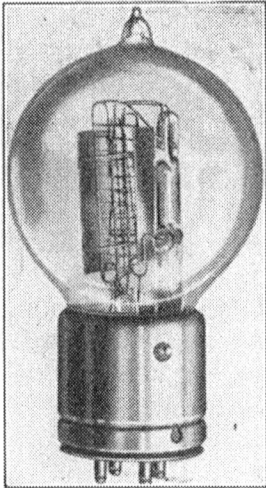

between cathode and anode of a three-electrode device is represented by means of lines of force. The anode is assumed to remain at a constant positive potential with respect to the cathode, the potential of which we can call zero. The three diagrams shown refer to the cases in which the potential of the grid is positive, zero and negative. Looking upon the intensity of the field as the number of lines of force passing through unit area, it will readily be seen in a general way how the potential of the grid affects the flow of electrons from the cathode.

But before considering the flow of electrons it must be pointed out that the diagrams in Fig. 69 represent the distribution of field intensity only for the case in which the space between cathode

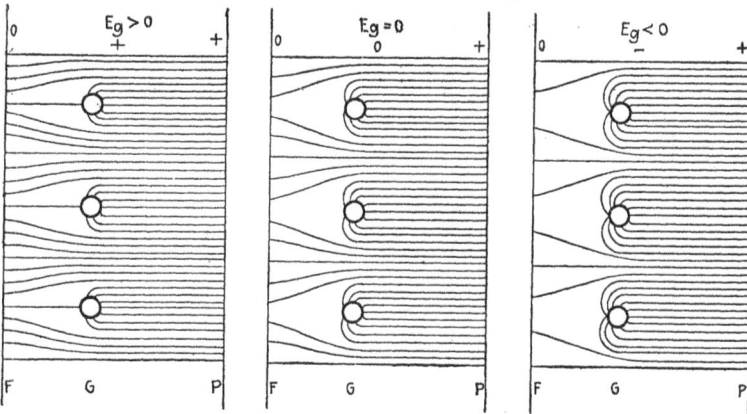

Fig. 69.

and anode is free from any dislodged electric charges. As soon as charges are introduced, such as electrons moving from cathode to anode, some of the lines of force proceeding from the anode will end on the electrons, and hence the density of the lines of force, i.e., the intensity of the field or the potential gradient, will be greater near the anode and less near the cathode than indicated in Fig. 69. This space charge effect can be made clearer by representing the field intensity as shown in Figs. 70 and 71. Fig. 70 shows the case in which there are no electrons in the space, such as would be the case if the cathode were cold. *BP* represents the potential of the anode, that of the cathode being zero. The potential gradient (or field intensity) is given by the slope of the lines *PaO*, etc. It is easily seen that the field between cathode and grid is the resultant of E_s and E_g. The lines PaO, PbO and PcO therefore represent the distribution of field intensity for the three cases in which E_s+E_g is greater than, equal to or less than zero.

If, now, the cathode be hot enough to cause a copious emission of electrons from it, the field intensity is no longer a linear function of the distance between cathode and anode, but can be represented in a rough way by the curves shown in Fig. 71. If $E_s+E_g>0$, the field distribution can be represented somewhat by the curve OaP. If $E_s+E_g \gtrless O$, the field between the cathode and the equivalent grid plane is negative and the emitted electrons are

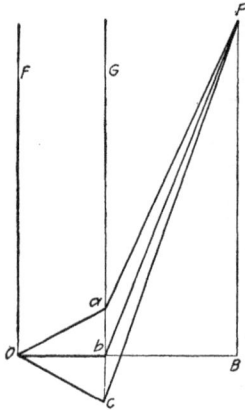

Fig. 70.

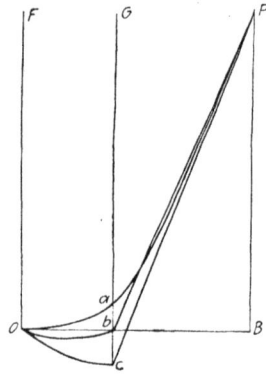

Fig. 71.

returned to the cathode. The curvature of the lines Ob and Oc is due to the initial velocities of the electrons (see Fig. 21).

The lines of force proceeding from the anode that reach through the grid represent the stray field due to E_s, which therefore tends to draw the electrons through the grid and throw them on to the anode. By varying the potential E_g of the grid the intensity of the field between the grid and cathode is varied in such a manner that the effect of E_g is similar to that of E_s, and whether or not electrons will flow away from the cathode depends on the resultant value of E_s and E_g. Now, E_s is always positive and therefore E_s+E_g will be positive (1) when E_g is positive, and (2) if E_g is nega-   .
tive and less than E_s.

(1) When E_g is positive some of the electrons moving away

from the cathode are drawn to the grid (see Fig. 69), while the rest are drawn through the openings of the grid to the anode under the influence of E_s. The relative number of electrons going to and through the grid depends upon the mesh of the grid, the diameter of the grid wires and the relative values of E_s and E_g. When, for example E_s is large compared with E_g the number of electrons going to the grid is comparatively small, but for any fixed value of E_s the grid current increases rapidly with increase in E_g. Hence, for positive values of E_g current will be established in the grid circuit FGE_g (Fig. 72).

(2) If, however, E_g is negative and less than E_s, as is generally the case, nearly all the electrons drawn away from the cathode pass to the plate, practically none going to the grid. In this case the resistance of the grid circuit is practically infinite for low frequen-

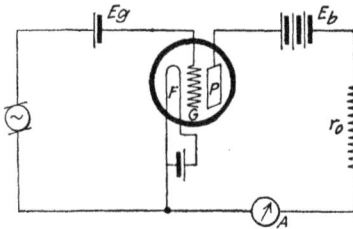

FIG. 72.

cies. The electrostatic capacities between the electrodes causes the impedance between filament and grid to have a value depending on the output circuit. For the present we shall neglect this effect, which is usually small, and later on investigate the conditions under which it can manifest itself to a marked extent.

If, now, an alternating E.M.F. be impressed on the grid circuit so that the grid becomes alternately positive and negative with respect to the filament or cathode, the resistance of the grid circuit FGE_g which is usually referred to as the input circuit, will, if the frequency is not too high, be practically infinite for the half cycle that the grid is negative and finite but variable for the positive half cycle. If, on the other hand, the alternating E.M.F. be superimposed upon a constant negative grid potential, which is so chosen with respect to the value of the impressed alternating voltage that the grid always remains negative with respect to the filament, the resistance of the input circuit is infinite.

It can now be seen in a general way how the device functions as a relay. Any variation in the grid potential changes the intensity of the field between filament and grid, resulting in a corresponding change in the number of electrons moving from filament to plate. Hence potential variations set up between filament and grid cause variations in current in the output circuit $PE_b r_0$, the power developed in the load r_0 being greater than that expended in the input circuit.

53. Current-voltage Characteristics of the Thermionic Amplifier. . Returning now to a consideration of the expression for the current

$$I_p = f\left(\frac{E_p}{\mu} + E_g + \epsilon\right). \quad \cdot \quad \cdot \quad \cdot \quad \cdot \quad \cdot \quad (2)$$

it is to be noticed in the first place that since this equation contains two independent variables, E_p and E_g, the three-electrode device possesses two families of characteristics, or the complete characteristic can be represented by a surface. The current as a function of the filament-plate voltage E_p can for various negative values of filament-grid voltage E_g be represented by a series of curves such as those shown in Fig. 73. It will be noticed that each of these curves is similar to the current-voltage characteristic of the simple two-electrode thermionic valve discussed in Chapter IV. The main difference is that in the three-electrode tube the current is limited not only by space charge and the voltage drop in the filament, but also by the grid. For the same potential on the plate the current in the three-electrode tube will therefore be smaller than in a simple valve. This follows directly from quation (2).

The relation between I_p and E_g for various values of E_p can be expressed by a set of curves similar to those shown in Fig. 73. Fig. 74 shows such a set of characteristics. The ordinates represent current to the plate and not necessarily the emission current, i.e., total current from the filament. When the grid becomes positive it takes current and so can distort the I_p, E_g curves. A set of grid current curves for various plate potentials is shown in Fig. 75. For the higher plate potentials these curves show a maximum. This is due to secondary electron emission from the grid by the impact of electrons coming from the filament (see p. 47). At the lower plate potentials the stray field given by $\dfrac{E_p}{\mu}$ is smaller, and fewer electrons are attracted to the plate. The secondary electron emission is then also less marked, so that the $I_g - E_g$ curve shows a rapid increase of I_g with increasing E_g.

So far it has not been possible to derive the equation of the whole characteristic theoretically with sufficient accuracy. For

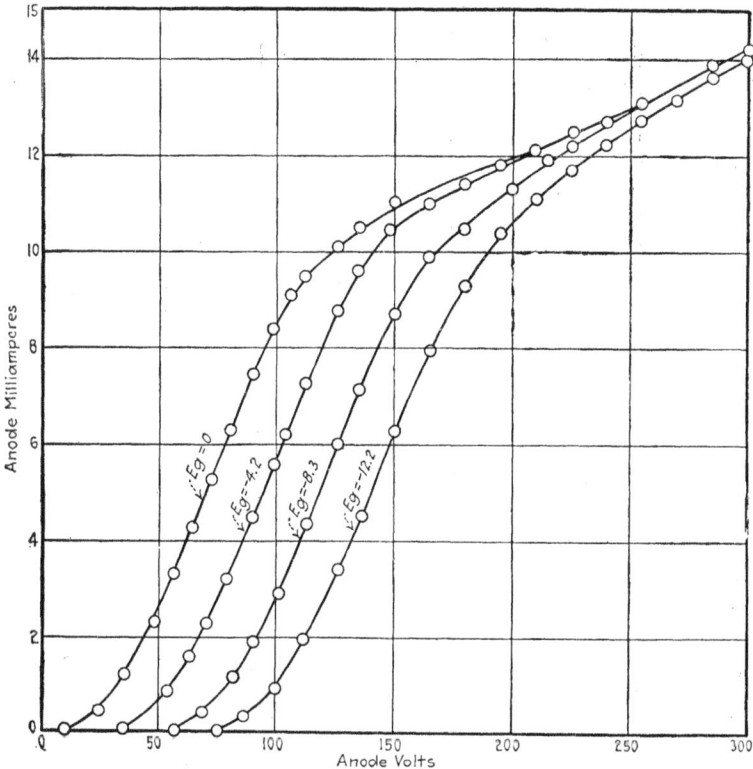

FIG. 73.

the operating range of the characteristic, when the tube functions as an amplifier, the plate current can be expressed by the equation.[1]

$$I_p = \alpha \left(\frac{1}{\mu} \Sigma E_p + \Sigma E_g + \epsilon \right)^2, \quad \cdot \quad \cdot \quad \cdot \quad \cdot \quad \cdot \quad (3)$$

where ΣE_p and ΣE_g are the filament-plate and filament-grid

[1] H. J. VAN DER BIJL, Phys. Rev., Vol. 12, p. 180, 1918.

voltages. If, for example, an alternating e.m.f. $e \sin pt$ be impressed on the grid circuit the equation takes the form:

$$I_p = \alpha \left(\frac{E_p}{\mu} + E_g + e \sin pt + \epsilon \right)^2. \quad \ldots \ldots \quad (4)$$

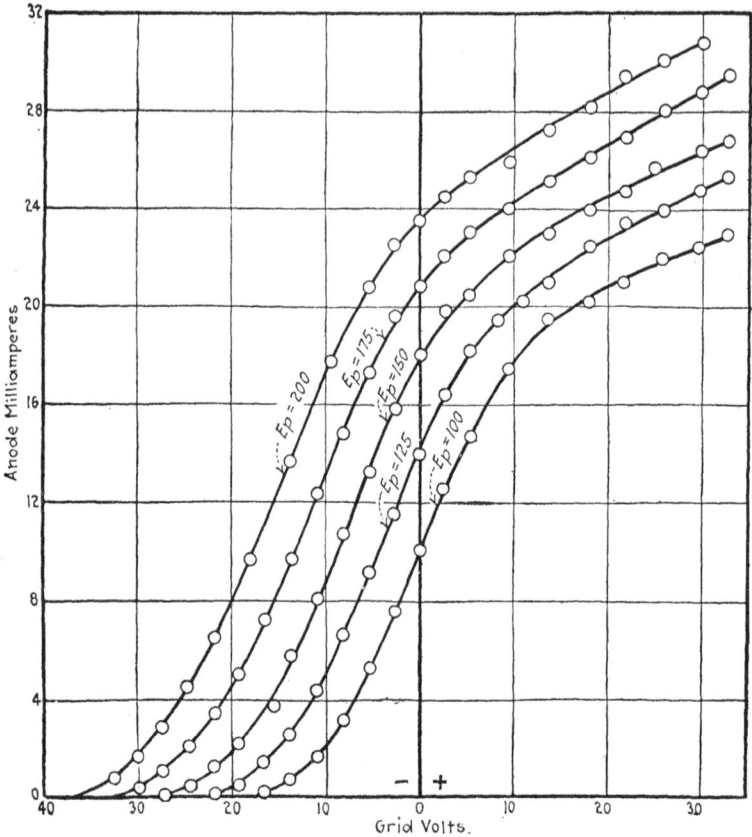

FIG. 74

This equation was determined empirically and is subject to certain limitations. In the first place, it does not apply to the horizontal part of the characteristic which gives the saturation current, but only to that part which obtains when the filament is hot enough to emit more electrons than are needed for the current convection through the tube. This is the condition under which the amplifier operates, because here the plate current can be varied by varying any of the applied voltages. Another condition for equation (4) is that the grid should not become sufficiently positive to distort the characteristic. Under these conditions I have generally found this equation to hold sufficiently well, to a first approximation at least, and have been using it in connection with work on the amplifier tube. The above equation does, however, not hold sufficiently accurately for purposes of radio detection, since this is determined by second order quantities.

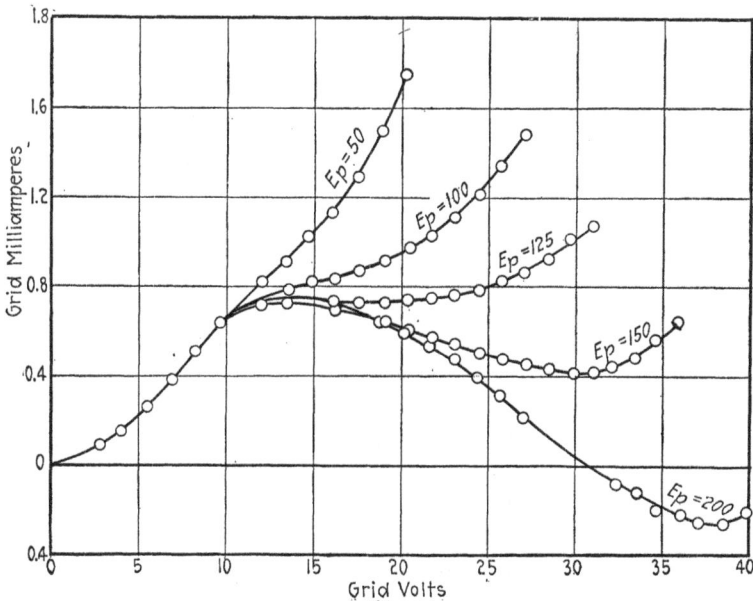

FIG. 75.

Latour [1] has derived some equations for the "relay effect" of audion tubes. He starts from the general functional expressions for the plate and grid currents: $I_f = F(E_p, E_g)$ and $I_g = f(E_g, E_p)$. In the expansion of these equations he neglects all quantities of the second and higher order, thus assuming that the current and voltage variations are very small, or that the characteristic is linear over the operating range.

[1] M. LATOUR, Electrician, December, 1916.

Vallauri[1] also assumes a linear characteristic by expressing the equation for the plate current in the form:

$$I_p = aE_g + bE_p + c.$$

It will be shown later that it is important to distinguish between the characteristic of the tube itself and that of the tube and external circuit combined. The latter can by taking special precautions be made practically linear. The characteristic of the tube itself can, however, not be regarded as linear over the range over which most amplifiers operate. The curvature of the characteristic cannot be neglected because it introduces distortion which, unless properly taken care of, makes it practically worthless as a telephone repeater, for example, on long telephone lines. When treating the tube as an oscillation generator the curvature of the characteristic can be neglected, because the oscillation current is established in an oscillation circuit which is usually tuned sufficiently sharply to eliminate the harmonics caused by the curvature of the characteristic.

Equation (3) gives the characteristic of the tube itself; that is, E_p and E_g are the potentials of the plate and grid with respect to the filament, and are not necessarily equal to the plate and grid battery voltages. E_p is, for example, only equal to the plate battery voltage E_b when the external resistance r_0 is zero (Fig. 72). When r_0 is not zero the potential drop established in r_0 by the current in the plate circuit causes a decrease in E_p, and it can readily be seen that if the current be varied, by varying the grid potential, E_p becomes a function of the plate current. This effect will be discussed more fully when we come to consider the characteristic of the tube and circuit (Section 58).

Langmuir[2] has expressed the equation for the characteristic as

$$I_p = A (E_p + kE_g)^{3/2}.$$

The extent to which the characteristics of practical tubes depart from the $\frac{3}{2}$-power relation was discussed in Chapter IV. In the case of the two-electrode tube the main cause of the deviation is the voltage drop in the filament. This has a greater

[1] G. VALLAURI, L'Elettrotecnica, Vol. 4, 1917, Electrician, Vol. 80, p. 470, 1917.

[2] I. LANGMUIR, Proc. I.R.E., p. 278, 1915.

effect at the lower than at the higher voltages. In three-electrode tubes the limitation of current by the grid accentuates this deviation. Thus, referring to equation (3), the constant μ is generally greater than unity and, therefore, although the plate voltage may be high, the effective voltage $\left(\dfrac{E_p}{\mu}+E_g+\epsilon\right)$ is low, so that the voltage drop in the filament has a relatively greater effect in causing a deviation from the $\frac{3}{2}$-power relation. As an example, suppose that $E_p = 100$ volts; $\mu = 5$; the voltage in the filament $E_f = 10$, and $E_g + \epsilon = 0$. Then the effective voltage is only twice the voltage drop in the filament. Under such conditions the deviation from the $\frac{3}{2}$-power relation is considerable. It is for this reason that the quadratic equation (3) is generally found to be more serviceable at least for that range of the characteristic over which the tube operates as an amplifier.

The quantity ϵ, which depends on the intrinsic potential difference between the filament and the system constituting the grid and plate, is usually small, but may, in some types of tubes, vary considerably. For tubes operating with high effective voltages ϵ can generally be neglected. But when the effective voltage is low, as in the detector and small amplifier tubes, variations in ϵ can, if not corrected for, cause deviations in the exponent of the effective voltage.

The important thing about the tube equation is that the current can be expressed as a function of $(E_p + \mu E_g)^1$.

Referring to equation (3) and Fig. 74, we see that the current is finite for negative values of the grid potential, and is reduced to zero only when

$$E_g = -\left(\frac{E_p}{\mu}+\epsilon\right) = E_s. \quad \ldots \ldots \ldots \quad (5)$$

This linear relation and equation (3) can be verified experimentally when the constants μ and ϵ are known. These constants

---

[1] This expression for the effective voltage in a three-electrode tube was established experimentally by the author and published in 1913 (Verh. d. D. Phys. Gesell., Vol. 15, p. 330, 1913). See also p. 44. The same expression has also been used by SCHOTTKY (Archiv. f. Elektrotechnik, Vol. 8, p. 1, 1919). BARKHAUSEN (Jahrb. d. drahtlosen Tel. & Tel., Vol. 14, p. 27, 1919) and others. See also W. H. ECCLES (Rad. Rev., Vol. 1, p. 69, Nov., 1919).

can be determined by methods which do not involve the exponent of equation (3). Let us assume a general exponent β, thus:

$$I_p = \alpha \left(\frac{E_p}{\mu} + E_g + \epsilon \right)^\beta.$$

Assuming the general case in which both E_p and E_g are variable, we have:

$$\frac{dI_p}{dE_g} = \frac{\partial I_p}{\partial E_p} \frac{dE_p}{dE_g} + \frac{\partial I_p}{\partial E_g}.$$

Now

$$\frac{\partial I_p}{\partial E_p} = \frac{\alpha \beta}{\mu} \left(\frac{E_p}{\mu} + E_g + \epsilon \right)^{\beta-1}$$

$$\frac{\partial I_p}{\partial E_g} = \alpha \beta \left(\frac{E_p}{\mu} + E_g + \epsilon \right)^{\beta-1}.$$

Hence

$$\frac{dI_p}{dE_g} = \alpha \beta \left(\frac{E_p}{\mu} + E_g + \epsilon \right)^{\beta-1} \left(\frac{1}{\mu} \frac{dE_p}{dE_g} + 1 \right). \quad . \quad . \quad (6)$$

Since the current can be varied by varying either one or both of the independent variables E_p and E_g, we can make these variations in accordance with the condition that the current I_p remains constant; for example, the current can be first increased by increasing E_p and then brought back to its original value by increasing the negative grid voltage E_g. The relation between the variations in E_p and E_g necessary to keep the current constant, can be obtained by putting $I_p =$ constant in equation (6). Then we have either

$$\frac{E_p}{\mu} + E_g + \epsilon = 0 \quad . \quad . \quad . \quad . \quad . \quad . \quad . \quad (5a)$$

or

$$\frac{dE_p}{dE_g} = -\mu. \quad . \quad . \quad . \quad . \quad . \quad (7)$$

These equations are therefore independent of the exponent of (3). Equation (5a) obviously states the condition that the current has the constant value zero, and shows that the stray field potential E_s is simply equal to the absolute value of the grid potential which is necessary to reduce the plate current to zero.

Referring to the above equations for the partial derivatives of I_p, it follows that a change in the grid potential produces μ-times as great a change in the plate current as an equal change in the plate voltage.

Equation (7) can be interpreted to mean that a potential variation $\delta E_g = e_g$ impressed between the grid and the filament is equivalent to introducing an E.M.F. in the plate circuit which is equal to μe_g.

This result is of fundamental importance and has been found of great value in the solution of many vacuum tube problems.

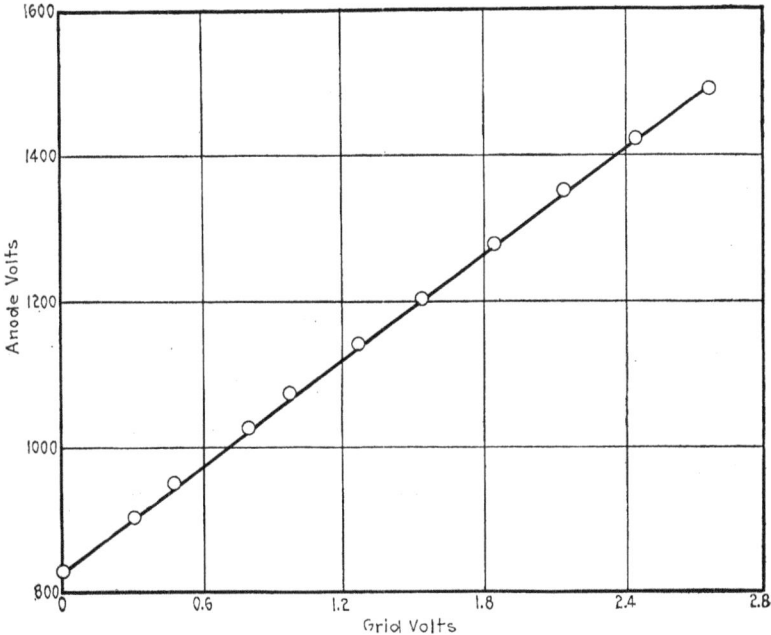

FIG. 76.

Integrating equation (7) we get

$$E'_p = E_p + \mu E_g. \quad \ldots \ldots \ldots (8)$$

While equation (5a) gives the relation between E_p and E_g necessary to neutralize the stray field and keep the current zero, equation (7) gives the relation necessary to keep the current constant at any convenient value. The verification of these relations is shown in Fig. 76.[1] The slope of these curves is equal to the constant μ.

[1] H. J. VAN DER BIJL, Phys. Rev., Vol. 12, p. 171, 1918. See also Fig. 14, p. 45.

The characteristic equation (3) was verified as follows: The tube was inserted in a circuit such as shown in Fig. 72, with the exception that the generator in the input circuit and the resistance r_0 were omitted. A convenient negative potential was applied to the grid, so that no current could be established in the grid

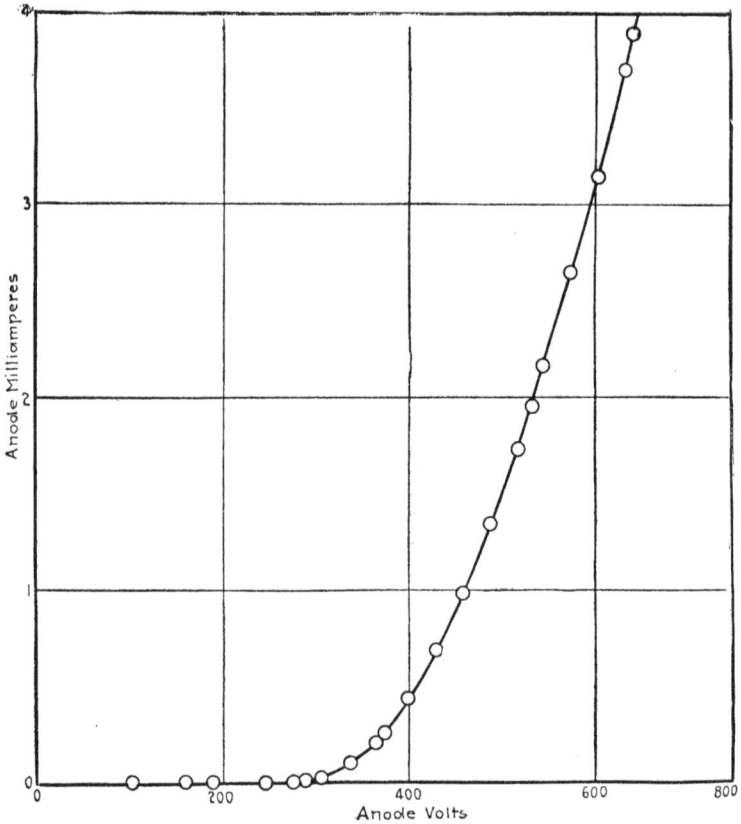

Fig. 77.

circuit, and the current in the plate circuit observed as a function of the plate voltage E_p. Since r_0 was zero E_p was always equal to E_b, the plate battery voltage. The grid being kept at a constant negative potential E_g with respect to the filament, current could not be established in the plate circuit until the $\dfrac{E_p}{\mu} + \epsilon$ became greater than E_g. The characteristic obtained is shown in Fig. 77.

From the value of the plate voltage for which the current is just reduced to zero we get

$$\epsilon = - \left(\frac{E_p}{\mu} + E_g \right),$$

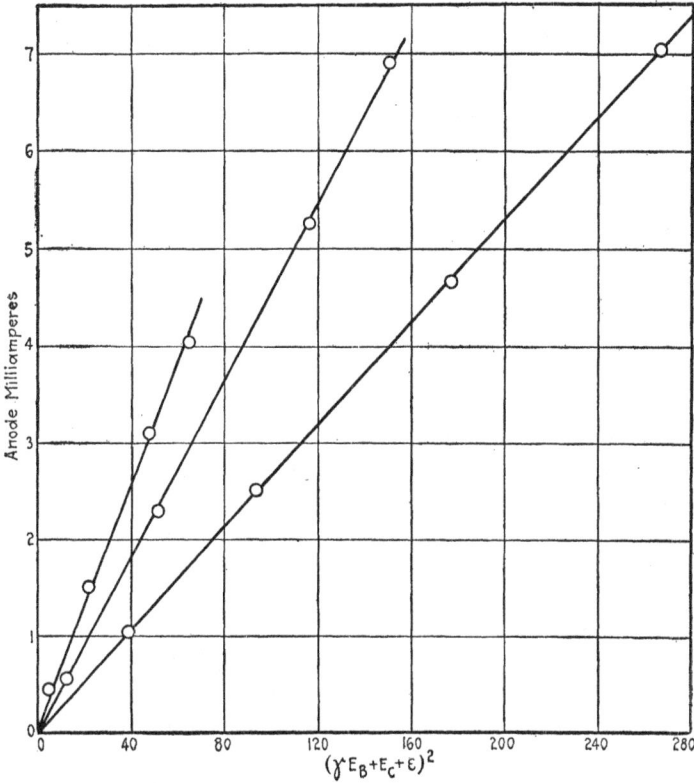

Fig. 78.

and since μ could be determined as explained above, this equation could be used to give ϵ. Once μ and ϵ are known the observed current can be plotted as a function of the expression

$$\left(\frac{E_p}{\mu} + E_g + \epsilon \right)^2$$

for arbitrary values of E_p or E_g. Some curves obtained in this way are shown in Fig. 78.

If we obtain a number of characteristics such as those shown in Fig. 74, which show the relations between the plate current and grid potential for a number of different plate potentials and plot the logarithms of I_p against the logarithms of the effective voltage $\left(\dfrac{E_p}{\mu}+E_g\right)$ the observed points for all the characteristics should, according to equation (3), lie on one straight line. This can be done by subtracting the applied grid potentials from the grid potential which is just necessary to reduce the current to zero, and plotting on logarithmic paper the values so obtained against the observed currents. (Note that the value of the grid potential necessary to reduce the current to zero is $\dfrac{E_p}{\mu}$.) The disadvantage of such a procedure lies in the uncertainty of the voltage at which the current becomes zero. However, the logarithmic plot of the curves of Fig. 74, and which is shown in Fig. 79 indicates a substantially good verification of equation (3). The slope of this lumped logarithmic line is almost exactly 2.

54. Amplification Constant. The constant μ appearing in the above equations is one of the most important constants of the audion or three-electrode tube. It will be shown later that μ is the maximum voltage amplification obtainable from the tube. This constant is also very instrumental in determining the current and power amplification and can therefore be referred to as the *amplification constant*. This constant plays an important part in all functions of the tube, as will be shown later when we come to consider its use as a radio detector, modulator, oscillation generator, etc. It will be noticed that since it appears in the stray field relation (equation (1)), which is a pure potential relation, the amplification constant is a function only of the geometry of the tube. It depends, for example, on the mesh of the grid, diameter of the grid wire and the distance between grid and plate. It can be determined from E_pE_g—curves shown in Fig. 76 and by methods which will be described later. In practice it is generally found that μ is not quite constant, its value decreasing somewhat at lower voltages. For the operating range of voltages commonly employed its value does, however, not vary much. (See Fig. 125.)

55. Plate Resistance and Impedance. The resistance of a tube is due to the work which the electrons emitted from the cathode must do in moving from cathode to anode. Let us consider the case of a single electron emitted from the cathode. In moving through the cathode surface it has to do an amount of work equivalent to the electron affinity and in moving from cathode to anode it has to do work in overcoming the contact potential difference between cathode and anode. This may sometimes assist the electron in moving from cathode to anode. (See Chapter III.) The total amount of work it has to do to overcome these

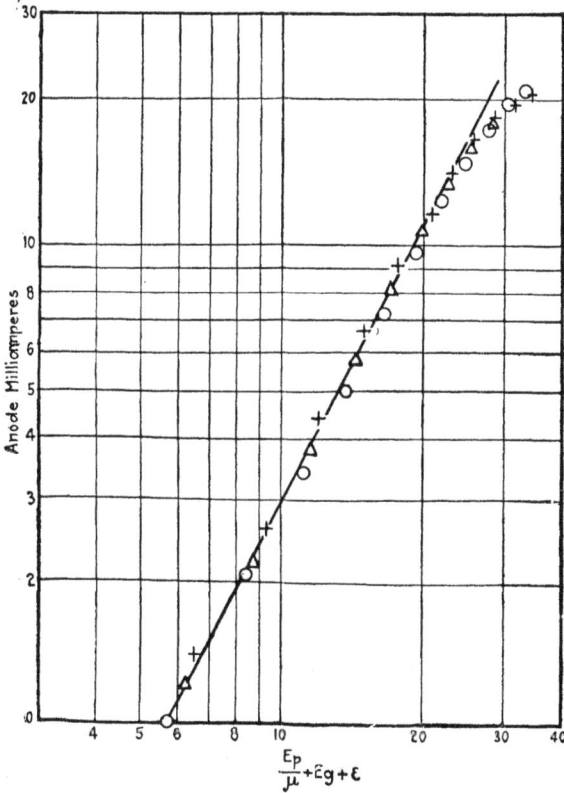

FIG. 79.

forces is generally small and never amounts to more than a drop of a few volts. If these were the only forces exerted on a large number of electrons escaping from the cathode the application of a small voltage between cathode and anode would almost immediately give rise to the saturation current, and the resistance of the tube would for all values of current less than the saturation current be very low. This is, however, not the case, since the electrons in the space exert a mutual repelling force on one another. This is the space charge effect explained in Chapters I and IV, and causes by far the greatest expenditure of energy on the part of the electrons in moving to the anode. This expenditure of energy causes the heating of the anode.

The true d-c. resistance of the tube is, of course, given simply by the ratio of the total amount of work done to the square of the current, i.e., by $\frac{E_p}{I_p}$. The a-c. resistance on the other hand, is given by the slope of the plate current characteristic, and since the characteristic is non-linear the a-c. and d-c. resistances are not the same. Referring to Fig. 80, the d-c. resistance at a voltage E_p is

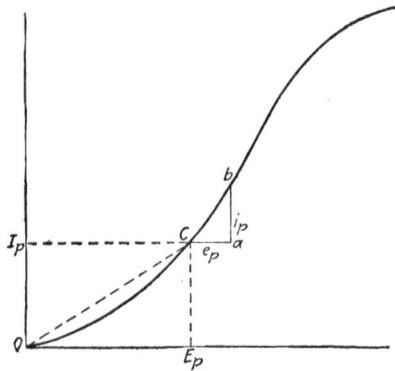

Fig. 80.

given by the reciprocal of the slope of the straight line OC, while the impedance of the tube is given by the ratio of the alternating voltage e_p between filament and plate to the alternating current i_p in the plate circuit. Now, the flow of electrons in the tube shows no lag, and for frequencies low enough to make the effect of the electrostatic capacity of the tube itself negligibly small, the condensive reactance thus being also practically infinite, the impedance is simply given by $\frac{e_p}{i_p} = \frac{ca}{ab}$ (see Fig. 80), and is then of the nature of a pure resistance. For most tubes used at present this approximation is satisfactory for frequencies up to the order of several hundred thousand cycles per second. For a tube like that shown in Fig. 68, for example the filament-plate capacity is of the order of a few micro-microfarads. Now we have $\dfrac{i_p}{e_p} = \dfrac{\partial I_p}{\partial E_p}$ when e_p and i_p are very small. But in practice we generally do not deal with very small current variations. To obtain an expression for the a-c. resistance for finite variations we must evaluate the partial derivative $\dfrac{\partial I_p}{\partial E_p}$ from the equation of the characteristic and integrate it over a complete cycle of variations, thus:

$$\frac{1}{r_p} = \frac{1}{2\pi} \int_0^{2\pi} \frac{\partial I_p}{\partial E_p}\, dt. \quad \cdots \cdots \quad (9)$$

For frequencies at which the electrostatic capacity of the tube cannot be regarded as negligibly small, we have in effect a condenser in shunt with the tube resistance. If x is the reactance due to the capacity of the tube the plate impedance Z_p can be obtained from the admittance Y_p:

$$\frac{1}{Z_p} = Y_p = \frac{1}{2\pi} \int_0^{2\pi} \frac{\partial I_p}{\partial E_p}\, dt - jb_p = g_p - jb_p \quad \cdots \quad (10)$$

where

$$\left.\begin{aligned} g_p &= \frac{r_p}{r_p^2 + x^2} \\[2mm] b_p &= \frac{x}{r_p^2 + x^2} \end{aligned}\right\} \quad \cdots \cdots \cdots \quad (11)$$

To evaluate expression (9) let us assume a general exponent for the characteristic equation:

$$I_p = \alpha \left(\frac{E_p}{\mu} + E_g + \epsilon + e \sin pt \right)^n. \quad \cdots \quad (12)$$

Then

$$\frac{\partial I_p}{\partial E_p} = \frac{\alpha n}{\mu} (E_\gamma + e \sin pt)^{n-1}$$

where

$$E_\gamma = \frac{E_p}{\mu} + E_g + \epsilon.$$

Hence

$$\frac{1}{r_p} = \frac{\alpha n E_\gamma^{n-1}}{2\pi\mu} \int_0^{2\pi} \left(1 + \frac{e}{E_\gamma} \sin pt \right)^{n-1}. \quad \cdots \quad (13)$$

Now the maximum value e of the input voltage is never greater than E_γ; for distortionless amplification e must always be less than E_γ (see Section 60).   Referring, for example, to Fig. 81, it will be seen that E_γ is the intercept cd when $E_g = 0$ or fd when $E_g = cf$.   Taking the latter case it will be seen that the maximum value of the input voltage e should not exceed the value fd otherwise we would be working beyond the point d, and then the lower peaks of the output current wave would be chopped off

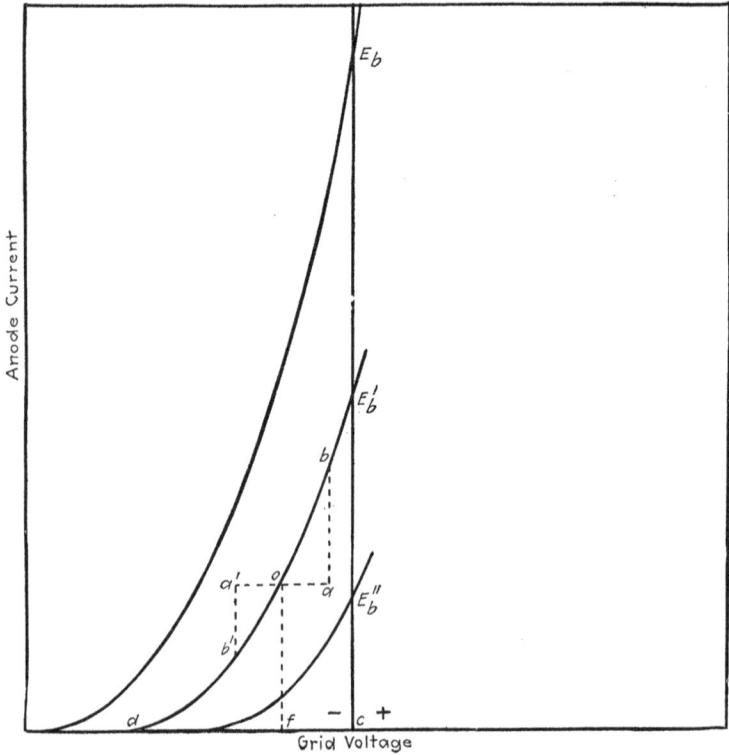

Fig. 81.

thus introducing harmonics.   Furthermore, since the maximum value of $\sin pt$ is unity and its odd powers vanish on integration the expression in the parentheses can be expanded into a series, the integral of which converges sufficiently rapidly to enable us to compute the resistance, for all practical values of n, from a few terms of the expansion.   The integrated series is:

$$\frac{1}{r_p}=\frac{\alpha n E_\gamma{}^{n-1}}{\mu}\left[1+\frac{(n-1)(n-2)}{(2)^2}\left(\frac{e}{E_\gamma}\right)^2+\frac{(n-1)\;\ldots\;(n-4)}{(2.4)^2}\left(\frac{e}{E_\gamma}\right)^4\right.$$

$$\left.+\frac{(n-1)\;\ldots\;(n-6)}{(2.4.6)^2}\left(\frac{e}{E_\gamma}\right)^6\ldots\right].\quad(14)$$

For the case of the amplifier $n=2$, and all the terms except the first vanish so that we get:

$$r_p=\frac{\mu}{2\alpha\left(\dfrac{E_p}{\mu}+E_g+\epsilon\right)}.\quad\ldots\ldots\quad(15)$$

The a-c. plate resistance is therefore to a first approximation independent of the a-c. input voltage.

It is customary to speak generally of the impedance of the tube, meaning thereby the plate impedance. It must, however, be remembered that unless the frequency is very high, the wattless component of the impedance is practically zero, and the impedance is then given by equation (14) in which case it is of the nature of a pure resistance. For a parabolic characterictic ($n=2$) equation (14) is simply the slope of the I_p-E_p characteristic.

56. Mutual Conductance. So far we have considered only the plate voltage-plate current characteristic from which we have deduced the plate impedance by obtaining an expression for the variation in plate current as a function of the plate potential variations. The usual thing in practice is to vary the plate current by varying the grid potential. As will be seen from the following the effects in this case can be deduced directly from the previous considerations by the introduction of the amplification constant μ.

Referring to the fundamental equation for the characteristic, we have

$$\frac{\partial I_p}{\partial E_g}=\mu\frac{\partial I_p}{\partial E_p}\quad\ldots\ldots\quad(16)$$

and

$$g_m=\frac{1}{2\pi}\int_0^{2\pi}\frac{\partial I_p}{\partial E_g}dt=\frac{\mu}{2\pi}\int_0^{2\pi}\frac{\partial I_p}{\partial E_p}dt.\quad\ldots\ldots\quad(17)$$

Equation (16) gives the slope of the plate current-grid potential characteristic and (17) gives the mutual conductance[1] g_m. It will be noticed that

$$g_m=\frac{\mu}{r_p}.\quad\ldots\ldots\quad(18)$$

[1] The expression " mutual conductance " for this quantity was suggested by HAZELTINE (Proc. I. R. E., Vol. 6, p. 63, 1918).

This is a very important quantity and is involved, as will be shown later, in all expressions giving the degree of merit of the tube when functioning as amplifier, detector, oscillator, etc. It is always desirable to have the mutual conductance as large as possible. While μ depends almost entirely on the structure of the grid and its position relative to the other electrodes, r_p depends upon μ and the the surface areas of cathode and anode as well.

The mutual conductance gives a measure for the effect of the grid potential on the plate current. The analogous expression for the effect of the plate potential on the grid current is given by

$$g_n = \frac{1}{2\pi} \int_0^{2\pi} \frac{\partial I_g}{\partial E_p} dt \quad . \quad . \quad . \quad . \quad . \quad . \quad (17a)$$

and may be called the reflex mutual conductance. At frequencies for which electrode capacities are effective to an appreciable extent we have to consider the mutual impedances $Z_m = g_m + jx_m$, etc., which cannot be obtained from the static characteristics of the tube.

57. Shape of Output Wave in Circuit of Low External Impedance. Consider the case of the tube circuit shown in Fig. 82

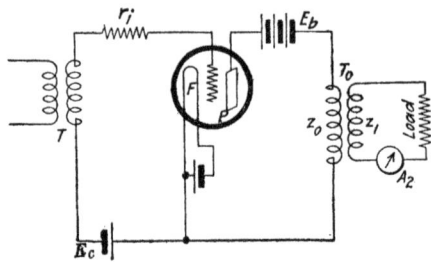

FIG. 82.

and let a voltage $e \sin pt$ be impressed between filament and grid. The resistance r_i may be that of the input transformer coil which is supposed to be wound to work into a practically open circuit. For the present we shall suppose that the external impedance Z_0 in the output is negligibly small compared with the plate-resistance of the tube, so that the characteristic of the circuit is practically the same as that of the tube.

If the constant grid voltage E_c is sufficiently negative to insure that the grid never takes current, the wave shape in Z_0 is deter-

mined by the characteristic equation (4). Thus if the voltage is
a sinusoid (Fig. 83, curve a) the output current is a lop-sided curve
shown in Fig. 83, curve bb'. This can readily be seen by referring
to Fig. 81, from which it will be seen that if the potential of the

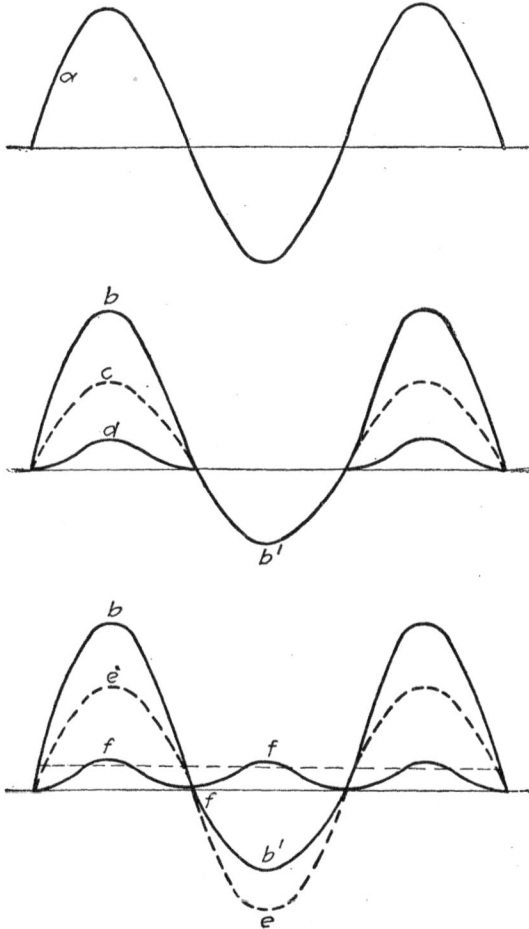

FIG. 83.

grid be varied about the value $E_c = cf$, the increase ab in plate current due to the decrease oa in the negative grid potential is greater
than the decrease $a'b'$ in current caused by an equal increase
oa' in the negative grid potential. This would produce distortion

since the output current is not an exact reproduction of the input. The curve b of Fig. 83, of course, shows the variation in the output current from its mean d-c. value which obtains when the input alternating voltage is zero.

Now, suppose the grid battery be so adjusted that whenever $e \sin pt$ is positive the grid takes current. Since the grid current characteristic is of the nature shown in Fig. 75, the grid current wave will be given by d (Fig. 83), providing the grid potential does not become sufficiently high to cause the emission of secondary electrons from it. Now this current in the grid circuit causes a voltage drop in r_i thus lowering the potential difference between filament and grid. The output current wave may therefore take the shape shown by curve cb'. This would therefore minimize the distortion if the quantities involved were correctly proportioned. On the other hand, once the grid current is established it increases so rapidly with further increase in the grid voltage that the increase in plate current during the half cycle when the grid is positive can become less than the decrease during the other half cycle, and this materially lowers the amplification.

Referring again to the case in which the grid is kept negative with respect to the filament, it will be seen on expanding equation (4) that the curve bb' (Fig. 83) consists of the following components:

$$I_p = \alpha\left(\frac{E_p}{\mu} + E_g + \epsilon\right)^2 + 2\alpha\left(\frac{E_p}{\mu} + E_g + \epsilon\right)e \sin pt$$
$$+ \frac{\alpha e^2}{2}\cos(2pt + \pi) + \frac{\alpha e^2}{2}. \quad (19)$$

The first term represents the steady direct current in the plate circuit which is maintained when the input e is zero, and is the value about which the plate current varies for finite values of e. The second term gives the alternating output current (ee, Fig. 83) which is in phase with the input voltage and which is the only useful current for amplification purposes. The harmonic represented by the third term and having double the frequency of the fundamental is present, as was to be expected from the parabolic shape of the characteristic. It is shown by the curve ff in Fig. 83. This is the undesirable term which causes distortion. The last term, which is proportional to the square of the input voltage is the change in the d-c. component due to the alternating input voltage (shown by the broken line in Fig. 83), and is the only effective component of the output current when using the device as a radio detector.

If the output transformer T_0 (Fig. 82) has a sharp frequency characteristic, and a pure note be impressed on the input of the tube, a current meter A_2 inserted in the load circuit would indicate a current which is proportional only to the second term of equation (19), that is, to the fundamental, and the distortion produced by the curvature of the characteristic would not be a serious matter. In telephony we have to deal, however, with frequencies ranging up to about 3000 cycles per second, and it is desirable to use a transformer with a flat frequency characteristic, so that all frequencies are transmitted with more or less equal facility. In such case the harmonic term would cause serious distortion of the speech wave. Distortion can, however, be reduced to a negligible quantity by properly choosing the impedance of the output transformer. We have so far assumed that the transformer impedance in the direction Z_0Z_1 is small compared with the plate resistance of the tube. In practice this is not so. In fact the best operation as power amplifier is obtained when this impedance is approximately equal to the plate resistance. In such case the characteristic of the circuit is different from that of the tube alone, as we shall now proceed to show.

58. Characteristic of Circuit Containing Tube and Resistance in Series. Let us first consider the simple case in which the output circuit is non-reactive and has a resistance r_0 (Fig. 72). Let the plate current be measured for different values of grid potential E_g, the plate battery voltage E_b remaining constant. When the grid is so much negative that the current in the plate circuit is reduced to zero, the plate voltage E_p is equal to the plate battery voltage E_b. But when current is established in the plate circuit there is a voltage drop across r_0, and E_p will be less than E_b, being given by

$$E_p = E_b - r_0 I_p, \quad \ldots \quad \ldots \quad \ldots \quad (20)$$

where I_p is the plate current, E_p thus becomes a variable. Substituting this value of E_p in the characteristic equation, thus:

$$I_p = \alpha \left[\left(\frac{E_b - r_0 I_p}{\mu} \right) + E_g + \epsilon \right]^2$$

we get, putting

$$\frac{E_b}{\mu}+E_g+\epsilon=E'_\gamma.$$

$$I_p=\frac{\left(\dfrac{2\alpha r_0 E'_\gamma}{\mu}+1\right)-\sqrt{1+\dfrac{4\alpha r_0 E'_\gamma}{\mu}}}{\dfrac{2\alpha r_0^2}{\mu^2}}. \qquad \ldots \quad (21)$$

This is the equation for the characteristic of the circuit consisting of the tube and resistance r_0, and it will be seen, if I_p be plotted against E'_γ, for various values of r_0, that the curvature of the characteristic is reduced as r_0 is increased, the characteristic becoming practically linear when r_0 is equal to or greater than the plate resistance. This is an important result for which I am indebted to my associate Dr. H. D. Arnold, and has an important bearing on the problem of distortionless amplification of telephonic currents.[1]

The effect of the resistance on the characteristic of the output circuit is shown graphically in Figs. 84 and 85. In the first the plate battery voltage E_b had a constant value equal to $\mu(E'_g+\epsilon)$, where E'_g is given by OO', while in Fig. 85 the plate battery was so adjusted for every value of r_0 as to keep E_p constant for zero grid voltage. It will be noticed that when $r_0=r_p$ (8000 ohms), the characteristic is substantially linear over a considerable range of input voltage.

59. Static and Dynamic Characteristics. So far we have considered only the static characteristics of the tube and its circuit. We have seen that the static characteristics of the tube itself, that is, the characteristics which are obtained when the external resistance is neglibibly small in comparison with the plate resistance of the tube, are given by equation (3), while those of the *output circuit*, containing an external resistance as well as the filament-plate resistance, are given by equation (21). The first mentioned set of characteristics is shown in Fig. 74, and the second set in Figs. 84 and 85. It is important to note that in Fig. 74 each characteristic is for a constant plate-filament voltage E_p, while in Figs. 84 and 85 each characteristic is for a constant plate battery voltage E_b. In this case, as was explained in the

---

[1] Other means for reducing distortion by using special circuit arrangements are discussed in Section 78.

previous paragraph, the filament-plate voltage E_p is variable, having a different value for each adjustment of the grid voltage E_g, due to the varying voltage drop in the external resistance. It is obvious that if the external plate circuit contains reactance

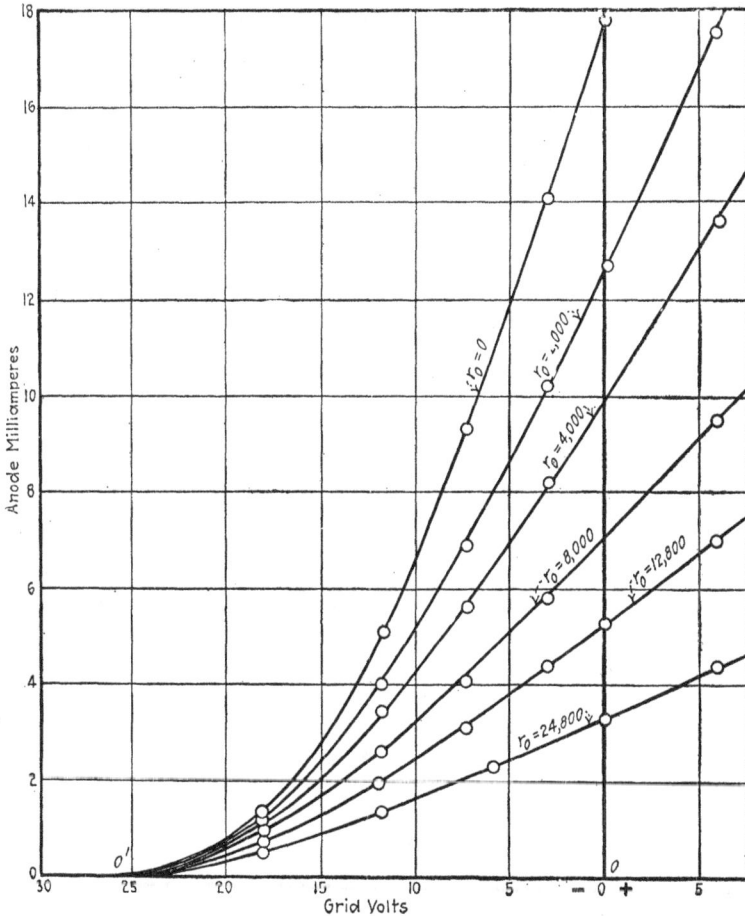

Fig. 84.

as well as resistance, the static characteristics take the same shape as when the reactance is zero, being determined only by the resistance component of the external impedance.

Now, the thermionic tube is used mostly in a-c. circuits. **It** is therefore necessary to know the shape of the characteristic that obtains when varying potentials are impressed on the grid, that is, it is necessary to know the shape of the dynamic characteristic. It is convenient to distinguish three cases: (1) the dynamic characteristic of the tube itself; (2) that of the plate circuit containing tube and non-inductive resistance, and (3) that of the circuit containing tube and impedance.

(1) The first can be determined with a circuit such as that shown in Fig. 72, provided that the resistance r_0 is zero and the current meter A has a resistance which is negligibly small com-

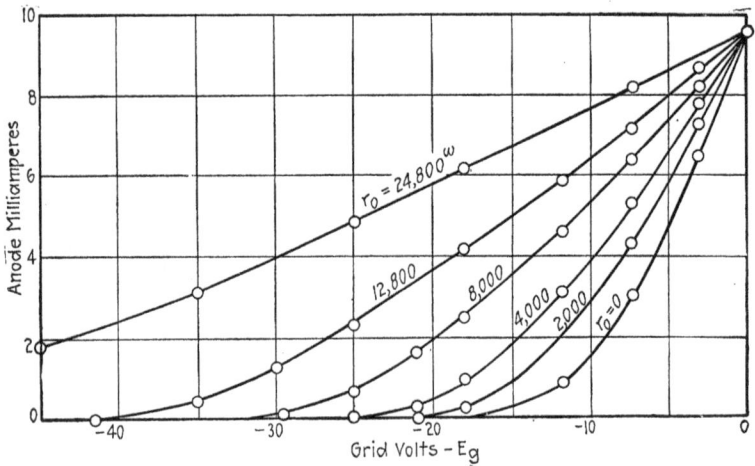

Fig. 85.

pared with the plate resistance. As far as the passage of electrons from cathode to anode is concerned, the thermionic tube, which operates with a pure electron discharge, shows no lag, such as is found to exist in an arc which depends for its operation on ionization by collision of the contained gas or vapor. The only reactance possessed by the tube is capacitive and is due to the electrostatic capacity between the electrodes. It is, therefore, in effect, a capacity shunted across the plate resistance. The capacity of ordinary tubes, is, however, so small (of the order of a few centimeters) that this parallel reactance can be regarded as practically infinite for frequencies ranging up to several hundred thousand cycles per second. Hence, for this range of frequencies the dynamic characteristic of the tube coincides with its static characteristic.

(2) If the external resistance r_0 (Fig. 72), instead of being zero, has a finite value and is non-inductive, the dynamic characteristic still coincides with the static characteristic, but they are different from the characteristic of the tube itself, being given by Figs. 84 and 85 instead of those shown in Fig. 74.

The effect of the external non-inductive resistance on the characteristic of the output circuit, when an alternating potential is impressed on the grid, can be explained as follows: Referring

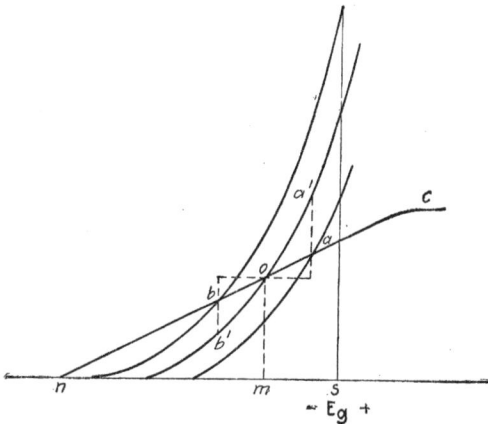

FIG. 86.

to Fig. 86, let the three parabolic curves represent the characteristics of the tube itself, the middle one of which, let us say, is the one obtained when the plate-filament voltage has a definite value E_p. The other two are the characteristics for higher and lower values of E_p. Let the tube be inserted in the circuit shown in Fig. 72. Let the constant grid battery voltage E_g be so adjusted that the direct current in the plate circuit, as measured with A, is mo. Now, on account of the voltage drop in r_0, due to the current I_p in it, the plate-filament voltage is $E_p = E_b - r_0 I_p$. If I_p be varied by impressing an alternating potential on the grid, E_p varies accordingly since E_b is constant. Thus, if the negative grid potential is decreased the plate current increases. This causes E_p to decrease to the value, say, corresponding to the lower characteristic shown in Fig. 86, and the current instead of increasing to a',

as it would if E_p remained constant, increases only to a. For the
same reason, when the negative grid potential is increased the current decreases only to b instead of to b'. The characteristic therefore straightens out and takes the shape given by boa, instead of
$b'oa'$.

Referring to equation (20), it will be seen that if we represent
the alternating plate voltage and current by e_p and i_p, respectively,
we have $e_p = -i_p r_0$. The plate current and plate voltage are therefore 180° out of phase. The plate current is, however, in phase
with the grid potential, so that the grid and plate potentials
differ in phase by 180°.

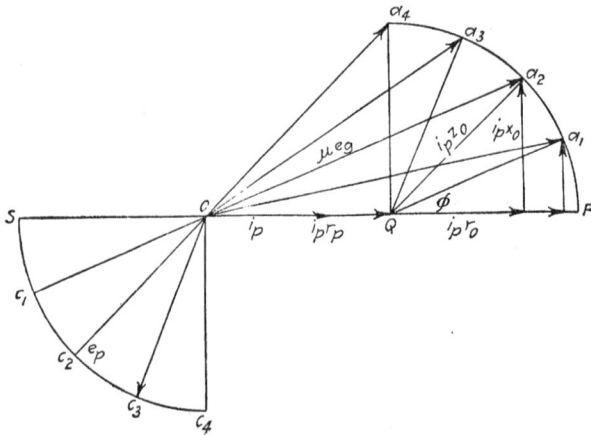

Fig. 87.

(3) Let the plate circuit now contain reactance as well as
resistance, that is, let it contain an impedance $Z_0 = r_0 + jx_0$. Here
we have $e_p = -i_p Z_0$, but on account of the reactance x_0 in the plate
circuit the phase difference between the plate and grid potentials
may differ from 180°. When this happens the dynamic characteristic of the plate circuit takes the form of a loop. To explain
this we can make use of the theorem stated on page 157, that a
voltage e_g applied between filament and grid is equivalent to an
electromotive force μe_g impressed on the plate circuit, where μ
is the amplification constant of the tube. The phase relations
are shown in Fig. 87 for various values of the angle $\phi = \tan^{-1}\dfrac{x_0}{r_0}$
of the external impedance. Let the plate current be represented

by i_p in the direction OQP. The voltage drop $i_p r_p$, in the tube, due to its plate resistance, is given by OQ. The drop $i_p Z_0$ in the external impedance Z_0 is given by Qa. Thus, in the case in which the angle ϕ is $45°$, $i_p Z_0 = Qa_2$, and is the vector sum $i_p r_0$ and $i_p x_0$, the total driving E.M.F., μe_g in the plate circuit is in this case given by Oa_2. Now e_p is equal to $-i_p Z_0$ and is given by Oc_2,

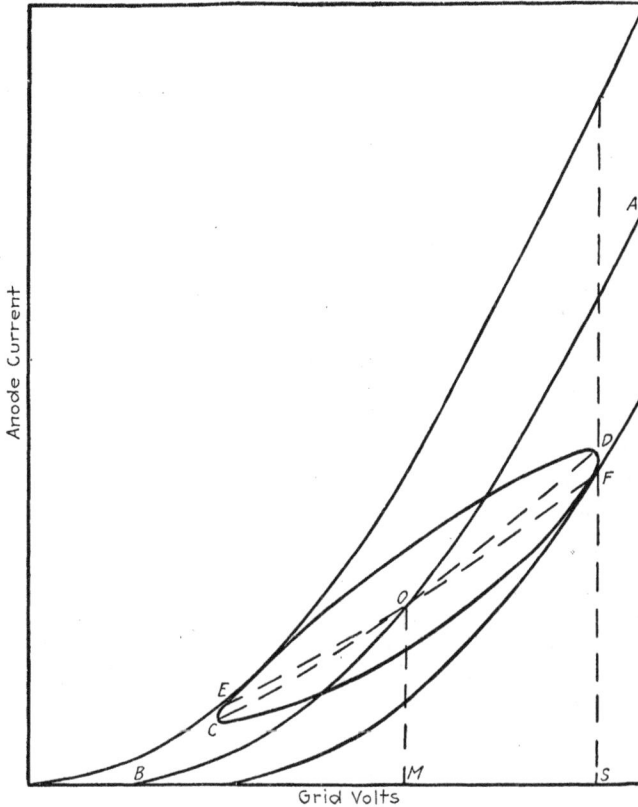

FIG. 88.

which is parallel to Qa_2. The phase difference between e_p and μe_g or e_g is therefore equal to the angle $a_2 O c_2$ which is $157.5°$. This is for the case in which the external impedance Z_0 is numerically equal to the plate resistance r_p ($OQ = Qa_2$), and has an angle of $45°$.

Referring now to Fig. 88, let the negative grid battery voltage be equal to MS, so that we operate around the point O of the tube characteristic AOB, which corresponds to the plate potential which obtains when the alternating potential impressed on the grid is zero. The other two tube characteristics are for the maximum and minimum potentials which the plate acquires when an alternating potential $e \sin pt$ is superimposed on the constant negative grid potential $E_g = MS$. If we now plot the plate current as a function of the varying grid potential e_g, considering at the same time that e_g and the alternating plate potential e_p are 157.5° out of phase, we obtain the loop shown in Fig. 88. The loop is, of course, due to the reactance in the external circuit, because there is no lag within the tube. This loop is not an ellipse, but has a curved axis CD, the general slope and curvature of which depends upon the angle between e_g and e_p, which in turn depends upon the angle ϕ of the external impedance. As ϕ decreases the loop narrows down, its axis straightens out and rotates in a clock-wise direction until, when ϕ is zero, that is, when the external circuit contains only non-reactive resistance, the loop degenerates into the line EF, which is the non-reactive dynamic characteristic *boa* shown in Fig. 86. It will be observed that if the angle θ between e_g and e_p is 157.5°, the axis of the loop very nearly coincides with the approximately straight line EF obtained when θ is 180°.

The angle θ depends not only on ϕ, the angle of the external impedance Z_0, but also on the value of this impedance compared with the plate resistance r_p. Thus if ϕ is 90° then $i_p Z_0$ is given by Qa_4, and if Z_0 is numerically equal to r_p, the angle $a_4 O c_4$ between e_g and e_p is $\theta = 135°$. But if $Z_0 = 3r_p$, the angle θ is about 160°. In this case also the axis of the dynamic characteristic coincides very nearly with the line EF which is obtained when $\theta = 180°$.

It is important to note the conditions that must be secured to make the axis of the dynamic characteristic approach a straight line. While the curvature of the characteristic enables the thermionic tube to perform certain very important functions, such as detection and modulation of oscillating currents, it is nevertheless an undesirable feature when the tube operates as an amplifier. It follows from the explanation given in Section 57 that unless the characteristic is straight the output current wave is not an exact enlarged reproduction of the wave impressed on the input. This causes distortion when amplifying telephonic currents, and to avoid it the amplifier must be operated under such conditions that its characteristic is substantially linear. Now, it will be shown later that when operating the tube as an amplifier, maximum power amplification is obtained when the external impedance is numerically equal to the plate resistance of the tube. If this equality is preserved and the angle of the external impedance is not greater than about 45°, the axis of the characteristic is, as we have seen, substantially linear over a considerable range of input voltage. In practice the conditions are often even better, because the angle of the external impedance is often much less than 45°. This is, for example, the case where the tube is operated as a telephone repeater; the secondary of the output transformer feeds into a long line of comparatively high resistance, so that the angle of the effective impedance into which the tube works is very small.

In cases where the angle of the external impedance is necessarily large, we can still secure a practically linear axis for the dynamic characteristic by making the external impedance larger than the plate resistance. We would therefore gain in quality of transmission at the expense of amplification. But the necessary sacrifice in amplification would not be large. Although maximum amplification is secured when the external impedance is equal to the plate resistance, the decrease in amplification is small even when the external impedance is twice as large as the plate resistance (see Fig. 112).

If the necessary precautions be taken to secure the conditions necessary to make the axis of the dynamic characteristic substantially linear, we can extend the theorem deduced on page 157 from the stray field relation: A voltage e_g applied between filament and grid establishes a current in the plate circuit which is given by

$$i_p = \frac{\mu e_g}{r_p + Z_0}, \quad \cdot \quad \cdot \quad \cdot \quad \cdot \quad \cdot \quad \cdot \quad (22)$$

where μ is the amplification constant of the tube, r_p its plate resistance and Z_0 the external impedance in the plate circuit.

If the conditions are not such as to make the characteristic linear this equation is still true as far as the fundamental frequency is concerned, but the curvature of the characteristic introduces harmonics which would necessitate the addition of terms of higher order of smallness to equation (22).

The theorem embodied in equation (22) is of fundamental importance and is instrumental in the solution of many vacuum tube problems. We shall have occasion to make extensive use of it in what follows.

60. Conditions for Distortionless Amplification. Distortionless amplification is obtained if the amplified current in the output circuit is, for the whole range of frequencies which it is desired to transmit, an exact enlarged reproduction of the input current. Distortion can be produced in two ways: (1) When currents of different frequencies are not amplified in the same proportion; (2) when the amplification is not independent of the input voltage.

(1) As far as the first effect alone is concerned, the amplification will be distortionless if the whole circuit is non-reactive. The circuits commonly used in connection with the tube are not non-reactive, but the necessary transformers and condensers can always be so chosen that for the operating range of frequencies the total impedance is not unduly affected by the frequency. As far as the tube itself is concerned it is to be noted that the capacities between the electrodes introduces a reactance effect. Of these we distinguish between the capacity between filament and plate, and the effective input impedance as measured between the filament and grid. When the amplification is expressed in terms of the potential actually applied to the grid, the only inter-electrode capacity that comes into consideration is the capacity between filament and plate. This is so small that when the amplification is expressed in this way it is found to be independent of the frequency for frequencies ranging up to several hundred thousand cycles per second. The power amplification is usually expressed in terms of the ratio of the power developed in the external output circuit to the total power impressed on the input. In this case the effective reactance due to the inter-electrode capacities depends on the circuit used, as will be explained in Sections 69 to 71. Under the conditions under which amplifiers are mostly operated, the electrode capacities usually have a very small effect. The general effect, however, is to decrease the amplification when the frequency becomes very high.

(2) The second condition for distortionless amplification will not be satisfied unless the axis of the dynamic characteristic of the output circuit is linear over the operating range of voltage. As was shown in the previous Section, this can be secured by

making the external impedance in the output circuit sufficiently large.

It is important to note that another condition for distortionless amplification is that the input voltage must be kept within certain limits determined by the d-c. plate and grid voltages and the structure of the tube. Let us assume that the external impedance is sufficiently large to straighten out effectively the characteristic. The question now is what range of input voltage can be employed without overtaxing the tube. If the input voltage is so large that the grid becomes sufficiently positive to take appreciable current, the positive halves of the output wave can be reduced in the manner explained in Section 57. This reduction is more marked the larger the external impedance in the output circuit, because the extent to which the grid can become positive without taking appreciable current depends on the potential difference E_p existing between filament and plate at the moment that the grid is positive and on the structure of the tube. Remembering that the stray field between filament and grid, due to the potential difference E_p, tends to draw the electrons *through* the openings of the grid, it will be seen that the larger E_p the higher must be the positive grid voltage to overcome the stray field and attract the electrons to the grid. Now the external impedance has the effect of decreasing the plate-filament potential difference when the flow of electrons from plate to filament through the impedance is increased, that is, during the half cycle when the grid is positive. This reduces the stray field and consequently increases the flow of electrons to the grid. This is the effect that gives rise to the bend C in the dynamic characteristic shown in Fig. 86. If we say that g is the positive potential with respect to the filament which the grid can acquire without taking appreciable current, we can state that one condition for distortionless amplification is

$$e \leqq - \mid E_g + \epsilon \mid + \mid g \mid \quad . \quad . \quad . \quad . \quad . \quad (23)$$

where E_g is the voltage of the grid battery, e the peak value of the input voltage, and ϵ the intrinsic potential difference between filament and grid.

Another condition is that the peak value of the input voltage must not exceed the value given by mn (Fig. 86), otherwise the negative peaks of the output current wave will be chopped off.

Now sn is given by $\dfrac{E'_p}{\mu}$, where E'_p is the potential difference between filament and plate at the moment when the grid has its maximum negative value, and sm is the voltage E_g of the grid battery.

We therefore have the two conditions:

$$\left.\begin{array}{l} e \leqq - \mid E_g + \epsilon \mid + \mid g \mid \\[2mm] e \leqq \left| \dfrac{E'_p}{\mu} + \epsilon \right| - \mid E_g \mid \end{array}\right\} \quad \cdots \quad (24)$$

or when the tube is working at full capacity, that is, when operating over the whole range of the characteristic.

$$c = - \mid E_g + \epsilon \mid + \mid g \mid = \left| \dfrac{E'_p}{\mu} + \epsilon \right| - \mid E_g \mid . \quad (25)$$

61. Amplification Equations of the Thermionic Amplifier. We shall now derive quantitative expressions for the amplification produced by the three-electrode thermionic tube. It will be recognized that when operating as a power amplifier the tube derives the extra power from the d-c. battery inserted in the plate circuit. The energy of the plate battery is released by the influence of the grid potential on the current in the plate circuit and the amount of power released depends almost entirely on the influence of the grid potential.

In deriving the following equations we assume that the grid is maintained sufficiently negative with respect to the filament to prevent any appreciable current convection between filament and grid; that is, the tube will be assumed to operate within the limits defined by equations (24) and (25). We shall also assume that the impedance conditions in the plate circuit are such as to make the characteristic of the plate circuit substantially linear over the operating range of voltages. These conditions can very nearly be satisfied in practice even when the circuit constants are so adjusted as to give a maximum degree of amplification. Under these conditions the alternating current i_p in the plate circuit is related to the alternating potential e_g, applied to the grid, by equation

$$i_p = \frac{\mu e_g}{r_p + Z_0}, \quad \cdots \quad \cdots \quad (22)$$

where r_p is the plate resistance and Z_0 the external impedance. This equation enables us to derive the amplification equations in a very simple manner.

62. Voltage Amplification. Consider first the case in which the tube is used as a voltage amplifier. The voltage developed in the impedance Z_0 is $e_0 = i_p Z_0$, which according to equation (22) becomes:

$$e_0 = \frac{\mu e_g Z_0}{r_p + Z_0},$$

and the voltage amplification is therefore

$$\mu' = \frac{e_0}{e_g} = \frac{\mu Z_0}{r_p + Z_0}. \quad \cdots \cdots \quad (26)$$

It must be noted that e_g is the a-c. potential difference actually established between filament and grid.

It will be seen that μ' increases as Z_0 is increased and asymptotically approaches the maximum value μ when Z_0 becomes infinitely large compared with r_p. The constant which depends on the structure of the tube and determines the stray field, is therefore simply the maximum voltage amplification obtainable from the tube. When a tube is to be used as a voltage amplifier it should therefore be designed to have as high a value of μ as possible. Fig. 89 shows a Western Electric voltage amplifier. The amplification constant μ of this tube is 40.

A voltage amplification of several hundred fold is not hard to obtain, it being simply necessary to design the tube accordingly, since μ is a structural constant. In using tubes as voltage amplifiers it is, however, necessary to consider also the other factors that

Fig. 89.

influence the voltage amplification. For example, it follows directly from equation (26) that the external impedance should be made several times as large as the plate resistance of the tube. Now, for the same amount of filament surface the plate resistance increases approximately as the square of μ (see equation 15) and may acquire such a high value as to necessitate an impracticably high external impedance. It is, therefore, often necessary, when increasing μ, to increase the amount of filament surface so as to reduce the plate resistance as much as possible. It is, of course, also possible to decrease the plate resistance by increasing the d-c. plate voltage, provided we do not operate beyond the minimum saturation voltage.

Referring now to equation (26) let $Z_0 = r_0 + jx_0$; the voltage amplification is then given by

$$\frac{e_0}{e_g} = \frac{\mu Z_0}{\sqrt{(r_p + r_0)^2 + x_0^2}}. \quad \cdots \cdots \quad (27)$$

Suppose the tube is inserted in the circuit shown in Fig. 90, and that it is desired to obtain the voltage developed between

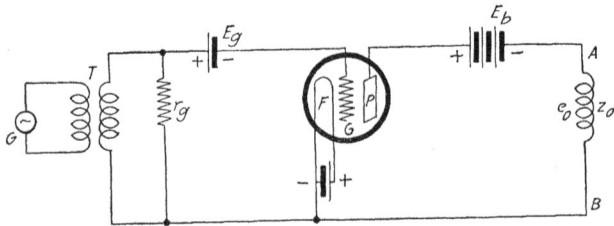

Fig. 90.

the ends A and B of the impedance Z_0. This voltage e_0 can be measured by connecting an electrostatic voltmeter between A and B.[1] The secondary of the transformer T can be wound to have as high an impedance as possible, thus impressing the highest possible voltage e_g on the grid for a given voltage in the primary of T.

Let us now consider the two extreme cases in which Z_0 is (1) a non-inductive resistance r_0 ($x_0 = 0$) and (2) a practically pure

---

[1] A thermionic tube can be used as an electrostatic voltmeter in the manner shown in Section 114.

reactance x_0 ($r_0=0$). In the first case the voltage amplification is given by

$$\frac{e_0}{e_g}=\frac{\mu r_0}{r_p+r_0}. \qquad \ldots \qquad \ldots \qquad (28)$$

The relation between $\frac{e_0}{e_g}$ and $\frac{r_0}{r_p}$ is shown by curve II of Fig. 91 from which it is seen that $\frac{e_0}{e_g}$ reacnes about 90 per cent of its maximum value μ when $r_0 = 10\ r_p$. (In computing these curves μ was taken equal to 10.)

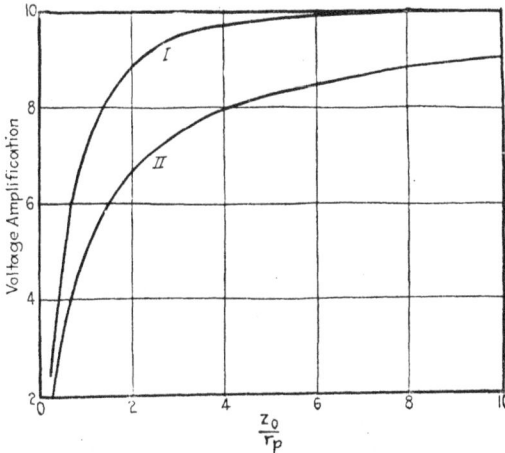

FIG. 91.

If Z_0 is a pure reactance x_0, the voltage amplification is given by

$$\frac{e_0}{e_g}=\frac{\mu x_0}{\sqrt{r_p^2+x_0^2}}. \qquad \ldots \qquad \ldots \qquad (29)$$

Curve I of Fig 91 shows the relation between $\frac{e_0}{e_g}$ and $\frac{x_0}{r_p}$. It is seen that there is a distinct advantage in making the tube work into a reactance, the voltage amplification rising to about 90 per cent of its maximum value when x_0 is numerically only twice r_p. It is, however, advisable to make the reactance as large as possible in order to minimize distortion due to the curvature of the characteristic. The use of a reactance instead of a resistance has another advantage. If Z_0 is a pure resistance and several times greater than the plate resistance, a considerable portion of the voltage of the plate battery is lost in Z_0, so that to secure the necessary potential difference between filament and plate it would be necessary to use a rather high plate battery voltage. This can be avoided by using instead of a pure resistance a choke coil which has a comparatively small d-c. resistance.

On the other hand, the value of the tube as a voltage amplifier lies in the fact that it can be operated in a non-inductive circuit, and in this respect it performs an important function, in that it serves the purpose of producing high degrees of amplification with very little distortion. Unlike the transformer, for example, it furnishes a voltage-amplifying means that is independent of frequency unless the frequency is very high. And, as a matter of fact, it can also be used to produce power amplification that is practically independent of frequency.

When several tubes are used in cascade formation in multi-stage non-inductive amplifier sets, all but the last tube should be used as voltage amplifiers, because the tube is a potential operating device. It works best as an amplifier when its grid does not take appreciable current; that is, when the tube operates within the limits defined by equations (24) and (25). The input power consumed by the tube is therefore usually very small, and all that is necessary is to make the input voltage applied between filament and grid as high as possible.

It must be pointed out that unless it is necessary to use a non-inductive circuit it is best to operate all tubes in a multi-stage amplifier set as power amplifiers, and use voltage step-up transformers between the tubes. Consider, for example, the circuit in Fig. 92. (The circuits shown here do not include details that are necessary to give best operation in practice. They are merely skeleton circuits intended to illustrate the points under consideration. Complete circuits will be discussed below.)

If the tube A were to be used as a voltage amplifier, it would be necessary to make Z_0 several times as large as r_p. This does not, however, give maximum total amplification, because when using transformers we have to consider the power, and maximum amplification is obtained when $Z_0 = r_p$, T_2 being used as a voltage step-up transformer. This can be shown as follows: The power in Z_0' will be a maximum for maximum voltage e'_g impressed on the input of the second tube B. Now, the voltage e_0 in Z_0 is given by:

$$e_0 = \frac{Z_0 \mu e_g}{r_p + Z_0},$$

where μ is the amplification constant of tube A. Now, the voltage ratio of the transformer T_2 is $\sqrt{\dfrac{Z_1}{Z_0}}$. Hence, the voltage impressed on tube B is:

$$e'_g = \frac{\mu e_g \sqrt{Z_1 Z_0}}{(r_p + Z_0)} \quad \cdot \quad \cdot \quad \cdot \quad \cdot \quad \cdot \quad (30)$$

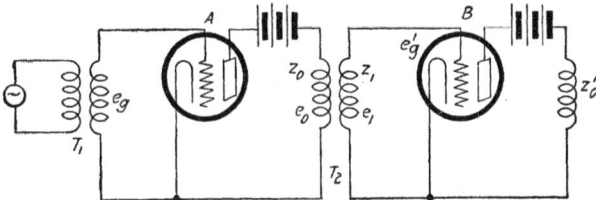

Fig. 92.

It is in all cases desirable to make Z_1 as large as can possibly be done in practice. Hence, regarding Z_1 as fixed and differentiating e'_g with respect to Z_0 and equating to zero, it will be seen that e'_g is a maximum when $Z_0 = r_p$, and this, it will be shown in the next paragraph, is the condition for maximum power in Z_0.

63. Power Amplification. The three-electrode thermionic tube can be used to amplify power, and in this property lies its great usefulness. It is its amplifying property that enables it to be used also as an oscillation generator. There are other types of amplifiers, such as, for example, the arc which amplifies in virtue of its negative resistance characteristic and therefore operates on an entirely different principle. But the thermionic amplifier, or audion, has certain marked advantages over other types. Unlike the arc it does not depend for its operation on ionization by collision of residual gas, and in fact operates satisfactorily only when the vacuum is so high that ionization by collision plays a negligibly small part in current convection in the tube. The discharge is therefore steady and reproducible. When using the device as a telephone relay, for example, steadiness and reproducibility are conditions that must be complied with, and this is also true of many other cases where amplifiers are used. It is furthermore capable of amplifying currents of frequencies ranging all the way up to several million cycles per second, and if properly designed it can be made to produce an extraordinarily high degree of amplification. I have for example, obtained with a specially designed tube, a power amplification of 3000-fold.

An equation for the power amplification can be obtained directly from the equations deduced above. Let us consider the circuit shown in Fig. 90. It is desired to amplify the power in the transformer T which may be at the end of a section of telephone line or may, for example, be connected in the output of the generator G. Let the a-c. potential impressed on the grid be e_g. Then the alternating current i_p in the output circuit $FPAB$ is

$$i_p = \frac{\mu e_g}{r_p + Z_0}.$$

where r_p is the plate resistance. The voltage e_0 in Z_0 is

$$e_0 = \frac{\mu e_g Z_0}{r_p + Z_0}$$

and hence the power in Z_0 is

$$e_0 i_p \cos \phi = \frac{\mu^2 e_g^2 Z_0 \cos \phi}{(r_p + Z_0)^2} \quad \cdot \quad \cdot \quad \cdot \quad \cdot \quad \cdot \quad (31)$$

where $\cos \phi$ is the power factor.

In order to get the power amplification it is necessary also to know the power expended in the input. The grid current does not bear a simple relation to the operating parameters, but to get an indication of how current in the grid circuit affects the amplification, we can expand the obvious functional relationship, $I_g = f(E_p, E_g)$ into a Taylor series, thus:

$$I_g + \delta I_g = f(E_p + \delta E_p, \ E_g + \delta E_g) = f(E_p + E_g) + \delta E_p \frac{\partial I_g}{\partial E_p} + \delta E_g \frac{\partial I_g}{\partial E_g},$$

second and higher order quantities being neglected. By making the following substitutions:

$$\delta I_g = i_g, \qquad \frac{\partial I_g}{\partial E_p} = g_n,$$

$$\delta E_p = e_p,$$

$$\delta E_g = e_g, \qquad \frac{\partial I_g}{\partial E_g} = g_g.$$

we get:

$$i_g = g_n e_p + g_g e_g{}^1.$$

Putting $\dfrac{e_p}{e_g} = \mu_1$, the input power becomes, if we neglect the power consumption in the input transformer:

$$e_g i_g = e^2{}_g(g_g + \mu_1 g_n).$$

Hence the power amplification becomes:

$$\eta = \frac{e_0 i_p \cos \phi}{e_g i_g} = \frac{\mu^2 Z_0 \cos \phi}{(r_p + Z_0)^2 (g_g + \mu_1 g_n)}. \quad \cdot \quad \cdot \quad \cdot \quad (32)$$

This equation shows how the power amplification is affected by the grid conductance g_g, the amplification factor μ and the reflex mutual conductance g_n. For a perfectly unilateral amplifier the output circuit has no effect on the input and then $g_n = 0$.

Conditions can readily be realized in practice which make both g_n and g_g negligibly small.[2] Conditions under which they become appreciable will be discussed in Section 69. If we neglect these quantities the input resistance is infinite and the power loss in the input indeterminate. We can, however, shunt the input

[1] This equation is equivalent to that derived by Latour. (Electrician, Dec., 1916.)

[2] When there are reactive effects, as, for example, when the output circuit is reactive, we should, strictly speaking, consider the mutual admittance and reflex mutual admittance instead of simply the mutual conductances, because under these conditions the grid potential and plate current are out of phase. The mutual admittances are then complex quantities involving the mutual conductances and the mutual susceptances. (See Fig. 87.) When the circuit constants are so proportioned that the axis of the dynamic characteristic is substantially linear, which is the condition for distortionless transmission, the angle of the mutual admittance is so small that we can, to a first approximation, neglect the mutual susceptances.

with a resistance r_g (Fig. 90), as was suggested by H. D. Arnold, and so proportion its value that the input transformer works most efficiently. The power expended in this resistance can then be taken as a measure of the input power. In telephone repeater circuits this resistance usually has a value of about 600,000 ohms.

Equation (32) then becomes:

$$\eta = \frac{\mu^2 Z_0 r_g \cos \phi}{(r_p + Z_0)^2}, \quad \cdots \cdots \quad (33)$$

or putting $Z_0 = r_0 + jx_0$, the power amplification can be expressed as:

$$\eta = \frac{\mu^2 r_g \sqrt{r_0^2 + x_0^2}}{(r_p + r_0)^2 + x_0^2} \cos \phi. \quad \cdots \cdots \quad (33a)$$

Let us first consider the case in which Z_0 takes the form of a non-inductive resistance ($x_0 = 0$). The power amplification is then simply given by

$$\eta = \frac{\mu^2 r_g r_0}{(r_p + r_0)^2}, \quad \cdots \cdots \cdots \quad (34)$$

and it will be seen by differentiating η with respect to r_0 and equating the derivative to zero, that the power amplification is a maximum when $r_0 = r_p$.

For the general case in which the reactance x_0 is not zero, we note that $\phi = \tan^{-1} \frac{x_0}{r_0}$ and $\cos \phi = \frac{r_0}{\sqrt{r_0^2 + x_0^2}}$. Substituting this in equation (33) the power amplification becomes:

$$\eta = \frac{\mu^2 r_g r_0}{(r_p + r_0)^2 + x_0^2} = \frac{\mu^2 r_g r_0}{(r_p + Z_0)^2}. \quad \cdots \cdots \quad (35)$$

This is also a maximum when Z_0 is numerically equal to r_p, as is shown in Fig. 112, page 220, where the curve represents the relation between the power amplification and the ratio $\frac{Z_0}{r_p}$ for the case in which angle of the external impedance Z_0 is 45°, that is, $\tan \phi = \frac{x_0}{r_0} = 1$. When the tube is used for the purpose of amplifying telephonic currents, the energy is translated into sound waves through the motion of the receiver diaphragm. In such cases, it will be shown later, the ratio $\dfrac{Z_0}{r_p}$ can deviate considerably from unity before resulting in any serious diminution of the effect produced upon the organs of hearing.

If we put $Z_0 = nr_p$ equation (33) becomes:

$$\eta = \frac{\mu^2}{r_p} \frac{r_g \cos \phi}{(1+n)^2}, \quad \cdot \quad \cdot \quad \cdot \quad \cdot \quad \cdot \quad \cdot \quad (36)$$

which clearly shows the importance of the mutual conductance $\dfrac{\mu}{r_p}$ or the steepness of the plate current-grid voltage characteristic. It will be seen later that this quantity plays an equally important role in the operation of the tube as oscillation generator and radio detector.

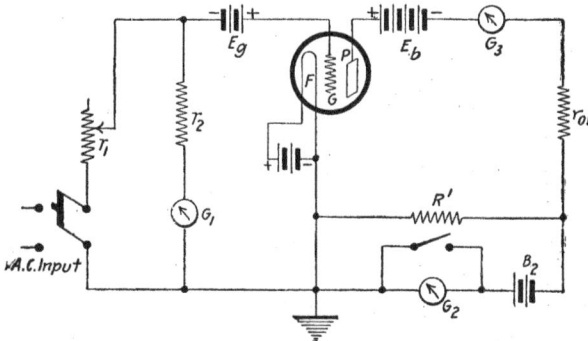

FIG. 93.

64. Experimental Verification of Amplification Equations. The amplification was determined experimentally as a function of the tube parameters with the circuit arrangement shown in Fig. 93.[1] This circuit was made non-inductive throughout. The input voltage could be varied by means of the resistance r_1 and measured with a Duddell thermo-galvanometer G_1 and resistance r_2. The grid battery E_g was inserted to insure that the tube was always operated within the limits given by equations (24). Now the amplifier is always operated with a battery in the plate circuit, so that there is a constant direct current in this circuit whether the a-c. input be applied or not. The application of the

[1] H. J. VAN DER BIJL, Phys. Rev., Vol. 12, p. 194, 1918.

a-c. input voltage establishes an alternating current in the plate circuit which is superimposed upon the constant direct current. This a-c. could not be measured accurately by simply inserting an a-c. meter in the plate circuit, because it was often small compared with the direct current that would constantly flow through the galvanometer. A galvanometer that would be capable of carrying the direct current would, therefore, not be sensitive enough to measure the increase in current due to the a-c. input with any degree of accuracy. On the other hand, it was not possible to separate the a-c. from the d-c. in the usual way with appropriate inductances and capacities, because then the amplification would be influenced in a large measure by the constants of the circuit. For these reasons the balancing scheme was used. The direct current was measured with the milliammeter G_3 and the alternating current with the thermocouple and milliammeter G_2. In series with G_2 was a battery B_2 so poled that when the input voltage was not impressed there was no current in G_2, the direct current being by-passed through R'. The resistance of G_2 was small compared with R' so that practically all the alternating current established in the plate circuit flowed through G_2. It is evident that the effective external resistance is r_0. The whole system was carefully shielded and care was taken to avoid any disturbing effects due to mutual and shunt capacity of the leads and resistances. Such precautions were necessary because the frequency at which measurements were made ranged from 200 to 350,000 cycles per second. The resistances r_1 and r_2 consisted, for example, of thin straight wires stretched on a board.

The amplification was found to be practically independent of frequency over the range mentioned above. The input voltage was varied from a few hundredths of a volt to several volts. Fig. 94 shows the relation between the voltage in r_0 (the output voltage) and voltage as measured with G_1 and r_2 (the input voltage). In these measurements r_0 was made equal to the plate resistance of the tube. The linear relation obtained shows that the amplification is independent of the input voltage, a result which justifies the use of equation (22).

Equation (28) was verified by measuring the output voltage for a constant input voltage and different external resistance r_0. The results are shown in Fig. 95 where the circles indicate the observed values and the curve was computed from equation (28).

The abscissæ give the ratio $\frac{r_0}{r_p}$ and the ordinates the voltage ampli-

fication $\frac{e_0}{e_g}$. The value of μ for this tube was 10.2, and the input

voltage in this particular experiment was 3.55 volts. In experiments like this it must be remembered that the plate resistance

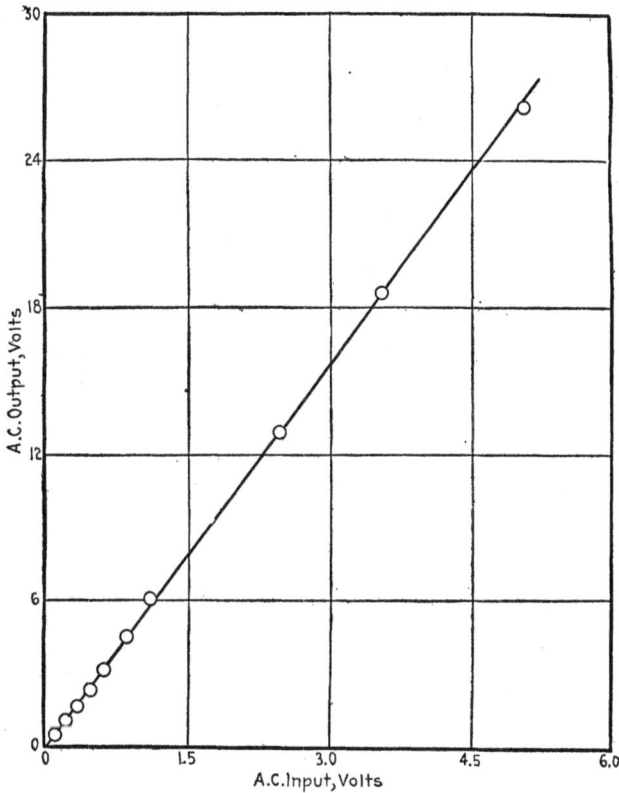

Fig. 94.

r_p of the tube depends upon the potential difference between filament and plate which, if the voltage of the plate battery remains constant, changes every time r_0 is given a different value. In order to operate the amplifier under the same conditions throughout the experiment, the plate battery voltage should always be adjusted to keep the plate resistance constant.

The power developed in the external resistance r_0 as a function of r_0 is shown in Fig. 96. The plate resistance of the tube was kept

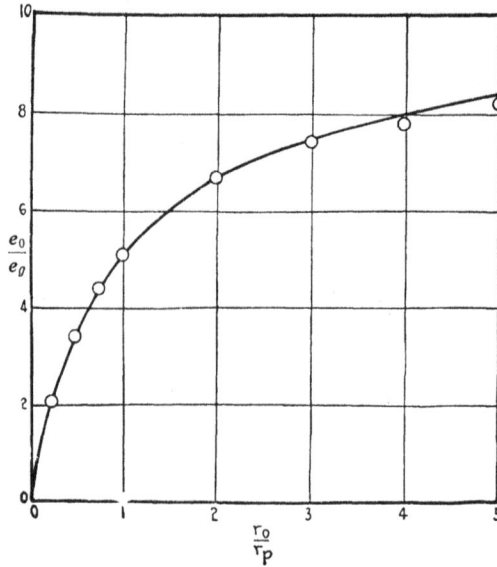

Fig. 95.

constant at 14,800 ohms. The power in r_0 is seen to be a maximum when $r_0 = 15,000$ ohms, which is in close argeement with equation

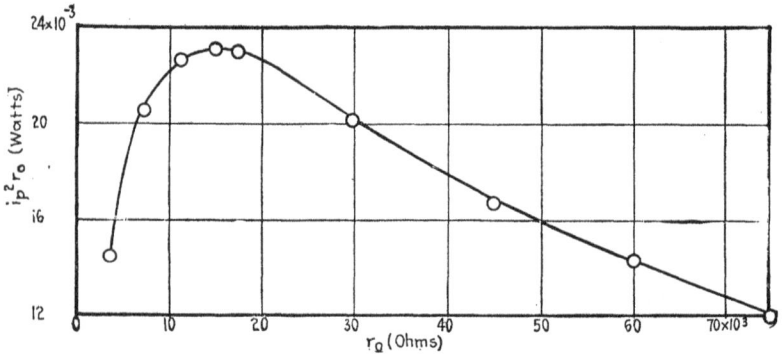

Fig. 96.

(34) which requires maximum power amplification when $r_0 = r_p$. The input in this experiment remained constant and corresponded

to an input voltage 3.55 volts. Putting the resistance r_2 and that of the galvanometer G_1 in series equal to r_g we may express the input power as $\dfrac{e_g{}^2}{r_g}$. Combining this with equation (34) the power developed r_0 is given by

$$P = \frac{\mu^2 e_g{}^2 r_0}{(r_0 + r_p)^2}. \quad \ldots \ldots \quad (37)$$

Now $e_g = 3.55$ volts and $\mu = 10.2$, and if we put $r_p = r_0 = 15,000$ ohms we find that the power in r_0 according to the above equation is 22×10^{-3} watt, which is in good agreement with the observed value, namely 23×10^{-3} watt.

65. Methods of Measuring the Amplification Constant. The amplification constant μ can be measured with considerable accuracy in a number of different ways. It is perhaps the most easily determined constant of the tube.

The first method we describe is not the most accurate nor the simplest, but furnishes a clear demonstration of the significance of μ. If the equation of the characteristic (equation 3) be differentiated partially first with respect to the plate voltage E_p and then with respect to the grid voltage E_g it will be found that

$$\mu = \frac{\dfrac{\partial I}{\partial E_g}}{\dfrac{\partial I}{\partial E_p}}, \quad \ldots \ldots \quad (38)$$

μ can therefore be obtained by measuring the slopes of the I_p, E_p and I_p, E_g characteristics, the slopes being taken at corresponding points of the two curves. In obtaining these curves care must be taken that the plate circuit does not contain an appreciable resistance which would influence the slopes of the characteristics.

Instead of going through the rather tedious process of taking the characteristic curves and then measuring their slopes at the desired points, we can determine μ by a method indicated by equation (8), which gives a linear relation between the plate and grid voltages necessary to keep the plate current constant. As Fig. 76 (p. 157) indicates, this method is quite accurate. It is of course not necessary to obtain more than two corresponding values of plate and grid voltages. For example, let the plate current for two definite values of E_p and E_g be I_p. Now let the plate voltage be increased to E'_p. This causes an increase in the plate current. In order to bring it back to its original value I_p, the grid must be made more negative with respect to the filament. If the necessary grid voltage is E'_g, μ is given by

$$\mu = \frac{E_p' - E_p}{E_g - E_g'}. \qquad \ldots \quad \ldots \quad \ldots \quad (39)$$

A convenient and rapid means of measuring μ is shown in Fig. 97.[1] E_1 is a battery of small dry cells of about 10 or 20 volts. By closing the key K opposite potentials are applied to the grid and

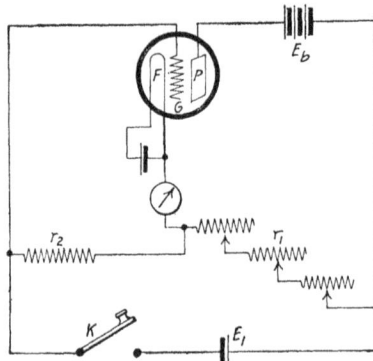

Fig. 97.

plate, their values depending upon those of r_1 and r_2. Since a potential applied to the grid produces μ-times the effect of a potential applied to the plate, it is evident that no change will be produced in the reading of the current meter by closing K if $\frac{r_1}{r_2} = \mu$. For convenience in measurement r_2 is given a fixed value of 10 ohms and r_1 consists of three dial rheostats of 1000, 100 and 10 ohms arranged in steps of 100, 10 and 1 ohms each. The rheostats are marked in tenths of the actual resistances, so that the setting of the dials gives μ directly.

A similar method has also been described by J. M. Miller,[2] who used a source of alternating current instead of the battery

[1] H. J. van der Bijl, Proc. I.R.E., Vol. 77, p. 112, 1919.

[2] J. M. Miller, Prcc. I.R.E., Vol. 6, p. 141, 1918.

E_1.   The meter is replaced by a telephone receiver and the resistances r_1 and r_2 are adjusted until the tone in the receiver is a minimum.   The use of an alternating current has the advantage that it also allows a simple determination of the plate resistance of the tube.

66. Measurement of the Plate Resistance.   We have seen that the characteristic of the amplifier can to a first approximation be given by equation (3), in which the exponent is 2.   For such a characteristic the plate resistance is the inverse slope of the I_p, E_p-curve and is given by equation (15):

$$r_p = \frac{\mu}{2\alpha\left(\dfrac{E_p}{\mu}+E_g+\epsilon\right)}. \quad \cdots \cdots \quad (15)$$

By multiplying numerator and denominator by the expression in the parentheses we can express r_p in the simpler form

$$r_p = \frac{E_p+\mu(E_g+\epsilon)}{2I_p}. \quad \cdots \cdots \quad (40)$$

or, neglecting the small quantity ϵ and putting $E_g = 0$,

$$r_p = \frac{E_p}{2I_p}. \quad \cdots \cdots \cdots \quad (41)$$

We can, therefore, obtain a fair estimate of the plate resistance by simply observing the plate current for the plate voltage at which it is desired to obtain the resistance.   It will be noted that the plate resistance, by which we mean the a-c. resistance, is half the d-c. resistance.   It is also to be noted that while the amplification constant μ is a geometrical constant, the plate resistance depends not only on the structure of the tube but also on the values of the plate and grid voltages.   If can, however, be fully specified for all operating plate and grid voltages by determining it as a function of the plate voltage, the grid voltage being kept zero.   The relation between r_p and E_p can be represented by a curve like that shown in Fig. 98.   Now, the resistance at any specified plate voltage E_p and a grid voltage E_g other than zero can be obtained by applying the stray field relation given in equation (1), from which it follows (neglecting ϵ) that the effective plate voltage is now $E_p + \mu E_g$.[1] All that is necessary, therefore, to obtain the plate resistance from the curve in Fig. 98 for any values E_p and E_g is to read off the resistance at an abscissa equal to $E_p + \mu E_g$.

In regard to Fig. 98 it should be noted that the resistance characteristic drops in virtue of the increase in slope of the plate current characteristic. Let us consider the curve shown in Fig. 99, which represents the relation between the plate current and the effective plate voltage $E_\pi = (E_p + \mu E_g)$. If this voltage has the value given by ob the direct current in the plate circuit will be represented by bb'. Let the grid voltage now be varied so that E_π oscillates between oa and oc, ab being equal to bc. The plate resistance is then the reciprocal of the slope of the line $a'c'$, and if

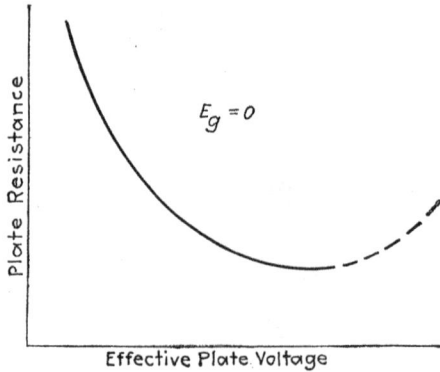

$$E_g = 0$$

Plate Resistance

Effective Plate Voltage

Fig. 98.

the characteristic is parabolic it follows directly from the properties of the parabola that $a'c'$ is parallel to slope of the curve at the point corresponding to the direct voltage $E_\pi = ob$. In the case of the parabolic characteristic the plate resistance is therefore simply given by the slope of the characteristic. If E_π oscillates between oc and od the plate resistance is smaller since the slope of $c'd'$ is larger. If now E_π is so large that it oscillates between od and of the resistance increases. This is shown by the broken part of the resistance characteristic in Fig. 98. In this case the resistance is no longer given by the slope of the curve at the point corresponding to the mean value of E_π. If E_π oscillates over the whole range oe the resistance is greater than in the case

[1] This applies for positive values of E_g only as long as the grid is not sufficiently positive to take an appreciable current.

where E_π oscillates over the range cd and the amplification will be less. This drop in amplification when the input becomes very large can of course always be avoided by operating at a higher plate potential E_p and increasing the saturation current by increasing the temperature of the filament.

Methods have been devised whereby the plate resistance can be measured dynamically with comparative ease. It is therefore

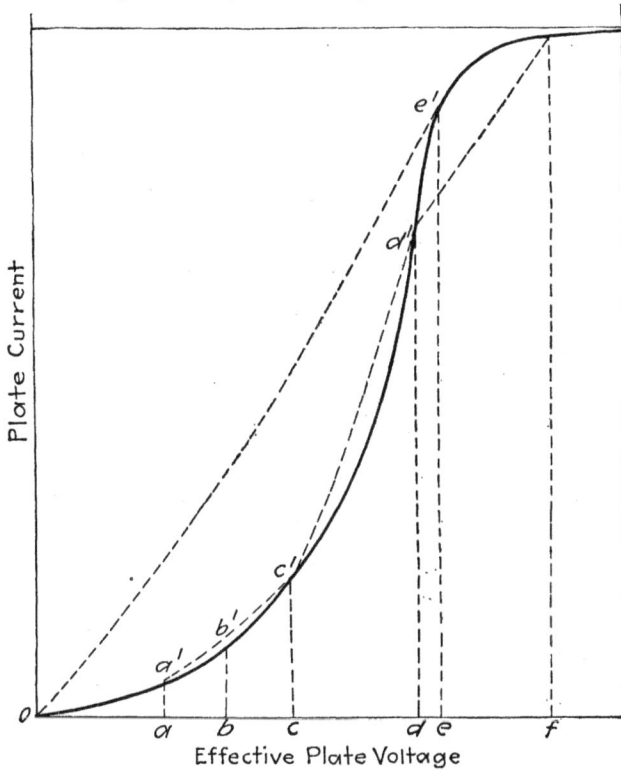

Fig. 99.

a simple matter to obtain a curve like that shown in Fig. 98. The following method was published by J. M. Miller.[1] Consider the circuit shown in Fig. 100. It will be recognized that with the key K_1 open and K_2 closed the circuit is the same as Fig. 97 except that the meter is replaced by the telephone receiver T and a source S of alternating current is used instead of the battery

[1] Loc. cit.

E_1. The circuit therefore gives a means of measuring μ, which can be done by adjusting r_1 until the tone in the receiver T vanishes. To measure the plate resistance r_p let the key K_1 be closed. If e_g be the alternating voltage applied between filament and grid the alternating current in the circuit FPr_0 is, by equation (22),

$\dfrac{\mu e_g}{r_p + r_0}$ and the voltage in r_0 is therefore $\dfrac{\mu e_g r_0}{r_p + r_0}$. Now it will be observed that if A is positive and B negative, the electron current to the plate will be increased if the effect of the applied grid voltage exceeds the opposite effect of the voltage simultaneously applied to the plate. The currents in r_0 and r_1 are therefore in phase. Hence, by adjusting r_0 until the potential drop in it is equal

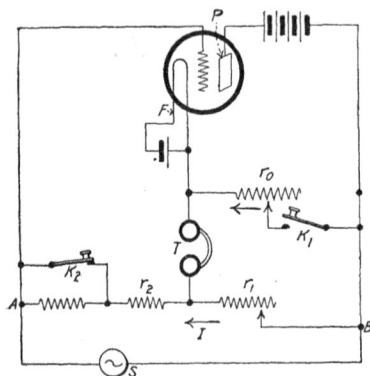

FIG. 100.

to that in r_1, the tone in the telephone receiver can be reduced to a minimum. If this is the case we have

$$\frac{\mu e_g r_0}{r_p + r_0} = I r_1.$$

But $e_g = I r_2$, hence the plate resistance is:

$$r_p = r_0 \left(\mu \frac{r_2}{r_1} - 1 \right), \quad \ldots \quad (42)$$

from which r_p can be computed. Miller puts $r_1 = r_2$ to obtain a simpler equation. But even so the method involves a calculation. A very valuable simplification hitherto unpublished was suggested by G. H. Stevenson. Suppose we adjust r_1 for minimum tone in

T when K_1 is open. Then $\mu = \dfrac{r_1}{r_2}$, and it will be seen from equation (42) that with this relation between r_1 and r_2 it would not be possible to obtain a balance with K_1 closed. But if r_2 be doubled, which can be done by opening K_2, thus adding a resistance equal to r_2, and r_0 be now adjusted, with K_1 closed, to give minimum tone in T, then $r_p = r_0$. This is the simplest method of measuring the plate resistance. By giving r_2 a fixed value of, say, 10 ohms and calibrating r_1 in the manner explained with reference to Fig. 97, we obtain a comparatively simple circuit which enables us to read the amplification constant and the plate resistance directly in terms of r_1 and r_0.

67. Direct Measurement of the Mutual Conductance. Once the amplification constant and plate resistance are known the mutual conductance can be obtained from equation (18)

$$g_m = \frac{\mu}{r_p}, \quad \cdot \quad \cdot \quad \cdot \quad \cdot \quad \cdot \quad \cdot \quad (18)$$

and it is therefore hardly necessary to measure it directly. However, since the mutual conductance is a good indication of the figure of merit of a tube, we shall briefly describe a few methods whereby it can be measured directly. Referring to equation (17) it can be seen that the principle of any method of direct measurement of the mutual conductance is to apply a potential difference between filament and grid by passing a current through a resistance shunting the grid and filament and balancing this current against the resulting current in the plate circuit. There are various ways in which this can be done. The circuit arrangement of a method proposed by S. Ballantine [1] is shown in Fig. 101.

The coils 1 and 2 are so connected that the currents i_1 and i_2 flowing in the directions of the arrows tend to neutralize each other's effect in the secondary of transformer T. If t_1 and t_2 be the inductance due to the coils 1 and 2, respectively, and R_1 be so adjusted that the tone in the receiver is a minimum, then

$$i_1 t_1 = i_2 t_2.$$

Ballantine assumes that $i_2 = \dfrac{\mu e_g}{r_p}$ from which he then obtains,

[1] S. BALLANTINE, Proc. I.R.E., Vol. 7, p. 134, 1919.

since $e_g = i_1 R_1$,

$$\frac{\mu}{r_p} = \frac{t_1}{R_1 t_2}. \qquad \cdot \quad \cdot \quad \cdot \quad \cdot \quad \cdot \quad \cdot \quad \cdot \quad (43)$$

The assumption made is, strictly speaking, justifiable only when the impressed oscillations are very small, because the current-voltage characteristic of the circuit is not linear unless the external impedance in the plate circuit (i.e., the impedance of coil 2) is large. On the other hand, if it is large the current i_2 cannot be

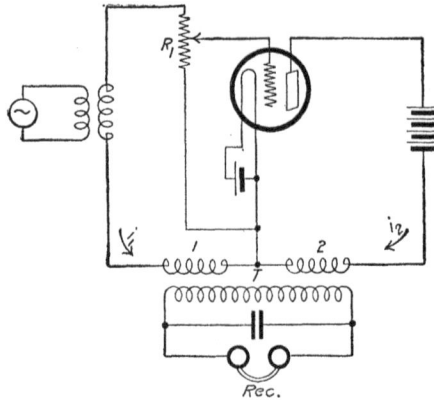

Fig. 101.

expressed by the above simple equation, but is given by equation (22), namely:

$$i_2 = \frac{\mu e_g}{r_p + Z_2}. \qquad \cdot \quad \cdot \quad \cdot \quad \cdot \quad \cdot \quad \cdot \quad (22)$$

where Z_2 is the impedance of coil 2. The mutual conductance of the tube is then given by:

$$\frac{r_p}{\mu} = \frac{e_g}{i_2} - \frac{Z_2}{\mu}, \qquad \cdot \quad \cdot \quad \cdot \quad \cdot \quad \cdot \quad \cdot \quad (44)$$

that is, by

$$\frac{r_p}{\mu} = R_1 \frac{t_1}{t_2} - \frac{Z_2}{\mu}. \qquad \cdot \quad \cdot \quad \cdot \quad \cdot \quad \cdot \quad (44a)$$

For a simplification of this dynamic method I am indebted to an hitherto unpublished suggestion of Mr. H. W. Everitt which is shown in Fig. 102.[1] It consists in replacing the transformer T by two non-inductive resistances r_1 and r_2, the telephone receiver being connected directly to them as shown. The effect of the

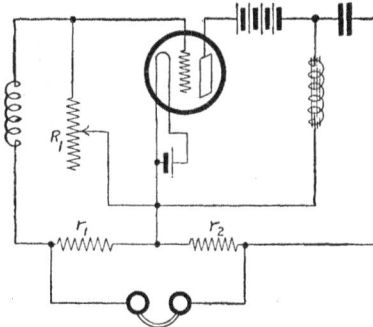

external resistance is shown in the following table, which gives observations obtained by Everitt. The second column gives the values that would be obtained with equation (43), and the last

r_2	$\dfrac{r_1}{r_2 R_1}$	$\dfrac{1}{\dfrac{r_2 R_1}{r_1} - \dfrac{r_2}{\mu}}$
100	1.33×10^{-3}	1.37×10^{-3}
1,000	1.09×10^{-3}	1.39×10^{-3}
2,100	0.86×10^{-3}	1.34×10^{-3}
10,000	0.37×10^{-3}	1.43×10^{-3}

column the corrected values according to equation (44a). The true value of $\dfrac{\mu}{r_p}$ computed from separately observed values of μ and r_p is, for the plate voltage used in these experiments, 1.31×10^{-3}. The values in the last column are not quite in agreement with this value, since the method is not very accurate, but they are grouped around a mean value. The values given in the second column are distinctly influenced by the external resistance, the deviation from

---

[1] This modification was also suggested by Ballantine and given in an addendum to his paper, which appeared about four months after the reading of the original paper at a meeting of the Institute of Radio Engineers.

the true value increasing with it.   When the external resistance is so large that it must be taken into consideration the method becomes tedious and has no advantage over obtaining $\frac{\mu}{r_p}$ from separate determinations of μ and r_p by the method explained with reference to Fig. 100.

A simple d-c. method of measuring the mutual conductance, due to E. V. Appleton [1] is of interest.   The circuit arrangement is shown in Fig. 103.   When the key K is open the galvanometer G indicates the normal plate current.   When K is closed the potential difference e_g applied between filament and grid is I_1R_1, where

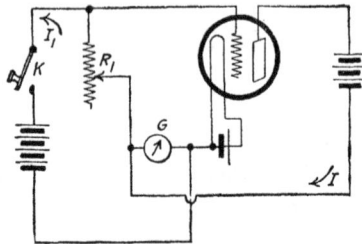

FIG. 103.

I_1 is the current in R_1.   This causes a change in the plate current equal to

$$I_1R_1\frac{\delta I}{\delta e_g}.$$

Now it will be seen that I and I_1 flow through the galvanometer in opposite directions.   Hence if R_1 be adjusted until the galvanometer reading shows no change, then

$$I_1 = I_1R_1\frac{\delta I}{\delta e_g}$$

or

$$\frac{\delta I}{\delta e_g} = \frac{1}{R_1}. \quad . \quad . \quad . \quad . \quad . \quad . \quad . \quad (45)$$

The plate circuit does not contain any external resistance, except that of the galvanometer, which is small.   This equation is,

[1] Wireless World, Vol. 6, p. 458, 1918.

strictly speaking, correct only when the potential applied to the grid is small, in which case we can put $\dfrac{\delta I}{\delta e_g} = \dfrac{\mu}{r_p}$ so that

$$\frac{\mu}{r_p} = \frac{1}{R_1}. \quad \cdots \quad \cdots \quad (46)$$

It will be evident from the foregoing that the mutual conductance is, like the plate resistance, a function of the d-c. plate and grid voltages. As in the case of the plate resistance the effect of the plate and grid voltages can be explained with reference to a curve like that shown in Fig. 99, except that here the abscissæ would represent the effective grid voltage, $E_\gamma = \left(\dfrac{E_p}{\mu} + E_g\right)$, instead of the effective plate voltage, $E_\pi = (E_p + \mu E_g)$. It is evident that $E_\pi = \mu E_\gamma$.

68. Circuit for Measuring Amplification Constant, Plate Resistance and Mutual Conductance. A set which makes possible the quick measurement of all three quantities, μ, r_p and g_m was devised by H. W. Everitt. It consists of the combination of three circuits shown in Fig. 104. For a certain setting of the keys on the box the circuit arrangement is that shown by circuit I (Fig. 104). As was explained above, when r_2 is so adjusted that the tone in the receiver is a minimum, then

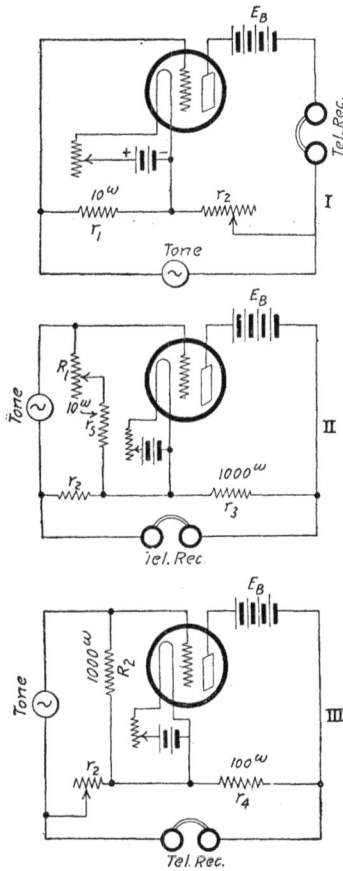

FIG 104.

explained above, when r_2 is so adjusted that the tone in the receiver is a minimum, then

$$\mu = \frac{r_2}{r_1}. \quad \cdots \quad \cdots \quad (47)$$

The resistance r_1 has a constant value of 10 ohms and r_2 is calibrated to read tenths of ohms, so that the reading of r_2 gives μ directly.

Now, to measure r_p the circuit is transformed into circuit II by the simple operation of throwing over a multiple-contact key. This is done without changing the setting of r_2 that gave the value of μ in circuit I. It is seen that r_2 is now transferred to the grid circuit and is replaced by a constant resistance $r_3 = 1000$ ohms.

Referring to this circuit and applying equations given in the previous sections, the voltage drop across r_3 is

$$e_3 = \frac{\mu e_g r_3}{r_p + r_3}$$

If i_1 is the current in the grid circuit,

$$e_g = i_1(R_1 + r_5).$$

Hence

$$e_3 = \frac{\mu r_3 i_1(R_1 + r_5)}{(r_p + r_3)}.$$

When R_1 is so adjusted that the tone in the receiver is a minimum, the voltage drop e_3 in r_3 is equal to the voltage drop $i_1 r_2$ in the resistance r_2. Hence, putting $e_3 = i_1 r_2$ we get:

$$r_p = \frac{r_3}{r_2}\mu(R_1 + r_5) - r_3.$$

Now since r_2 has the same value that it had in circuit *I* and $r_1 = 10$ ohms, we have from equation (47)

$$r_2 = \mu r_1 = 10\mu.$$

Hence

$$r_p = \frac{r_3 R_1}{10} + \frac{r_3 r_5}{10} - r_3.$$

But since r_5 is a constant resistance of 10 ohms the last two terms vanish. Furthermore, $r_3 = 1000$ ohms, so that the plate resistance is given directly by

$$r_p = 100 R_1. \quad \ldots \quad \ldots \quad (48)$$

The dials of R_1 are marked 100 times their actual ohmic resistance so as to make the set direct reading.

Next, to measure the mutual conductance $g_m = \dfrac{\mu}{r_p}$ a second multiple-contact key is operated which transforms the circuit into III (Fig. 104). The resistance r_2 is the same as that used in circuits I and II.

It was shown in Section 67 that if r_4 is small compared with the plate resistance of the tube then

$$\frac{\mu}{r_p} = \frac{r_2}{r_4 R_2}. \qquad \ldots \ldots \ldots \quad (49)$$

By making $r_4 R_2$ an even multiple of 10, we have

$$\frac{\mu}{r_p} = r_2 \times 10^n.$$

For the chosen values of r_4 and R_2, namely 100 and 1000 ohms, $n = -5$, and since the dials of r_2 are marked in tenths of ohms:

$$g_m = \frac{\mu}{r_p} = [r_2] \times 10^{-4}, \qquad \ldots \ldots \quad (50)$$

$[r_2]$ being the reading indicated on the dials.

The measurement of the tube constants with this set is a very quick and simple operation. The complete set is shown in Fig. 105. It includes a tone source, such as that described on page 223. The tube to be tested is inserted in the socket as indicated. The transformation of the circuits is accomplished with the keys 2 and 3. A and B represent the resistances r_2 and R_1 of Fig. 104.

69. Influence of the Electrode Capacities. The amplification equations derived in sections 61 to 63 express the quantities considered in terms of the potential variations actually applied to the grid. When considering the power supplied to the input circuit it is necessary to determine to what extent the electrode capacities can influence the results. The potential variations impressed on the grid when the power is supplied to the input circuit can be influenced by the electrostatic capacities between the electrodes of the tube. The capacity between grid and plate effects a coupling between the output and input circuits, so that the tube is not a perfect unilateral device. The extent to which the output circuit reacts on the input depends on the constants of the circuits.

The solution of the network involving the electrode capacities was given by H. W. Nichols [1] and by J. M. Miller.[2]

[1] H. W. NICHOLS, Phys. Rev., Vol. 13, p. 405, 1919.
[2] J. M. MILLER, Bureau of Standards, Bulletin No. 351.

Fig. 106 represents the equivalent network of the tube and circuit. *G, F* and *P* denote the grid, filament and plate. This circuit represents the condition that the grid is kept at a negative potential with respect to the filament, so that there is no convection current between them. The resistance to the convection current between filament and plate is represented by r_p and is in

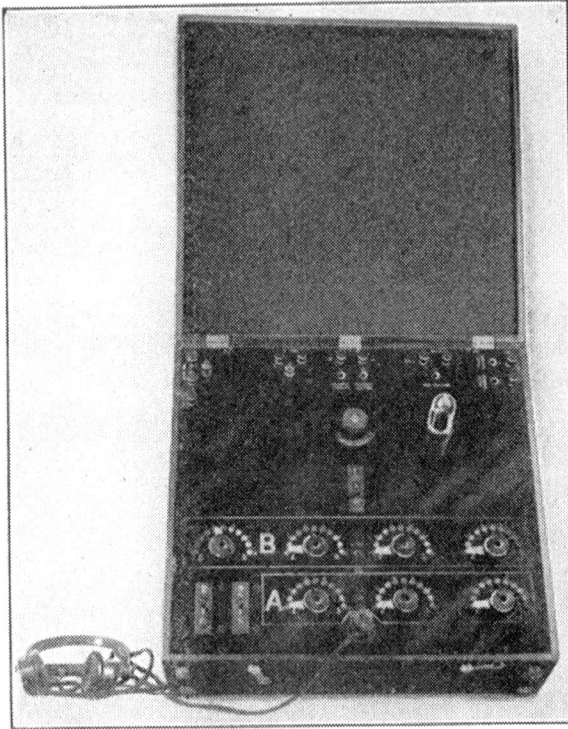

FIG. 105.

shunt with the capacity C_2 between the filament and plate. Z_g represents the impedance as measured between filament and grid and is the effective input impedance. Remembering that a potential e_g impressed on the grid introduces an E.M.F. equal to μe_g in the plate circuit, the input impedance Z_g can be obtained by including in the plate circuit a fictitious generator giving μe_g as indicated in the diagram and solving the Kirchoff equations for the network.

Unless the frequency is very high (over a million cycles per second) we can neglect the capacity C_2 between filament and plate, since it is shunted by the plate resistance which is then low compared with the impedance due to C_2. The equation given by Nichols for the effective input impedance is:

$$Z_g = \frac{1}{jC_1\omega} \frac{1+j\omega C_3 W}{1+j\omega C_3 W + \frac{C_3}{C_1}\left(1+\frac{\mu W}{r_p}\right)}. \quad \dots \quad (51)$$

where $W = \dfrac{r_p Z_0}{r_p + Z_0}$; ω is $2\pi \times$ frequency, and j is the imaginary unit $\sqrt{-1}$. The other quantities are indicated in Fig. 106.

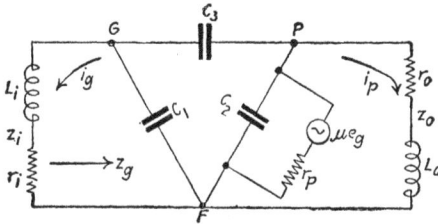

FIG. 106.

For most tubes used at present this equation is applicable for frequencies up to about a million cycles per second.

Let the external output impedance take the general form $Z_0 = r_0 + jx_0$. Then equation (51) can be transformed into:

$$Z_g = -\frac{ac+bd}{c^2+d^2} + j\frac{ad-bc}{c^2+d^2}. \quad \dots \quad (52)$$

$$= -r_g + jx_g$$

where the coefficients have the values

$$\left. \begin{aligned} a &= r_p + r_0 - \omega r_p r_0 C_3 \\ b &= \omega r_p r_0 C_3 + x_0 \\ c &= \omega^2 r_p r_0 C_1 C_3 + \omega x_0 (C_1 + C_3 + \mu C_3) \\ d &= \omega^2 r_p x_0 C_1 C_3 - \omega r_p (C_1 + C_3) - \omega r_0 (C_1 + C_3 + \mu C_3) \end{aligned} \right\} \quad \dots \quad (53)$$

It will be seen from inspection that the effective input impedance will generally comprise a resistance r_g which may be positive or negative, and a reactance x_g which is capacitive.

70. Case 1. Low Frequencies: $\omega < 10^6$. In this case we can neglect ω-terms where they occur in the same expression with terms containing ω in a lower order, e.g., neglect ω^2 in comparison with ω.

Let the output impedance be inductive $x_0 = L_0\omega$. Evaluation of the coefficients (53) gives for the *input resistance*:

$$r_g = \frac{r_p C_3(r_p r_0 C_3 + r_0^2 \mu C_3 + r_0^2 C_3 - \mu L_0)}{[r_p(C_1+C_3)+r_0(C_1+C_3+\mu C_3)]^2}, \quad \cdots \quad (54)$$

and for the *input reactance*:

$$x_g = \frac{1}{\omega\left[C_1+C_3+\dfrac{\mu r_0 C_3}{r_p+r_0}\right]}, \quad \cdots \cdots \quad (55)$$

and therefore the effective input capacity is:

$$C_g = C_1 + C_3\left(1+\frac{\mu r_0}{r_p+r_0}\right). \quad \cdots \cdots \quad (56)$$

Now, C_1 is the electrostatic capacity between filament and grid. The effective input capacity is greater than the electrostatic capacity by the amount shown by equation (56). The increase depends on the electrostatic capacity between grid and plate, and on the *resistance* in the output load. It also depends on the amplification constant μ. It will be recognized that this equation contains the expression for the voltage amplification as a function of the external output resistance (see equation 28). The effective input capacity therefore increases with the output load resistance in the manner indicated by Fig. 95, page 192. Miller [1] has measured the effective input capacity as a function of the external output resistance. The following table shows the agreement between his observed and computed values for the case of a *VT*-1 tube.

	Input Capacity.	
r_0, Ohms.	Computed.	Observed.
0	....	27.9
8,000	51.4	49.0
16,000	64.5	61.5
49,400	78.9	76.1
97,000	84.2	84.3
139,000	86.1	87.6

[1] Loc. cit.

Equations (54) and (55) show that the effective input resistance is for low frequencies ($\omega < 10^6$) independent of the frequency, but depends, among the other circuit constants, on both resistance r_0 and inductance L_0 in the external output circuit. The input reactance, on the other hand, is inversely proportional to the frequency and depends on the resistance r_0, but not on the inductance in the output.

From equation (55) it follows that the amplification given by the tube would decrease as the frequency is increased. But this tendency to distort is in itself not due to power consumption in the input, but is occasioned by the decrease in the input grid potential, due to the lowering of the input reactance.

The power consumed in the input is determined by the equation (54). This equation contains a negative term in the numerator and therefore if the output inductance L_0 is large enough, the input resistance can be negative. Under these conditions the tube would tend to produce oscillations or " sing " through its internal capacities. This tendency to sing is frequently a source of annoyance in amplifier circuits.

Miller has computed the relation between r_g and the output inductance L_0. In amplifier circuits we are usually more interested in the external output impedance than in the output inductance. Fig. 107 shows the relation between the effective input resistance r_g and the ratio $\dfrac{Z_0}{r_p}$ of external output impedance to plate resistance for various angles $\phi = \tan^{-1} \dfrac{L_0}{r_0}$ of the output impedance $Z_0 = r_0 + jL_0$. For a pure inductance in the output ($r_0 = 0$), equation (54), reduces to the simple form

$$r_g = -\frac{L_0 C_3}{r_p (C_1 + C_3)^2}, \quad \cdots \cdots \quad (57)$$

and is therefore always negative. As the angle ϕ decreases, the negative value of r_g decreases and finally becomes positive.

The curves in Fig. 107 were computed with the following values of the constants:

$$r_p = 5 \times 10^3 \text{ ohms}$$
$$C_1 = 5 \times 10^{-12} \text{ farad;}$$
$$C_3 = 15 \times 10^{-12} \text{ farad;}$$
$$\mu = 5;$$
$$\omega = 2 \times 10^4.$$

These are approximately the constants of a type of tube that is commonly used for amplifying telephonic currents.

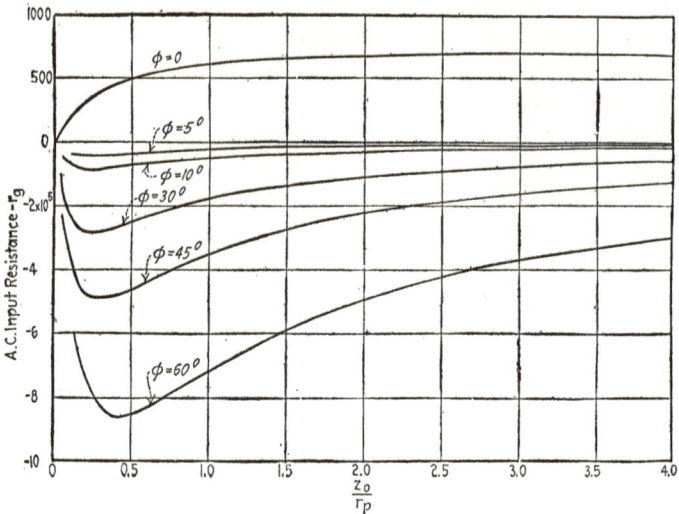

Fig. 107.

The corresponding input reactances x_g are shown in Fig. 108. The input reactance is therefore generally much larger than the input resistance.

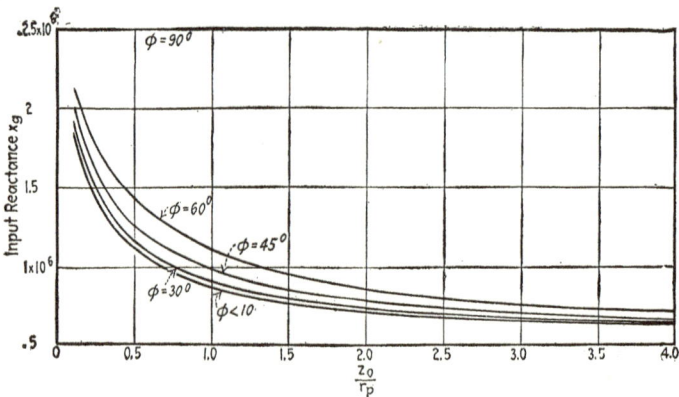

Fig. 108.

It can be seen from the above equations that if the output reactance is capacitive, then the input resistance is always positive This is also the case when the output is a pure resistance ($x_0 = 0$).

Under these conditions the tube would absorb power from the input. But for ordinary frequencies this power absorption is negligibly small.

The power absorbed in the input can be obtained as follows: Let us impress an alternating potential e_g on the grid. Then the grid current is

$$i_g = \frac{e_g}{Z_g} = \frac{e_g}{r_g + jx_g},$$

the condition being that the grid is at all times negative with respect to the filament, so that there is no convection current between filament and grid. The power absorption is therefore due entirely to the reaction of the output on the input circuit. This power is:

$$P_g = i_g{}^2 r_g = \frac{e_g{}^2 r_g}{(r_g + jx_g)^2}$$

or

$$P_g = e_g{}^2 \frac{r_g}{r_g{}^2 + \dfrac{1}{\omega^2 C_g{}^2}}.$$

For $\omega < 10^6$ we can neglect $r_g{}^2$ in comparison with $\dfrac{1}{\omega^2 C_g{}^2}$ and write

$$P_g = \omega^2 e_g{}^2 r_g C_g{}^2. \qquad \ldots \ldots \ldots \quad (58)$$

Substituting the values of r_g and C_g from equations (54) and (55), for the case of a pure resistance in the output ($L_0 = 0$), we obtain

$$P_g = \omega^2 e_g{}^2 \, r_p C_3{}^2 K(1 + \mu K), \quad \ldots \ldots \quad (59)$$

where $K = \dfrac{r_0}{r_p + r_0}$. This shows that the only inter-electrode capacity that is effective in causing input power loss is C_3, the capacity between grid and anode.

In order to obtain the order of magnitude of P_g for a common type of tube, we can insert the values for the constants given on page 209 into equation (59). When $r_0 = r_p$ (the condition for maximum power output), $K = \tfrac{1}{2}$. Putting $e_g = 5$ volts, we get:

$$P_g = 4.8 \times 10^{-17} \times \omega^2 \text{ watt.}$$

For a tube of the type considered, and for telephonic frequencies ($\omega < 2 \times 10^4$), the ratio of output to total input power is about 300, and for an input voltage of 5 volts the power in the output is about 30×10^{-3} watt. The total input power is therefore about 1×10^{-4} watt. For $\omega = 2 \times 10^4$, the power P_g consumed in the effective grid resistance is about 2×10^{-8}, which is still very small compared to the total input power which, in normal operation of a tube, is consumed in the input transformer, and in the high resistance usually bridged across filament and grid, as shown in Fig. 90, page 182. Of course, the power consumption by the effective grid resistance can become quite large when the frequency is high, since it increases with the square of the frequency within the frequency range considered. At extremely high frequencies the effective grid resistance again becomes negligibly small, as will be seen from the following.

71. Case II. High Frequencies. When the frequency is very high, we cannot neglect the capacity C_2 between filament and plate because the impedance due to C_2 can obviously become comparable with or even lower than the plate resistance r_p with which it is in parallel. (See Fig. 106.) In this case the effective input impedance is given by

$$Z_g = \frac{ac+bd}{c^2+d^2} + j\frac{bc-ad}{c^2+d^2},$$

where the coefficients now have the values:

$$\left.\begin{array}{l} a = \omega r_p r_0 (C_2 + C_3) + x_0 \\ b = \omega r_p x_0 (C_2 + C_3) - r_p - r_0 \\ c = \omega r_0 (C_1 + C_3 + \mu C_3) + \omega r_p (C_1 + C_3) \\ \qquad\qquad - \omega^2 r_p r_0 (C_1 C_3 + C_1 C_2 + C_2 C_3) \\ d = \omega x_0 (C_1 + C_3 + \mu C_3) + \omega^2 r_p r_0 (C_1 C_3 + C_1 C_2 + C_2 C_3) \end{array}\right\}, \quad (60)$$

and x_0 can be $L_0\omega$ or $\dfrac{1}{C_0\omega}$. Since ω is large we can neglect the ω-terms of lower order in comparison with those of the succeeding and higher orders. This gives:

$$\left.\begin{array}{l} r_g = 0 \\ C_g = \dfrac{C_1 C_3 + C_1 C_2 + C_2 C_3}{C_2 + C_3} \end{array}\right\}, \quad \cdot \quad \cdot \quad \cdot \quad \cdot \quad (61)$$

independent of the constants of the external output circuit. At very high frequencies, therefore, the grid does not absorb power, but the amplification is lowered because the input current is practically short-circuited by the electrode capacities.

The voltage amplification as a function of the frequency, for frequencies ranging from 3×10^5 to 4 million cycles per second, can be measured with the circuit arrangement shown in Fig. 109, for which I am indebted to my associate Dr. J. B. Johnson. A_1 is the tube with which the high-frequency current was amplified. The a-c. output voltage from this tube was measured by means of a bridge arrangement shown on the right of the figure, in which the plate-filament resistance of the tube A formed one arm of the bridge. The other three arms are indicated by B, C and D. The output from the tube A_1 is impressed on the grid

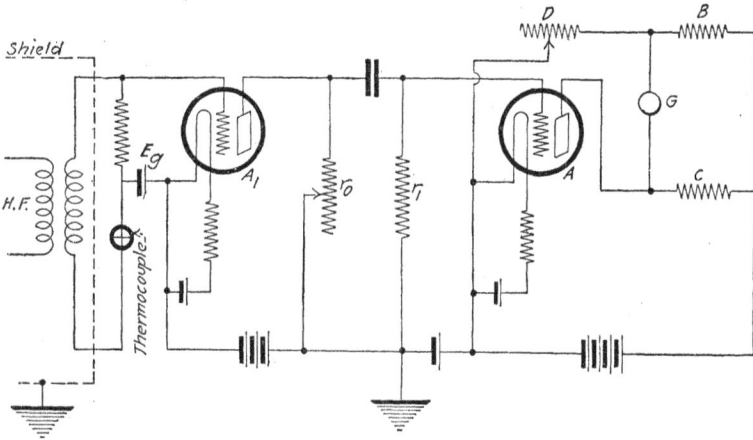

FIG. 109.

of the tube A, which acts as a detector, so that small potential variations applied to its grid unbalances the bridge.

The bridge was calibrated by applying known a-c. voltages to the grid of tube A, and noting the deflection of the galvanometer G. The known voltage is applied, as is commonly done, by connecting between the grid and filament of tube A, a small resistance and a thermocouple in series. The high-frequency current is passed through the resistance and thermocouple, and the input voltage is given by the known value of the resistance, and the reading is indicated by the thermocouple. In order to measure the amplification, the tube A_1 is inserted as shown, and the input voltage measured with resistance and thermocouple as indicated. In such case we therefore measure the voltage actually applied to the grid of the tube A_1, so that the capacity between filament and grid of this tube does not affect the results. Now, the actual voltage amplification depends on the load in the output of the tube. This load is constituted by the resistance r_0 and r_1 in parallel. This is, however, not the only impedance in the output of tube A_1, because there is parallel with r_0 and r_1, the bridge circuit connected through the capacities of tube A, which forms one arm of the bridge. Hence, unless these capacities are very small the values of the voltage in the output of tube A_1 (and which is impressed on the input of tube A) will be smaller than the output voltage that would actually be given by the tube A_1 if its output were not connected to the bridge. The galvanometer would therefore indicate read-

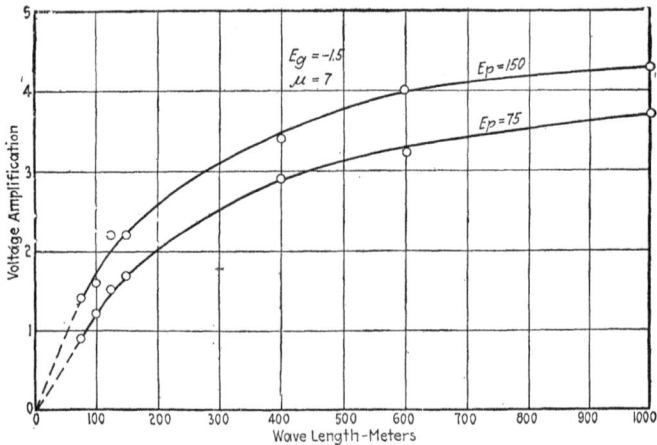

FIG. 110.

ings that are too small. In order to avoid such an effect, the tube A was specially constructed to have as small a capacity as possible between its electrodes. This capacity should be small compared with the capacity between the elements of the tube A_1. Fig. 110 shows two curves obtained by Johnson which indicate the relation between the voltage amplification and the wave length for a standard type of tube such as is shown in Fig. 68. The two curves are for different plate potentials. The higher plate potential of course gives a higher amplification because the plate resistance of the tube is lower. It is seen that the amplification at 1000 meters is about three times as large as the amplification at 100 meters, the amplification at both plate potentials dropping **to** zero when the frequency comes infinitely high.

The reduction in the amplification can for a given frequency be avoided, as Nichols suggested, by shunting the grid-plate capacity with a inductance of such a value as to make the impedance between grid and plate infinite at the given frequency. The value of this inductance would, of course, depend on the other reactances in the circuit. Making the grid-plate impedance infinite is equivalent to putting $C_3 = 0$ in equation (61). Then $C_g = C_1$, and this is in parallel with the tuning capacity in the input, which is generally shunted across the input inductance. By properly adjusting the tuning capacity the filament-grid impedance can be made large, thus increasing the input potential applied to the grid. This scheme has now been in use for several years and found to give satisfactory results.

72. Practical Measurement of Amplification. It will be evident that the amplification that can be produced by a tube depends not only upon its structural parameters, but also upon the constants of the circuit in which it is used. When a tube is designed for a specific purpose and is to be operated in a specific circuit, the amplification can be measured quickly and conveniently by a method that has been in use in telephone practice for a number of years. The desirability of a method of testing tubes rapidly for amplification becomes apparent where they are manufactured in comparatively large quantities. These tubes are, for example, used extensively on the long distance telephone lines of the Bell System and elsewhere as telephone repeaters. Tubes used for this purpose are made with great care and are designed to operate in circuits of definite constants, the variation of the tube constants being kept within close limits.

The method of measuring amplification consists in amplifying a current of a given frequency with the tube and then passing the amplified current through an artificial line of variable and known attenuation before it passes through a telephone receiver. By means of a switch the current can also be transmitted directly from the source to the receiver. When the attenuation of the artificial line is so adjusted that the intensity of the current in the receiver is the same whether it passes directly to the receiver or through the tube and line to the receiver, then the current is attenuated by the line as much as it is amplified by the tube, and the amplification can therefore be obtained from the known constants of the line.

The circuit arrangement used for testing audio frequency amplifiers is shown in Fig. 111. U is a source of alternating current which will be described later. The current is preferably passed through a wave filter to obtain a pure note of about 800 cycles per second. By means of the switch W this current can be transmitted directly to the telephone receiver or be impressed on the input circuit of the tube. S represents the artificial line and takes the form of a shunt to the receiver. It is, however,

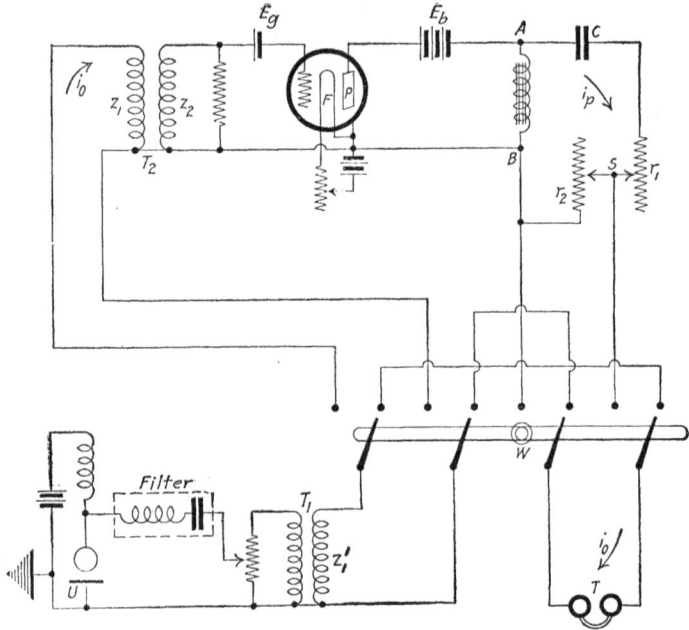

Fig. 111.

not a simple shunt, but is so designed that the total impedance to the output current is always constant for all values of the branch current in the receiver. This is done, as will be explained below, by the addition of the series resistance r_1. In this form the shunt is known as a " receiver shunt." The direct current in the plate circuit is supplied through the choke coil, as shown, to insure that the plate resistance remains constant. Although the total output impedance remains constant for all adjustments of the shunt the resistance does not. Hence, if the choke coil were omitted the potential difference between filament and plate would change with the shunt adjustment and this would result in a change in the plate resistance. The secondary of transformer T_2 is wound to have as high an impedance as possible, so as to impress the highest possible voltage between the grid and filament. The grid battery E_g is inserted to keep the grid sufficiently negative with respect to the filament to prevent it from taking current. It is important to note that the current drawn from the generator U must remain the same for both positions of the switch W. The primary of transformer T_2 is therefore wound so that the impedance as measured across the terminals of the primary is the same as that of the telephone receiver.

Now, if i_p be the total alternating current in the plate circuit, and i_0 the branch current in the receiver, then we can put

$$\frac{i_0}{i_p} = e^{-\alpha d}, \quad \cdots \cdots \cdots \quad (62)$$

where d represents the length of artificial line of which α is the attenuation constant per unit length.

The value of d depends of course upon the relative values of the series and shunt resistances r_1 and r_2 of the receiver shunt. If these are so adjusted that the intensity of the note in the telephone receiver is the same for both positions of the switch W, then the expression $e^{\alpha d}$ gives a measure of the current amplification produced by the tube provided the following conditions are satisfied. First, the current i_0 in the telephone receiver must be equal to the current in the primary of transformer T_2. This can be done by making the impedance Z_1 of T_2 equal to that of the receiver T. Second, for " amplification " to have any meaning the impedance to the input current i_0 must be equal to the impedance to the output current i_p. Now the total impedance to the input current is that of the circuit $Z'_1 Z_1$ (which is equal to that of the circuit $Z'_1 T$) and the impedance to the output current is that of the circuit $FPATB$. These two impedances must be equal. The receiver shunt must therefore be so designed that for all the necessary adjustments of its shunt and series resistances r_2 and r_1 the total impedance to the output current remains constant. If the plate resistance differs markedly from the impedance of the receiver a transformer can be inserted between the coil AB and the receiver shunt S.

Writing equation (62) in the form

$$d = K \log \frac{i_p}{i_0}, \quad \cdots \cdots \quad (63)$$

where

$$K = \frac{2.303}{\alpha}, \quad \ldots \ldots \quad (64)$$

we can express the current amplification in terms of d instead of $\frac{i_p}{i_0}$, that is, we express it on the logarithmic scale.

The constant α is arbitrary and can be given any convenient value, provided it be definitely specified when the amplification is expressed in terms of d. Unfortunately the value of α commonly used in telephone practice is not a convenient value, but it has already found its way into extensive vacuum tube practice and we shall adopt it here. As commonly used, α is the attenuation constant per mile of the so-called " standard No. 19 gauge cable," which has a capacity of 0.054 mf., and a resistance of 88 ohms per mile. It is to be noted that this reference cable has neither inductance nor leakance. The constant α is determined by these two quantities and the frequency of the current. If the frequency is 800 cycles per second, $\alpha = 0.1091$. This makes $K = 21.13$ and the amplification can thus be expressed in terms of miles (d) of standard cable. And this simply means that if the amplification is d miles, it would take a standard No. 19 gauge cable d miles long to reduce the current to its original value. When expressing power amplification instead of current amplification the constant K in equation (63) must be divided by two.

There is an advantage in expressing amplification on the logarithmic scale which is in accordance with our lack of conception of absolute intensity of sound. By using the logarithmic scale and taking length of cable as the standard of reference, we obtain a definite idea of what the energy means when it is to be used for operating a telephone receiver. The value of the attenuation constant α happens to be such that steps of 1 mile of the chosen standard cable afford a very convenient unit of measurement. For many purposes it is, however, sufficient if the receiver shunt is calibrated in steps of two miles of the standard cable. A little practice then makes it possible to estimate amplification to within 1 mile.

For convenience of reference the following table is attached, giving the relation between miles of the standard No. 19 gauge cable and the current and power ratios as computed from equations (63) and (64).

Miles of Std. Cable.	Current Ratio.	Power Ratio.
1	1.11	1.24
2	1.24	1.54
3	1.38	1.92
4	1.54	2.39
5	1.72	2.97
6	1.92	3.70
7	2.14	4.60
8	2.39	5.72
9	2.66	7.12
10	2.97	8.85
11	3.31	11.02
12	3.70	13.7
13	4.12	17.0
14	4.49	21.1
15	5.12	26.37
16	5.71	32.7
17	6.94	40.7
18	7.11	50.6
19	7.92	62.8
20	8.83	78.3
21	9.84	97.4
22	11.0	121.1
23	12.25	150.7
24	13.65	187.1
25	15.2	233.4
26	17.0	289.8
27	18.9	359.8
28	21.1	447.8
29	23.5	559.2
30	26.3	693.5
32	32.6	1.07×10^3
34	40.5	1.65×10^3
36	50.5	2.57×10^3
38	62.6	3.96×10^3
40	78.1	6.15×10^3
42	97.2	9.44×10^3
44	120.8	1.47×10^4
46	150.6	2.27×10^4
48	186.6	3.51×10^4
50	232.3	5.43×10^4
55	400.0	1.61×10^5
60	691.9	4.79×10^5
65	1.18×10^3	1.43×10^6
70	2.05×10^3	4.265×10^6
75	3.54×10^3	1.285×10^7
80	6.09×10^3	3.75×10^7
85	10.55×10^3	1.13×10^8
90	18.16×10^3	3.32×10^8
95	31.27×10^3	9.95×10^8
100	53.83×10^3	2.95×10^9

Fig. 112 shows the relation between power amplification and the ratio of external impedance Z_0 to the plate resistance r_p. The curves were computed from equation (35) with the following values: $\mu = 5$, $r_p = 5 \times 10^3$ ohms, $r_g = 6 \times 10^5$ ohms, and the impedance Z_0 was assumed to have an angle of 45°. The lower curve gives the amplification expressed in ratio of output to input power, while the upper curve gives the power amplification expressed in miles of standard No. 19 gauge cable. It is seen that while the power ratio varies considerably from the maximum value when the ratio $\dfrac{Z_0}{r_p}$ deviates from unity, yet the change in the effect produced on the ear, which is more in accordance with the upper curve

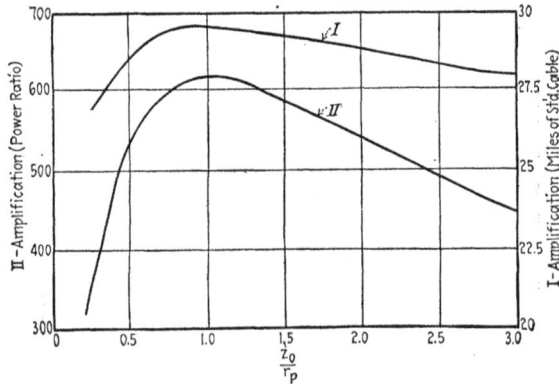

FIG. 112.

expressed on the logarithmic scale, is quite small. A change in amplification of one standard cable mile is not serious. The external impedance into which the tube works can therefore have values ranging from about one-half to two and one-half times the plate resistance without producing any marked change in effect as heard in the telephone receiver.

Now, it was stated above that the condition to be satisfied by the receiver shunt is that its insertion in the circuit must not change the total impedance of that circuit, and this must be true for all adjustments of its shunt and series resistances. The receiver shunt is, of course, so arranged that the shunt and series resistances are only inserted in definite pairs, the object being that the change in the total impedance, due to the insertion of a shunt resistance must be compensated for by the addition of a corresponding series resistance.

Referring to Fig. 113, let r_1, r_2 represent the receiver shunt, r_p the plate resistance of the tube which is non-reactive and constant, and Z_0 the impedance of the receiver, the angle of which is $\tan^{-1} \dfrac{x_0}{r_0}$, and this is the angle of the external impedance when the receiver shunt is cut out ($r_1 = 0$, $r_2 = \infty$). But when the shunt is inserted the angle of the effective external impedance depends on the values of r_1 and r_2, and changes with every adjustment of r_1 and r_2. When the angle of the receiver is large and if accuracy is desired, this must be taken into consideration in com-

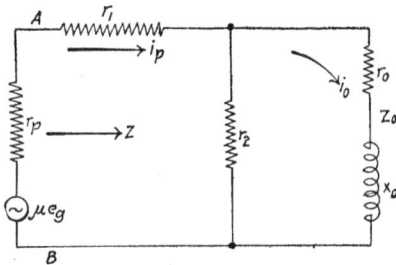

Fɪɢ. 113.

puting receiver shunts. The necessary values of r_1 and r_2 can be determined graphically or in the following simple manner by first neglecting the effect of the change in angle and then applying a simple correction method.

The first condition to be satisfied is that the current i_0 in the impedance Z_0 must be related to the total output current i_p in the plate circuit by the equation

$$\frac{i_0}{i_p} = e^{-\alpha d}, \quad \cdots \cdots \cdots \quad (62)$$

where the current attenuation produced by the shunt is then equivalent to d miles of cable of which α is the attenuation constant per mile at the frequency of the current, which we shall take to be 800 cycles per second, thus making $\alpha = 0.109$ for the " standard cable."

Now the alternating current established in the circuit (Fig. 113) is due to the alternating potential e_g applied to the grid. The impressed E.M.F. is therefore μe_g and is applied in the branch $r_p r_1 r_2$.

If Z be the total impedance of the circuit and R and X the resistance and reactance, respectively, of the circuit to the right of AB, we have as a second condition to be satisfied by the receiver shunt:

$$Z = \frac{\mu e}{i_p} = r_p + R + jX = \text{constant}. \quad . \quad . \quad . \quad (65)$$

When the shunt is cut out ($r_1 = 0$, $r_2 = \infty$) then $R = r_0$ and $X = x_0$.

Summing the E.M.F.'s in the two branches we have

$$e = i_p r_p + i_p r_1 + i_p r_2 - i_0 r_2 \quad . \quad . \quad . \quad . \quad . \quad (66)$$

$$0 = i_0 Z_0 + i_0 r_2 - i_p r_2 . \quad . \quad . \quad . \quad . \quad . \quad . \quad (67)$$

From (67) and (62) we obtain directly

$$r_2 = \frac{Z_0}{e^{\alpha d} - 1},$$

or

$$r_2 = \frac{Z_0}{\sinh \alpha d + \cosh \alpha d - 1}. \quad . \quad . \quad . \quad . \quad (68)$$

The values of r_2 necessary to give any desired attenuation d can thus be obtained directly from a table of hyperbolic functions.

In deriving equation (68) we made use of equation (62), which holds for a circuit of zero reactance, while in the receiver shunt circuit the reactance is not zero. The values of r_2 obtained will therefore not be correct unless the angle of the impedance Z_0 is small or the attenuation large. If the angle of Z_0 is not greater than about 45° the values of r_2 given by equation (68) are sufficiently accurate for current attenuations greater than those produced by 6 miles of standard cable ($d > 6$). For smaller attenuations or if the impedance Z_0 has a large angle the values of r_2 obtained from (68) can be corrected as follows: Suppose it is desired to compute a receiver shunt giving a maximum attenuation of 30 miles of standard cable and allowing the attenuation to be varied in steps of 1 mile each. This shunt, we shall suppose, is to operate with a receiver having a large angle, say, 70°. We can use equation (68) to obtain an idea of the range of values of r_2 that would be necessary to give attenuations ranging from 1 to 30 miles, by computing r_2 for d equal to 2 and 30, say. By choosing five or six convenient arbitrary values of r_2 covering this range, we can compute the corresponding current attenuations $\frac{i_p}{i_0}$, or d, that would be produced by the chosen values of r_2 in parallel with $Z_0 = r_0 + jx_0$, and by plotting a curve between these values of d and the chosen values of r_2, the correct shunt resistances can be read from the curve for all the desired values of d from 1 to 30 miles. It will be recognized that by this method the shunt is computed by the quick and simple process of determining d for chosen values of r_2 instead of the lengthy and tedious way of determining r_2 from the desired values of d.

Once the values of the shunt resistances r_2 are known, the corresponding series resistances r_1 can be obtained from the condition stated by equation (65) that the total impedance must remain constant.

The generator U (Fig. 111) can be of any type that gives a constant note of about 800 cycles per second. It may, for example, be an audion oscillator or a microphone generator, as shown in Fig. 111. This type of generator is very convenient when compactness is desired. Its principle of operation is like that of an interrupter or " buzzer," although it is much superior to the interrupter, which is for this purpose practically useless because of its inconstancy and need for constant adjustment. The

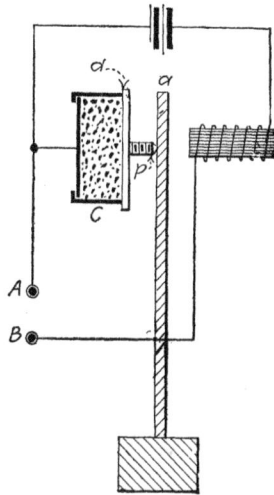

FIG. 114.

microphone generator is shown diagrammatically in Fig. 114. C is a carbon button, the diaphragm d of which is under tension, due to the pressure of the armature a against the pin p. The current from the battery causes the armature to be attracted to the coil. This releases the pressure on the diaphragm and increases the resistance of the carbon. The resulting decrease in current reverses the process and an alternating current is established in a circuit connected to AB. It is seen that the current is never broken as in the case of an interrupter, but the current strength is merely varied by the varying pressure exerted on the carbon. This device is therefore free from troubles attending the interrupter, such as corroding of contacts, etc. Fig. 115 shows a photograph of the microphone generator. It operates on a voltage of 3 to 5 volts and gives an alternating current output of several milliamperes in a resistance equal to its own resistance which varies approximately between 50 to 100 ohms.

The use of this generator makes it possible to include the whole amplification test circuit shown in Fig. 111, in compact form in a portable box. Fig. 116 shows such an amplifier test set used for measuring the amplification of tubes in the laboratories of the Western Electric Company. The dial of the receiver shunt is shown at the lower right-hand corner of the box.

FIG. 115

73. Amplification as a Function of the Operating Parameters. In the early part of this chapter it was pointed out that the operating range of the characteristic of the tube is characterized by the condition that the filament temperature must be so high that the effective grid voltage $\left(\dfrac{E_p}{\mu}+E_g+\epsilon\right)$ is not high enough to draw all the electrons to the anode as fast as they are emitted from the cathode, for if this were so, a change in the applied voltage would not produce any appreciable change in the anode current. In other words, the temperature of the filament must be high enough to comply with the condition of " temperature saturation " which is indicated by the horizontal part of the plate current-filament current characteristic (see Fig. 18, page 51). The effect of filament temperature is indicated in Fig. 117, which gives the relation between the filament voltage and the amplification measured at a constant plate voltage in a circuit like that shown in Fig. 111. For this particular type of tube, it is seen from the curve, the filament voltage must never drop below 3.0 volts. On the other hand, it should not be increased more than is absolutely necessary, for this would shorten the life of the filament. Satisfac-

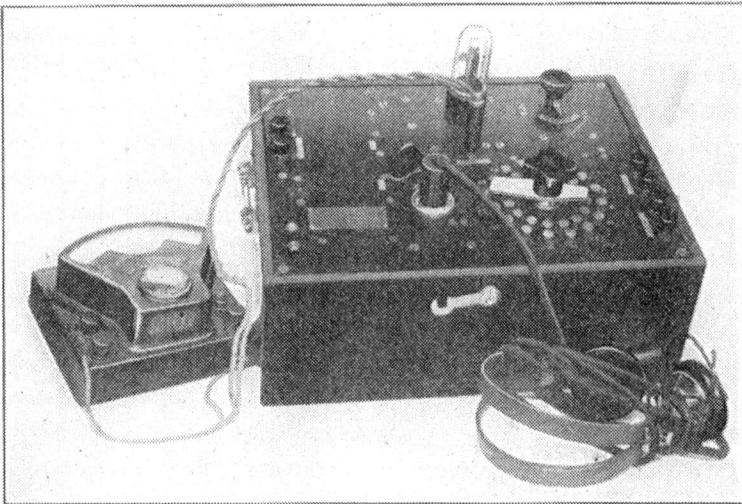

FIG. 116.

tory operation and long life can therefore be obtained by suitably adjusting the filament current.

Referring to equation (36) it is evident that the amplification increases as the ratio $\dfrac{\mu^2}{r_p}$ is increased provided the plate resistance r_p always remains equal to the external impedance of the plate circuit. For a given type of tube μ remains practically constant, but r_p can be decreased by increasing the plate potential provided the filament temperature remains high enough to insure " temperature saturation." (It will be remembered that the filament temperature necessary for this condition increases as the plate voltage is increased.) It is therefore to be expected that the amplification would increase with increase in the plate voltage. If the amplification is measured for increasing plate voltages E_p in the circuit of Fig. 111, which contains a fixed external impedance, the amplification generally tends toward a maximum value, as shown in Fig. 118. This is due to the counter balancing effect occasioned by the deviation of the ratio $\dfrac{Z_0}{r_p}$ from unity. (See Fig. 112.)

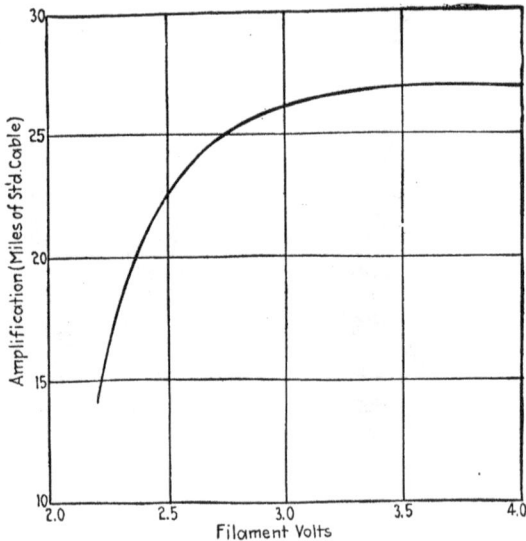

Fig. 117.

74. Tube Constants as Functions of the Structural Parameters.

It will be recognized from the foregoing that the amplification constant μ and the plate resistance r_p play an important part in the operation of three-electrode tubes. They are the two main constants that figure in the design of such tubes. Except at low effective voltages, μ is practically independent of the applied grid and plate potentials, and is determined by the dimensions of the grid and the distance between grid and plate. The plate resistance depends not only on the dimensions and disposition of the electrodes, but also on the applied voltages. It is important to know for the purpose of designing tubes, how these constants depend on the structural dimensions of the tube.

75. Calculation of Amplification Constant. The constant μ is due to the electrostatic screening effect of the grid. An expression for this effect had been derived by Maxwell long before the audion came into existence.[1] The problem that Maxwell set himself was to determine the extent to which a wire grating or gauze could protect apparatus enclosed by it from external electrostatic disturbances. His solution is, however, directly applicable to the audion. It was applied and extended to include cylindrical tubes by Abraham, King, Schottky, and v. Laue.[2]

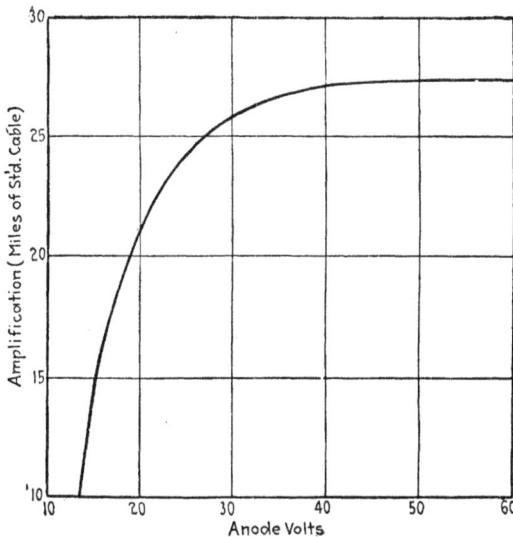

FIG. 118.

Maxwell's results can be expressed as follows: Let F and P (Fig. 119) be two infinitely large parallel planes with a grating G of parallel wires interposed between them. Let the potentials of F, G and P be V_f, V_g and V_p, respectively. The wires of the grid are of the same thickness and have a radius r, the distance between the wires being a. Let f and p be the distances of F and P from the grid. The assumption is made that f and p are large

---

[1] J. C. MAXWELL, "Electricity and Magnetism," Vol. 1, p. 310.

[2] MAX ABRAHAM, Archiv. für Elektrotechnik, Vol. 8, p. 42, 1919. R. W. KING, paper read at October meeting of Am. Phys. Soc., Philadelphia, 1919. SCHOTTKY, Archiv. für Elektrotechnik, Vol. 8, p. 12, 1919. v. LAUE, Ann d. Phys., Vol. 59, p. 465, 1919.

compared with a, which again is large compared with r. If σ and σ' are the charge densities induced on F and P, respectively, then

$$4\pi\sigma\left(p+f+\frac{pf}{\alpha}\right)=\left(1+\frac{p}{\alpha}\right)V_f-V_p-\frac{p}{\alpha}V_g \quad . \quad . \quad . \quad (69)$$

$$4\pi\sigma'\left(p+f+\frac{pf}{\alpha}\right)=-V_f+\left(1+\frac{f}{\alpha}\right)V_p-\frac{f}{\alpha}V_g, \quad . \quad (70)$$

where

$$\alpha=-\frac{a}{2\pi}\log_e\left(2\sin\frac{\pi r}{a}\right). \quad . \quad . \quad . \quad (71)$$

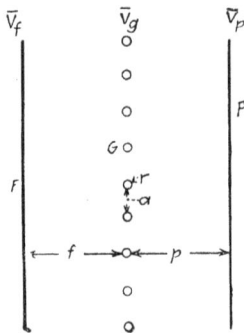

Fig. 119.

The quantity σ gives a measure of the intensity of the field near F, which affects the flow of electrons from F. These equations apply, of course, to the case in which there are no electrons in the space between the electrodes to cause a distortion of the field. As far as the determination of the screening effect of the grid for most practical purposes is concerned, it is found that the presence of the electrons in the space usually does not materially influence the results. From these equations we can obtain the following interesting results:

Suppose, first, that the grid wires are infinitely thin, so that $\frac{r}{a}$ is zero. Then α is infinitely large and equation (69) reduces to the simple form applicable to two plates without the grid, namely,

$$4\pi\sigma_1(p+f)=V_f-V_p. \quad . \quad . \quad . \quad . \quad . \quad (72)$$

If, on the other hand, the grid is of finite dimensions, but G is connected to F, i.e., $V_g=V_f$, then

$$4\pi\sigma_2\left(p+f+\frac{pf}{\alpha}\right)=V_f-V_p. \quad . \quad . \quad . \quad . \quad (73)$$

From these two equations it follows that the charge induced on F, when the grid is interposed and connected to F, is to the charge induced on F by the same potential on P when there is no grid, as 1 is to

$$1+\frac{pf}{\alpha(p+f)}.$$

This result expresses the screening effect of the grid in its simplest form.

Another case that is of interest is the relation between the stray field acting through the grid when the latter is connected to F and the field obtained when grid and plate P are connected together, the potential applied to the plate P being the same. This is readily obtained by putting $V_p = V_g$ in equation (69). Thus:

$$4\pi\sigma_3\left(p+f+\frac{pf}{\alpha}\right)=\left(1+\frac{p}{\alpha}\right)(V_f-V_p). \quad \ldots \quad (74)$$

Comparison with equation (73) gives

$$\frac{\sigma_3}{\sigma_2}=1+\frac{p}{\alpha}. \quad \ldots \ldots \ldots (75)$$

What we are interested in when using the tube in an a-c. circuit is the relation between the variation in the field produced at F by a variation of the grid potential to that produced by an equal variation in the potential of the plate, both fields acting at the same time. This can be obtained directly as King[1] has done, by evaluating the partial derivatives of σ with respect to V_g and V_p (equation 69):

$$\frac{\partial\sigma}{\partial V_g}\bigg/\frac{\partial\sigma}{\partial V_p}=\frac{p}{\alpha}. \quad \ldots \ldots \ldots (76)$$

Let F be the filament of a vacuum tube. Then, applying the usual notation, $V_f-V_p=E_p$ and $V_f-V_g=E_g$, and substituting the value of α from (71), we obtain directly from equation (76):

$$\frac{dE_p}{dE_g}=\mu=-\frac{2\pi p}{a\log_e\left(2\sin\frac{\pi r}{a}\right)}. \quad \ldots \ldots (77)$$

[1] Loc. cit.

When the diameter of the wire is small compared with the distance between adjacent wires, $\frac{\pi r}{a}$ is small, and we can write approximately:

$$\mu = \frac{2\pi p}{a \, \log_e \dfrac{a}{2\pi r}} , \quad . \quad . \quad . \quad . \quad . \quad . \quad . \quad . \quad (78)$$

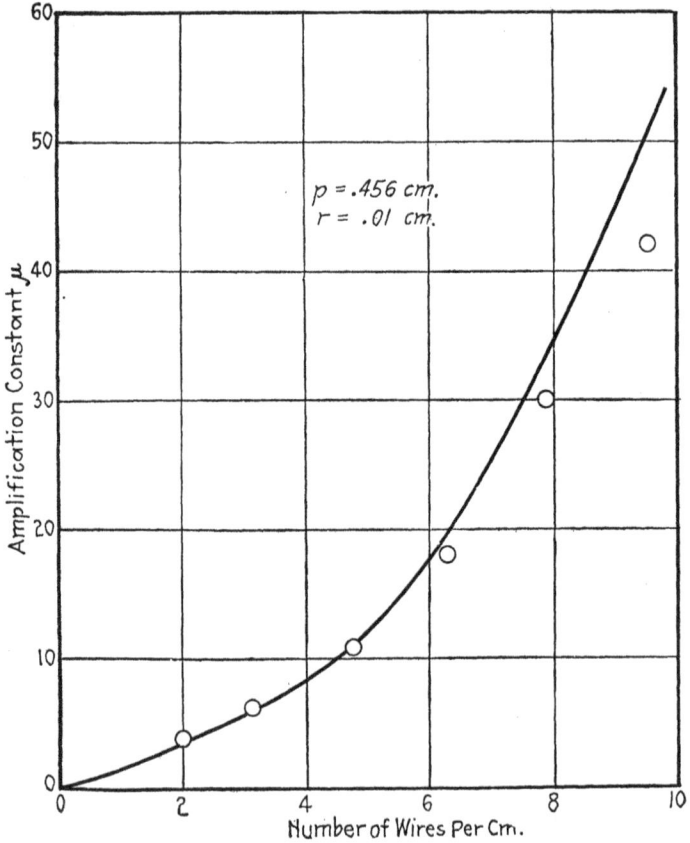

$$p = .456 \; cm.$$
$$r = .01 \; cm.$$

Amplification Constant μ

Number of Wires Per Cm.

Fig 120.

where

$p =$ distance between grid and plate;
$a =$ distance between adjacent grid wires;
$r =$ radius of grid wire.

This equation does not give as good results as the empirical equation that will be given below (equation 79). But for values of μ ranging from about 2 to 20, equation (78) can be used for designing tubes with a sufficiently high degree of accuracy for most practical purposes. The extent of the agreement between calculated and observed values is shown in Figs. 120 and 121. The points indicate observed values, while the smooth lines represent equation (78). Each point represents the average of a number of tubes. The deviation at the higher values of μ where the wires are close together, is inherent in the equation which was derived on the assumption that the distance between successive wires is large compared to the thickness of the wires.

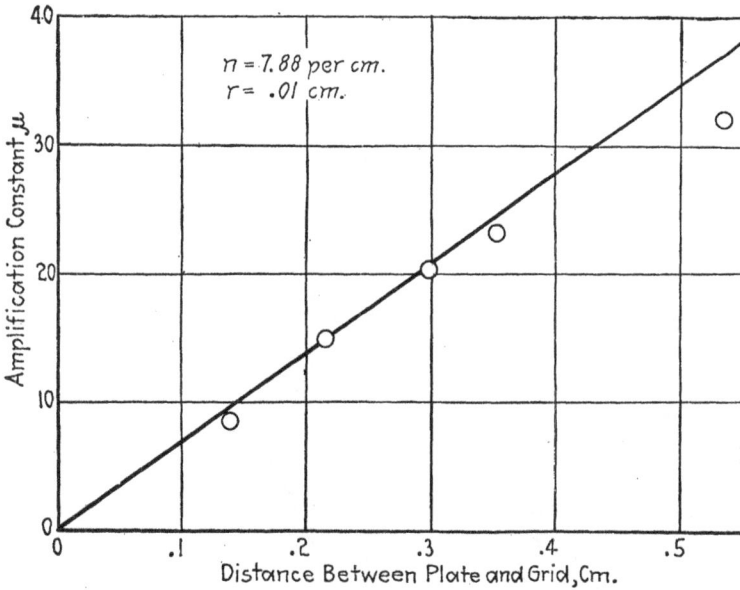

n = 7.88 per cm.
r = .01 cm.

Amplification Constant μ

Distance Between Plate and Grid, Cm.

FIG. 121.

On account of the accuracy with which tubes must be designed for telephone repeater purposes, the author carried out an extensive series of measurements in 1914, to establish an empirical formula relating the tube constants with its structural parameters. The equation which was found to give the best results is:

$$\mu = Cprn^2 + 1, \quad \ldots \ldots \ldots \quad (79)$$

where

p = distance between plate and grid;
r = diameter of grid wires;
n = number of wires per unit length.

C is a constant which for the parallel-plane type of tube (see Fig. 68) has a value of 80. Since this equation is non-dimensional, C is independent of the system of units used in expressing the tube dimensions and is independent of the size of the tube structure.

It will be recognized that this equation is the same as equation (25a) given on page 46 where $d = p$ and $k = \dfrac{1}{Crn^2}$.

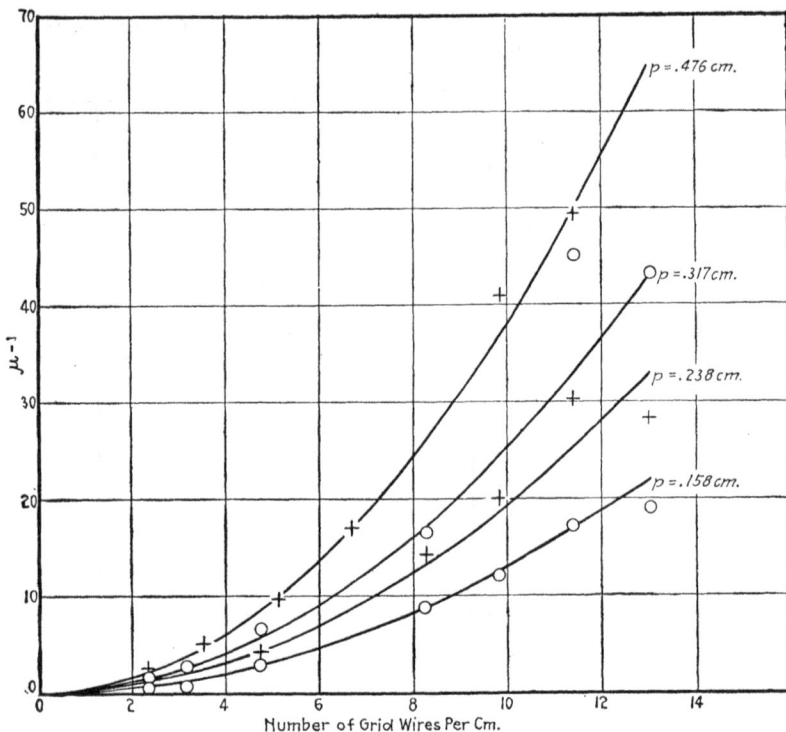

Fig. 122.

Equation (79) has been determined from measurements made on a large number of carefully constructed tubes in which not only the quantities given in the equation were varied, but also the distance between filament and plate, and the distance between filament and grid varied over a wide range. The constant is, however, independent of these latter two distances, as the equation shows. This is in accordance with Maxwell's result, which also states that the stray field between filament and grid is independent of the distance between them. (See Equation 75.) Equation (79) is more accurate than the theoretical equation (78) and therefore has been used by the Western Electric Company for the design of substantially all its tubes. The accuracy with which this equation holds is shown in Fig. 122, where $\mu-1$ is plotted as a function of n, the number of wires per centimeter length of the grid for various distances p between grid and plate. The curves are computed from equation (79), while the circles and crosses represent the observed values of $\mu-1$.

The radius r of the grid wires in these tubes was 1.02×10^{-2} cm. The relation between μ and r is shown in the following table, which also contains values to indicate the range over which distances of the grid and plate from the filament were varied. The agreement between observed and computed values is, as will be seen from the table, quite good for values of $\frac{r}{a}$ ranging to about 0.3. The thickest grid wire used in these tubes had a radius of 2.54×10^{-2} cm. It is usually desirable to use thin wires, unless

$p+f$	f	p	n	r	Amplification Constant μ.	
					Observed.	Calculated. (eq. 79)
.635	.158	.475	5.12	.0102	10.8	11.1
.635	.158	.475	5.12	.0191	18.0	19.0
.635	.158	.475	5.12	.0254	25.8	26.1
.318	.158	.158	8.26	.0102	8.4	9.8
.397	.158	.238	8.26	.0102	14.7	14.5
.476	.158	.317	8.26	.0102	20.2	18.7
.556	.158	.397	8.26	.0102	23.2	23.0
.635	.158	.476	9.84	.0102	42.0	38.5
.635	.158	.476	6.7	.0102	18.1	18.3
.635	.238	.397	8.26	.0102	24.6	23.1
.635	.317	.317	8.26	.0102	16.5	18.6
.635	.397	.238	8.26	.0102	14.0	14.5
.635	.476	.158	8.26	.0102	11.0	9.8
.635	.158	.476	11.4	.0102	50.5	51.5

p = distance grid and plate in centimeters;
f = distance grid and filament in centimeters;
n = number of wires per centimeter;
r = radius of grid wire in centimeters.

requirements of rigidity necessitate the use of heavy wires, such as is the case when the grid is in the form of a helix, supported only at the ends, or sometimes even at one end only.

King in the paper referred to above has also given an equation for μ for the three classes of cylindrical structures shown in Fig. 123. His argument applies particularly to the case in which the grid wires are parallel to the axis of the structure, but the resulting

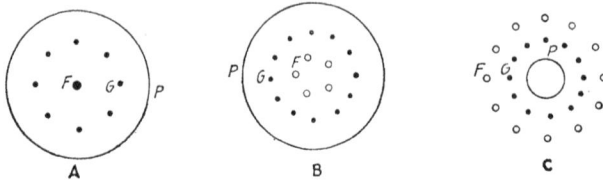

Fig. 123.

equation applies almost equally well when the grid is a helix. The equation is:

$$\mu = \frac{\pm 2\pi n \dfrac{\rho_g}{\rho_p}(\rho_p - \rho_g)}{\log_e \dfrac{1}{2\pi n r}}, \quad \cdots \quad \cdots \quad (80)$$

where n = number of grid wires per unit length;

   r = radius of grid wires;

   ρ_p, ρ_g = radii of anode and grid.

As in the case of parallel-plane structures, μ does not depend on the distance between filament and grid. The negative sign in equation (80) is to be used for the type of tube shown in Fig. 123C.

Equation (80) gives a reasonably good agreement with observed values of μ. With the help of the equations given above, it is possible to determine beforehand the tube dimensions required to give the desired value of μ.

76. Calculation of Plate Resistance. The plate resistance can in general not be determined with such simple equations as those which make possible the calculation of μ. But in designing tubes it is necessary to make the ratio $\dfrac{\mu}{r_p}$, where r_p is the plate resistance, as large as possible. For any given value of μ it is therefore desirable to make r_p as small as possible.

Now r_p can be decreased by decreasing the distance $p+f$ between *filament* and plate. On the other hand μ increases with increasing distance p between *grid* and plate, but is independent of the distance f between filament and grid. Hence, in order to keep μ large and r_p small, f should be kept as small as possible, i.e., the grid should be close to the filament.

The plate resistance depends, furthermore, on the size of the electrodes. It is within certain limits inversely proportional to the area of the anode as well as that of the cathode. It will be evident that there are limitations to increasing the area of the anode. For example, if the cathode is a single straight filament and the anode a plane parallel to the filament, there would be a limit to the size of the anode beyond which any further increase in its size would not contribute appreciably to a reduction in the resistance. On the other hand, the resistance can be reduced very much by using two plates, one on either side of the filament, as is mostly done. If the cathode consists of more strands of filament, the anode area can, of course, be further increased to advantage.

If the anode is cylindrical, an increase in its diameter would increase the distance between filament and anode in the same proportion as the anode area is increased. Considering a surface element of the anode, the resistance is proportional to the square of the distance between the cathode and the anode element. But for a cylindrical anode the area can be increased only by increasing the radius in the same proportion, so that the resistance *increases* linearly with the area of the anode; or, what is the same thing, it increases linearly with the radius.

Fig. 124.

In this connection, we may note an interesting relation between cylindrical and parallel-plate tubes, which was pointed out by R. W. King. Suppose a cylindrical structure, having a thin filament stretched along the axis of the anode (Fig. 124) be unfurled so that the anode becomes a plate having a width equal to $2\pi\rho_p$. Let the filament be replaced by a surface equal to the anode area and at a distance ρ_p from it. Assuming the cathode to be an equi-potential surface, the space current per unit area of the

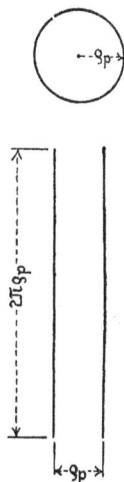

parallel plane structure is given by equation (9) Chapter IV. Hence the current for this tube is given by

$$I = 2.33 \times 10^{-6} \frac{2\pi \rho_p l E^{3/2}}{\rho_p{}^2},$$

where l is length of the structure (perpendicular to the paper). That is

$$I = 14.65 \times 10^{-6} \frac{l}{\rho_p} E^{3/2}.$$

This is the same equation that applies to the cylindrical tube (see equation (14), page 60). The two structures, therefore, give the same space current.

As regards the effect of the area of the cathode, there are also certain limitations. If the area of the cathode be increased by increasing its diameter, the resistance will not be reduced proportionately because of the density of the space charge of the electrons in the neighborhood of the filament. The total saturation current will, of course, be greater, but will only be obtained at a higher voltage. The better way to increase the area of the cathode is to increase its length. However, this generally means an increase in the voltage drop in the filament, due to the heating current, and this in itself increases the plate resistance, due to the limitation of the current by the filament voltage. (See Fig. 20, page 61.)

King has also derived equations for the space current as a function of the structural parameters for both cylindrical and parallel-plane structures, on the assumption that the current can be taken to vary as the $\frac{3}{2}$-power of the effective voltage. Since the effective voltage $\frac{E_p}{\mu} + E_g$ is generally not large compared with the voltage drop in the filament, the limitation of space current by the latter must be taken into account, in which case the current will be governed by equations (18) and (19) of Chapter IV. If the voltage drop in the filament be neglected the computed current will in general be considerably larger than the observed current.

77. Types of Thermionic Amplifiers. The number of different types of thermionic tubes now in use has become so large that no attempt will be made to describe or even mention all. The purpose in describing any is merely to give the reader a quantitative idea of characteristics of tubes used in practice. At the outset it may be stated that tubes are used ranging from a type that consumes for its operation a small fraction of a watt, to types that give several thousand watts' output in the form of alternating current. This widely varying range is occasioned by the widely differing conditions to be satisfied, depending on the purpose for which the tube is to be used. If the tube is to operate as an amplifier with the telephone receiver connected directly in its output circuit, the output power necessary need not be more than a very small fraction of a watt—one millionth of a watt is quite sufficient to give a very loud tone in most well-constructed receivers. If, on the other hand, the tube is to be used as a telephone repeater, inserted at a point on the telephone line about midway between the sending and receiving stations, the tube must give a sufficient amount of power to give clearly audible speech in the receiver after the telephone currents have been attenuated by the line between the repeater and the receiver. Then, again, if the tube is used to amplify modulated high-frequency oscillations, for example, before being impressed on an antenna for radio transmission, it must obviously be capable of giving a much larger output power, the magnitude of which depends upon the distance over which transmission is to take place, and can range all the way up to several kilowatts. When the necessary power is too large to be handled by one tube, a number of tubes can be used in parallel.

In designing tubes for amplification purposes several factors have to be taken into consideration. It is, for example, necessary to consider the output power that is necessary. To obtain best operation the plate resistance of the tube should be made equal to the impedance into which the tube works. If this is not possible or desirable from the point of view of tube construction, a transformer could be used in the output circuit to match the tube resistance on the one side and the line impedance or the impedance of the recording apparatus on the other. It is also possible to use two or more tubes in parallel, thus reducing the total plate resistance. Referring to equation (31) it will be seen that the output depends also upon the input voltage e_g and the amplification constant μ. The input voltage must be kept within the limits defined by equations (24) and (25). Furthermore, μ and the plate resistance must be so chosen that the amplification has the desired value. This is usually as large as possible.

There are thus a number of requirements to be satisfied and they differ with different operating conditions. The following tubes represent a few standard types. The tube characteristics are specified sufficiently fully by giving the value of μ, the filament constants, the relation between plate current and plate voltage,

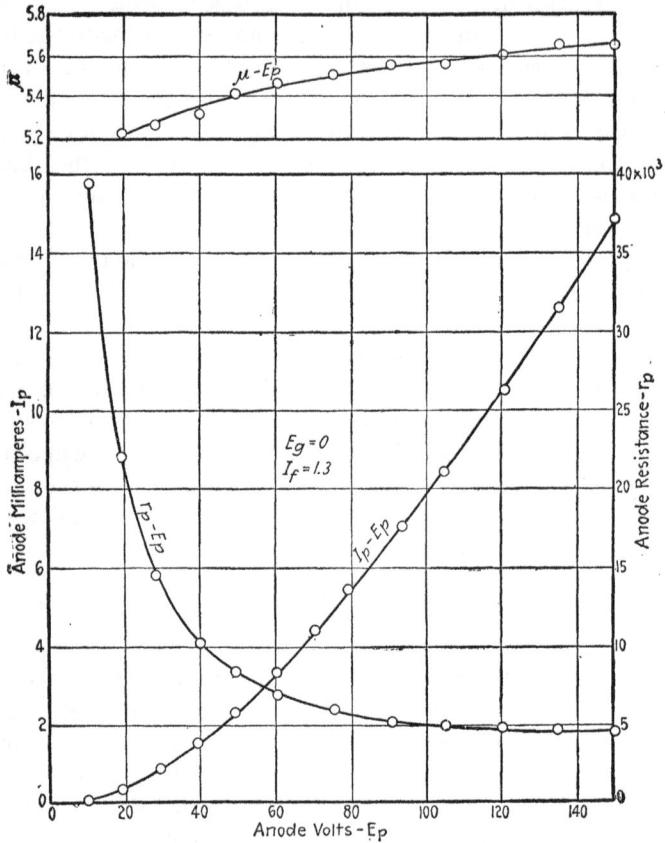

Fig. 125.

and the relation between plate resistance and plate voltage, over a range of operating plate voltages. The slope $\dfrac{\mu}{r_p}$ of the plate current-grid potential characteristic can then be obtained directly from the known values of μ and r_p.

The tube shown in Fig. 68, page 146 represents a modern type of telephone repeater manufactured by the Western Electric Company and used on the lines of the Bell Telephone System. The overall length of the tube and base is about 4 inches. The plates are of nickel and their edges are turned up to prevent warping due to the high temperature to which they rise when bom-

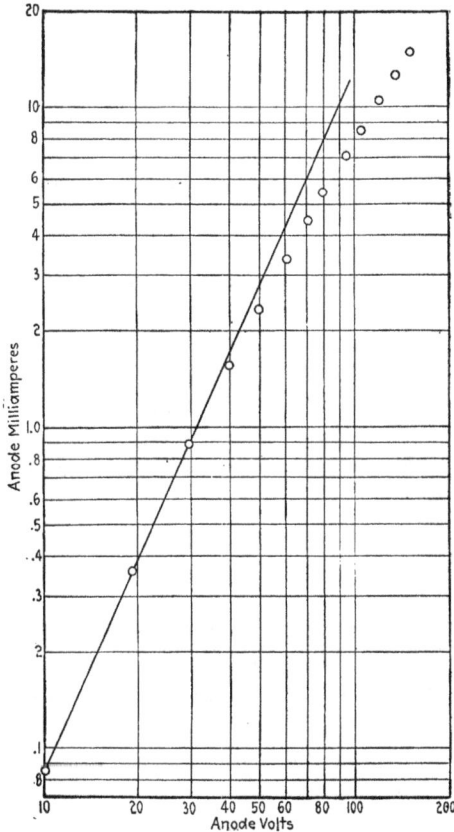

FIG. 126.

barded by the electrons during the process of evacuation. It contains an oxide-coated platinum filament operating on a normal filament current 1.3 amperes, the voltage being 7 volts. When operated as a telephone repeater the d-c. plate voltage is 160 volts, and the grid is maintained negative with respect to the filament by a battery of 9 volts. The characteristics are shown in Fig. 125. The plate resistance is shown here and in the following curves as a function of the plate potential, the grid potential being zero. To obtain the plate resistance for any grid potential other than zero, all that is necessary is to add μE_g to the plate potential and read the resistance from the curve at the value of plate potential equal to the value so obtained. Thus, since μ is about 5.6, the plate resistance at a plate potential $E_p = 160$, and grid potential $E_g = -9$ is that corresponding to an abscissa of $160 - 9 \times 5.6 = 100$ volts, namely, 5000 ohms. The minimum amplification required of this tube is 25 miles of standard cable, which corresponds to a power amplification of 230. (See table on page 219.) The logarithmic plot of this tube's characteristic is shown in Fig. 126. The slope of the line is close to 2, indicating a parabolic relation between current and voltage over the operating range.

Fig. 127 shows a type of tube, commonly known as the VT-1, that is suitable for use either as detector or amplifier and is designed to operate on a plate voltage of about 30 volts when delivering power directly to a telephone receiver. Its operating filament current and voltage are 1.1 amperes and 2.5 volts, and its amplification constant is 6. Its other characteristics are shown in Fig. 128. The logarithmic plot of the characteristic is shown in Fig. 129. The slope of this line is also close to 2. The minimum amplification at 30 volts on the plate is 24 miles of standard cable. This tube was manufactured by the Western Electric Company for use as an aeroplane radio receiver. The aeroplane radio transmitter tube resembles in its structural features the one shown in Fig. 68 but was designed to operate on plate voltages ranging from 275 to 350 volts instead of 160 volts, the voltage of the telephone repeater. The evident rugged construction of these tubes was found necessary to enable them to withstand the rather severe vibration to which they are subjected on an aeroplane. In the case of the receiver tube (Fig. 127) the filament, plate and grid are sup-

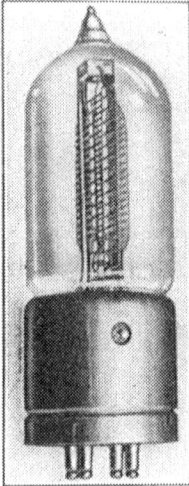

FIG. 127.—Western Electric Receiving Tube.
Length: 10 cms.

ported from the top by means of a block of lavite. The lower part of the plate forms a collar which fits tightly round the reentrant tube. The double plates and grids are each stamped in one piece from sheet metal, thus facilitating quantity production.

A tube of simple construction manufactured by the General Electric Company is shown in Fig. 130. The anode consists of a nickel cup about $\frac{9}{32}$ inch in diameter and $\frac{9}{32}$ inch high, and is

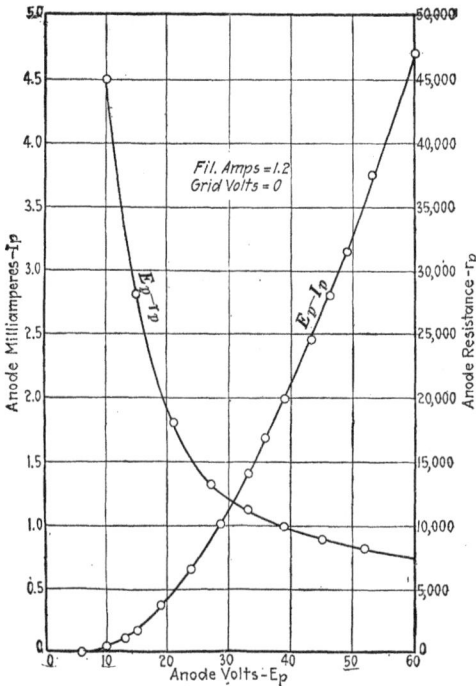

Fig. 128.

placed over the grid and filament. The grid is in the form of a helix enclosing the filament which is likewise helical and consists of tungsten wire.

In one type of tube of this construction the normal operating filament current is 1.1 amperes and the filament voltage 3.6 volts. The amplification produced by the tube is equivalent to about 20 miles of standard cable.

Tubes that are used as amplifiers and radio detectors and in-

tended to deliver power directly to a telephone receiver need not
be capable of handling more power than is necessary to operate
the receiver and in most cases this corresponds to a very small
plate current.   This makes it unnecessary and undesirable to

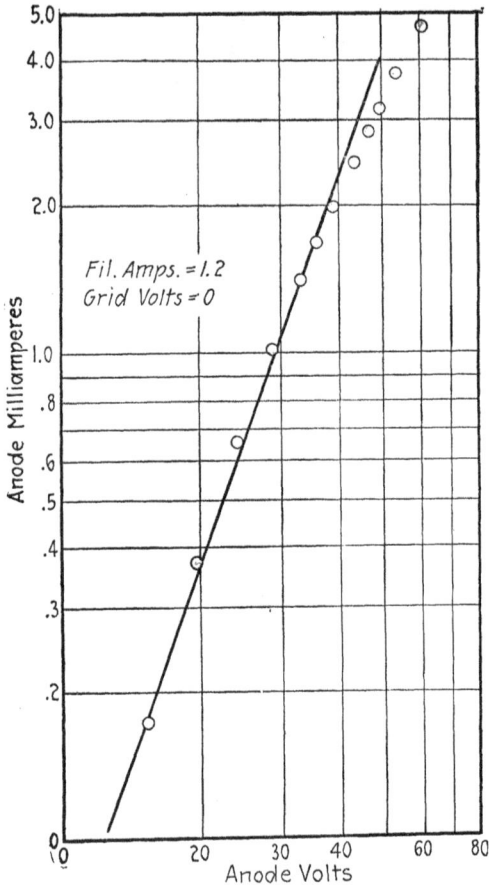

FIG. 129.

dissipate more than a fraction of a watt in heating the filament.
The desirability of reducing the power necessary for heating the
filament is apparent.   A tube in which this has been done is shown
in Fig. 131.   The filament is of the oxide-coated type, 2×10^{-3}
inch in diameter and is designed to operate on one dry cell, the

operating voltage ranging from 1.0 to 1.5 volts with a normal filament current of about 0.2 ampere. The normal power dissipated in the filament is therefore about one-quarter of a watt This tube is a good detector and gives about 20 miles amplification with a plate voltage of about 30 volts. It was designed not only for the purpose of reducing as much as possible the power consumption in the filament, but also to facilitate the construction. The four stout wires which connect to the electrodes, are moulded in a glass bead 11, or other insulating substance. This forms a unit on which the rest of the structure is built. The grid is sufficiently rigid to be supported only from the lower end. The plate is likewise supported by a single wire going through the glass bead. This means of supporting gives a sufficiently rugged structure on account of its small size and light weight. The structure thus mounted on the bead 11 is inserted into a tube, the lower end of which is pressed, as shown at 2, Fig. 131, thus making an air-tight seal. This method of constructing the tube eliminates the making of flares and presses.

Fig. 130.—General Electric type Receiving Tube. Length: 10 cms.

Another type of tube of simple construction is shown in Fig. 132. This drawing was made from one of several tubes designed and used by the author in his work on vacuum tubes, about six years ago. The same type was subsequently developed in Europe and used extensively in military operations during the war. The simplicity of construction of this tube lies in the horizontal mounting of the electrodes.

Turning now to a consideration of the high-power type of tube, it must be noted that this type of tube belongs to a class that is entirely different from the small types discussed above, and presents problems that make its construction more difficult. For example, on account of the large amount of power dissipated in such a tube, the electrodes and walls of the vessel must be more thoroughly freed of gas during evacuation than is required of the low-power type of tube.   The heating of the plates during operation makes it necessary to compute their size on the basis of the power dissipated at them by electron bombardment.   This can be done in the manner explained in Chapter IV.   On

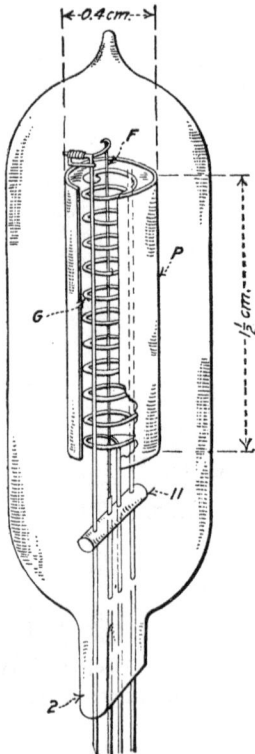

FIG. 131.

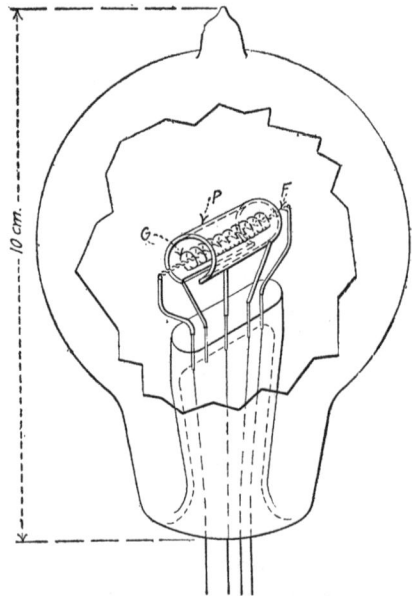

FIG. 132.

account of their low vapor pressure, the refractory metals such as tungsten and molybdenum are very suitable materials for use in high-power tubes. Tungsten can, for example, be heated during evacuation to about 2500° K. to drive out occluded gases, while during operation the tungsten anodes can safely rise to a temperature of about 1500° K., thus making it possible to use a smaller anode area than is necessary when less refractory metals are used as anodes.   On the other hand, high-power tubes that are to operate

on comparatively low plate voltages often require so much filament that when the filament strands are properly spaced and
supported in a mechanically convenient
manner, they occupy a surface which
is so large as to necessitate a large
plate area. It is also necessary, in designing power tubes, to consider the
size of the bulb and make it large enough
to prevent overheating of the glass.
The radiating area of the glass bulb,
expressed in square inches, should be at
least equal to the number of watts dissipated inside the bulb.

Fig. 133 shows a type of power tube
that was used in 1915 by the American
Telephone & Telegraph Company and
the Western Electric Company to transmit speech from Arlington, Va., U. S. A.,
to Paris and to Honolulu, a distance in
the latter case of about 5000 miles.
The anode, it will be observed, consists
of metallic ribbon so arranged on glass
frames that it can be heated during
evacuation by passing a current through
it. This is of considerable help in driving out occluded gases. A glass frame
situated midway between the frames on
which the anodes are mounted serves to
support the grid and filament.

A General Electric Company type of
tube which also enables the anode to be
heated by passing a current through it is
shown in Fig. 134.[1] Here the anode consists of tungsten wire stretched back and
forth on a glass frame. The grid consists
of fine tungsten wire wrapped around
another glass frame.

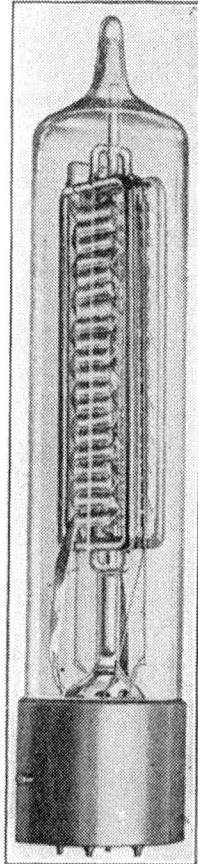

Fig. 133.—Western Elec-
tric Type Power Tube,
used in early Long-
distance Radio-phone
Experiments. Length:
30 cms.

A type of power tube of British design is shown in Fig. 135.
The electrodes are supported, as indicated, by springs fitting

[1] I. Langmuir, General Electric Review, Vol. 18, p. 335, 1915.

tightly in the side elongations of the bulb. This type of tube operates on a plate voltage of several thousand volts.

A high-voltage tube of more rugged construction is the General Electric type shown in Fig. 136. The anodes are stout tungsten plates, about $2 \times 2\frac{1}{2}$ inches, and the grid consists of fine tungsten wire wrapped around a molybdenum frame, thus encasing the filament. This tube operates normally on about 1500–2000 volts, and is capable of delivering several hundred watts a-c. output, the space current ranging from 150 to 250 milliamperes.

The power obtainable from a tube can be greatly increased by designing the tube to operate on high voltages. This, however, requires ordinarily a d-c. source of high voltage. This can be

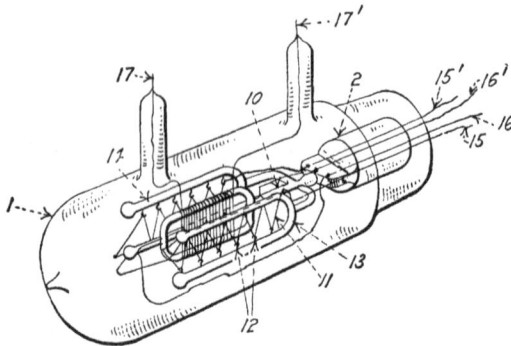

Fig. 134.

obtained either by using a valve rectifying system, as explained in Chapter VI, or by using d-c. generators. Such sources of high voltage, direct current are usually not efficient. But the power obtainable from a tube can, of course, also be increased by increasing the space current; that is, by increasing the area of the cathode or its thermionic efficiency. (See page 78.) The coated type of cathode has a high thermionic efficiency and therefore lends itself well to the construction of high-power, low-voltage tubes. Such a tube is shown in Fig. 137. This tube is capable of giving an a-c. output of several hundred watts when operating on a plate voltage of only about 700–1000 volts.

In order to avoid using large containers, we can use cooling means to carry off the heat dissipated at the anode, instead of depending only on radiation. Cooling can be effected in several ways: The device can, for example, be immersed in a liquid bath or a blast of air can be d rected against it. Or the anode can be so constructed that it can be cooled by water circulation. A Western Electric type of tube in which this is done is shown in Fig. 138. The anode consists of a metal tube about half an inch

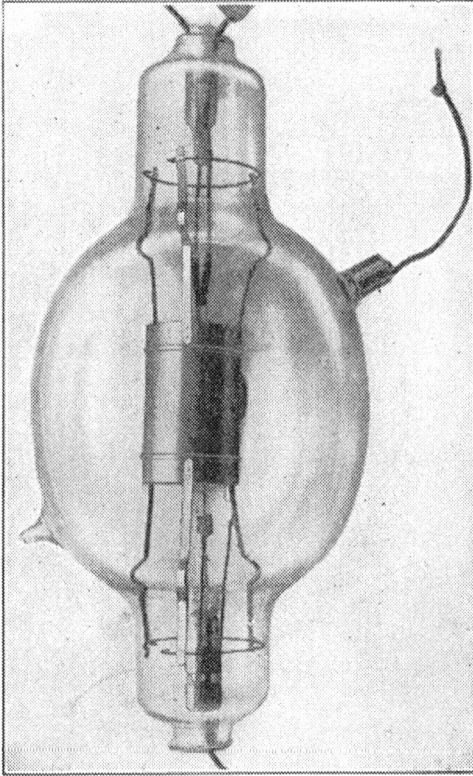

Fig. 135.—British Type Power Tube.

in diameter. The bottom of this tube is closed and the top sealed hermetically to the re-entrant glass tube, thus allowing water to be circulated inside of it. The grid forms a helix round the anode, and the cathode consists of a number of filaments arranged outside the grid to form a cylindrical system concentric with the grid and anode. Tubes of this type have been made to dissipate several kilowatts.

Other means have been used for preventing the power dissipation at the anode from raising its temperature too high.   One way is to make the area of the anode as large as possible, but to this there are limitations.   For example, increasing the area of

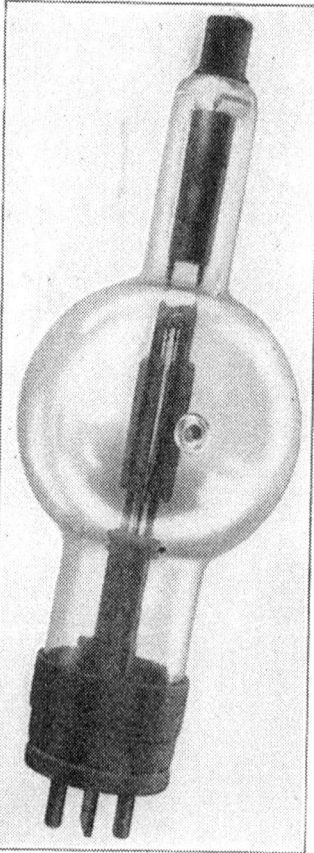

FIG. 136.—General Electric Type Power Tube.

FIG. 137.—Western Electric Type Power Tube.   Length: 30 cms.

a cylindrical anode by increasing its diameter causes an increase in the resistance of the tube.   This is avoided in a type of tube of General Electric design.   The anode is shaped in the form of a cylinder having four radial flanges, which help to increase the radiating area.

An obvious way of keeping the temperature of the anode low is to increase its thermal emissivity by blackening its surface. (Equation 23, Chapter IV.)

78. Amplification Circuits. A great variety of circuit arrangements have been devised in which the thermionic tube can be operated as an amplifier. We shall, however, discuss only a few

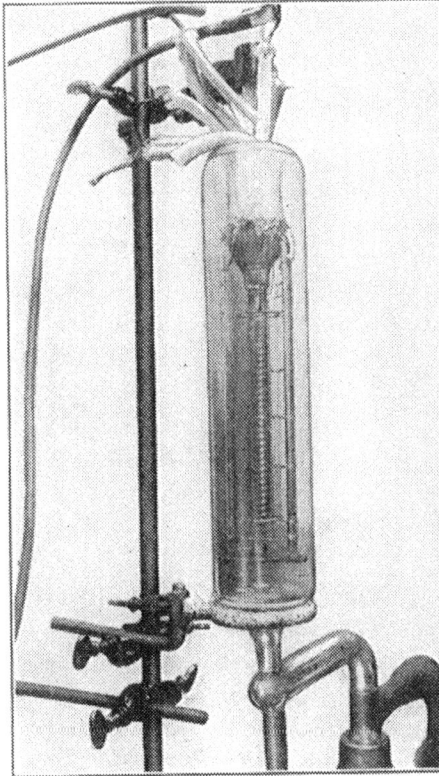

Fig. 138.—Power Tube with Water-cooled Anode.

types of circuits for the purpose of illustrating the points that must be considered in designing efficient circuits. The necessary considerations follow directly from the theoretical discussions in the foregoing. For example, it was shown that the current in the circuit can be controlled by potential variations impressed on the grid. It is, therefore, necessary to make the potential variations impressed on the grid as large as possible. For this purpose the step-up transformer T_1 (Fig. 139) is inserted in the input circuit. The secondary of this transformer should be wound to have the highest possible impedance. The input transformers that are used in connection with telephone repeater tubes on the Bell telephone lines, have a secondary impedance of 600,000 ohms at a frequency of 800 cycles per second. It is desirable to shunt the secondary of the input transformer with a resistance r_o about equal to the secondary impedance. The output transformer T_2 is so wound that the tube works into an impedance equal to its plate resistance, the secondary impedance of T_2 being equal to the impedance of the line or device to which energy is delivered. If the line impedance is equal to the plate resistance of the tube, this transformer can be omitted. The choke coil L and condenser C

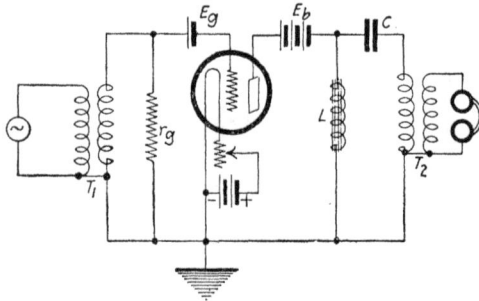

FIG. 139.

are inserted if the direct current established in the plate circuit by the battery E_b is too large to pass directly through the winding of the output transformer without causing undue heating of its coils. The grid is maintained at a negative potential, with respect to the filament, by the battery E_g, which should be so adjusted with respect to the plate voltage, the constants of the tube, and the peak value of the input voltage that the limit equations (24) and (25) are satisfied.

Instead of using a separate grid battery, the grid can be maintained negative with respect to the filament by making use of the voltage drop in the filament rheostat. The arrangement is shown in Fig. 140. In this case the plate circuit is connected to the positive terminal of the filament battery, thus increasing the potential difference between the filament and plate by an amount equal to the voltage of the filament battery. This is sometimes desirable since the amplification increases with the plate voltage.

It was shown in Chapter III, that there exists an intrinsic potential difference between the filament and grid, the value of which depends largely upon the nature of the surfaces of these electrodes. This quantity was designated in the equations by ϵ. Now, ϵ can be either positive or negative, and is seldom greater than 1 volt. If it is negative, the grid will be intrinsically at a negative potential with respect to the filament. If this is the case and the alternating input voltage is small no extra provision need be made to keep the grid negative. If ϵ is very small or has a positive value, the grid battery should be inserted or a scheme can be resorted to which is illustrated in Fig. 141 and for

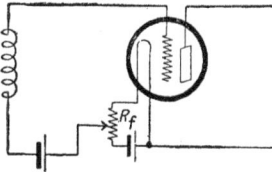
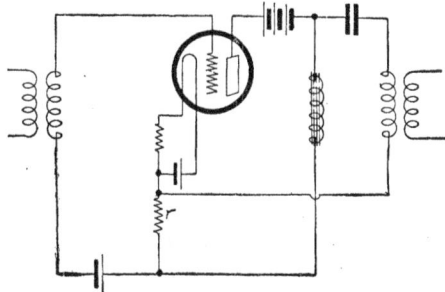

Fig. 140.   Fig. 141.

which I am indebted to my associate Mr. R. H. Wilson. The scheme consists essentially in the addition of a resistance r in the d-c. branch of the output circuit. A positive value of ϵ causes electrons to be attracted to the grid and this results in a decrease in amplification due to a reduction in the effective alternating grid voltage, as explained on page 167. But a positive value of ϵ also increases the direct-plate current. This increases the voltage drop in the resistance r which consequently makes the grid more negative with respect to the filament. The more negative ϵ is, the smaller will be the space current flowing in r and this tends to decrease the negative potential on the grid. In this way the circuit tends to effect a balance, so that tubes having different values of ϵ will not, when used in this circuit, give widely varying degrees of amplification. It will be recognized that there is an optimum value for the resistance r, because if r be made too large the voltage drop in it would make the grid so much negative with respect to the filament and consequently the impedance of the tube so high that the amplification drops.

The effect of the resistance is illustrated in the following table, which gives the amplification in miles of standard cable obtained from a number of tubes having different values of ϵ. The best value of resistance according to this table is about 1000 ohms.

	Amplification in Miles of Standard Cable.					
	$r=0$	$r=400$	$r=800$	$r=1200$	$r=2000$	$r=4000$
+0.16	16	22	24	24	25	24
+0.11	14	20	24	25	25	24
−0.13	14	22	26	26	26	24
−0.85	21	21	21	21	20	19
−0.69	22	21	21	21	20	18
−0.27	24	24	24	23	23	22

Tubes are frequently used in cascade formation to secure higher amplification in the so-called multi-stage amplifier sets. There are numerous circuit arrangements whereby this can be

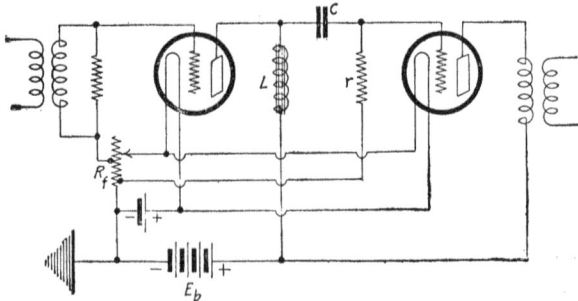

Fig. 142.

done. Fig. 142 shows a two-stage set in which the first tube is operated as a voltage amplifier. The inductance L has an impedance which is large compared with the plate resistance of the first tube, thus permitting the largest possible voltage amplification. It was shown in Section 62 that if $L\omega$ is twice as large as the plate resistance of the tube to which it is connected, the voltage amplification is about 90 per cent of its maximum value μ. If, however, L is a practically pure reactance, it is desirable to make it more than twice as large as the plate resistance, in order to make the phase difference between grid and plate potentials as near as 180° as possible, thereby straightening out the tube characteristic and minimizing distortion. (See Section 59.) The circuit shown in Fig. 142 is so arranged that both tubes can be operated from the same plate battery. The grids can be maintained negative with respect to their adjacent filaments by means of grid batteries (not shown) or by connecting them to convenient points on the

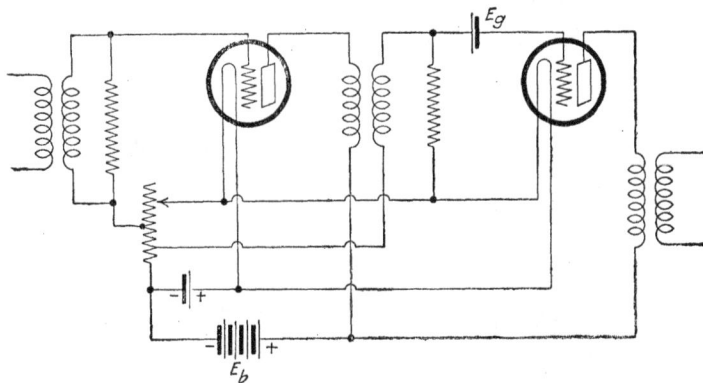

FIG. 143.

filament rheostat R_f. The resistance r should be large, preferably of the order of one or two megohms, and merely serves the purpose of maintaining the grid of the second tube at the desired d-c. potential.

Instead of using the inductance L and condenser C, a step-up transformer can be inserted between the tubes, as shown in Fig. 143. When this is done both tubes should be operated as power amplifiers (see page 184). The primary impedance of the inter-tube transformer should therefore be equal to the plate resistance of the first tube and its secondary should be wound to impress the highest possible voltage on the grid of the second tube.

The thermionic tube makes it possible to obtain high degrees of amplification with non-inductive circuits by means of an arrangement suggested by H. D. Arnold.[1] Fig. 144 shows a

[1] U. S. Patent 1129943, 1915.

non-inductive amplifier. Instead of the inductance L a non-inductive resistance r is used. If the grid battery E_g were omitted the grid of tube B would be at the same potential as the plate of tube A. The grid of B would therefore be positive with respect to its filament by an amount equal to the potential difference between filament and plate of tube A. To avoid this the negative voltage E_g is applied to the second grid to give it the appropriate negative potential with respect to its filament.

This non-inductive type of amplification circuit is a very important contribution made possible by the thermionic tube, because it enables us to produce almost any degree of amplification without the use of transformers. In many instances transformers are undesirable. This is, for example, the case when dealing with

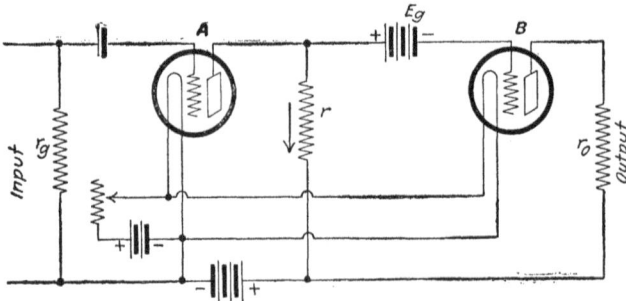

Fig. 144.

currents of very low frequency, such as are used on telegraph lines and especially on submarine telegraph cables. Transformers for such frequencies are unpractical, being costly and inefficient. Besides, they distort the wave form which it is very desirable to preserve. Even when dealing with currents of frequencies covering the audible range, transformers produce distortion which in some cases is very serious. This can, for example, happen in the transmission of music. When speech is transmitted through a system having a transmission band ranging from a few hundred to about 2000 cycles per second, the speech is still perfectly intelligible and, in fact, a smaller frequency range often suffices. Transformers that are used on telephone lines have a fairly flat frequency characteristic and are very satisfactory for speech transmission. But for the transmission of music a much wider range of frequencies is necessary. It is known that to preserve the quality of many musical tones, the system must be capable of transmitting with equal facility frequencies ranging up to several thousand cycles per second. In all such cases the non-inductive amplification circuit is of value. Care must, of course, be taken to eliminate the distortion produced by the curvature of the characteristic of the tube itself.

In ordinary circuits, such as that shown in Fig. 139, transformers and coils are used for convenience to serve definite purposes. The output transformer T_2 is, for example, inserted to match the impedance of the line or device into which the tube works with the impedance of the tube. This secures maximum power amplification. The input transformer likewise matches the impedances and has a very high secondary impedance because of the high input impedance of the tube. Referring to equation (33) (page 188), it will be noticed that the power amplification η is directly proportional to the input resistance r_g. Hence, if the tube is to amplify currents from a low impedance line and the input transformer T_1 were omitted, the amplification would be very small. To overcome this Arnold suggested using voltage amplifier tubes to step up the input voltage. These tubes then feed into tubes having an impedance sufficiently low to be connected directly to the output circuit. This can be done by designing tubes to have a low plate resistance or using a number of tubes in parallel.

Telephone transformers that are commonly used on the input side of vacuum tubes have voltage step-up ratios ranging from about 18 to 40. Voltage amplifier tubes having an amplification constant $\mu = 40$ are commonly used. It follows from equation 28, page 183) that if the resistance r (Fig. 144) is made five times as large as the plate resistance r_p of the tube A, the voltage amplification produced by this tube is 33. It can, therefore, take the place of the input transformer. The voltage can, of course, be amplified still more by increasing the number of tubes in the cascade series.

Fig. 144 shows the filaments connected in parallel to a common battery. They can, of course, also be operated in series from a common battery. Which ever is the more desirable depends upon the filament battery available. Both arrangements are used where tubes are operated in parallel to give increased output. But when the filaments are connected in series and the grids in parallel, provision must be made to counteract the difference in potential between the successive grids and filaments, due to the voltage drop in the preceding filaments. This is done by the insertion of appropriate grid batteries.

In designing multi-stage amplifier circuits, it is very important to make sure that all the tubes operate in accordance with the limit equations (24) and (25). In all cases the voltages impressed on the inputs of the tubes must be as high as possible, irrespective of what the input power may be, because the power developed in the output depends primarily on the input voltage. Now, when using a tube as a voltage amplifier it must be designed to have a large amplification constant μ. This, according to equation (27) produces a large voltage amplification. But, referring to equation (25), it is seen that the larger the value of μ the smaller is the input voltage e_g that can for constant plate battery voltage E_b be impressed on the input without causing distortion. When it is necessary to use a multi-stage amplifier set the input voltage on the first tube is generally so small that μ for the first tube can be quite high and the plate voltage not large, because $\dfrac{E_b}{\mu}$ need not be larger than about $2e_g$, where E_b is the voltage of the plate battery. But the voltage impressed on the second tube after being amplified by the first tube is then very much larger and the second tube must be capable of handling this increased voltage. If the first tube, operating on a definite plate battery, is just capable of handling the voltage impressed on its input, then, in order to handle the amplified voltage, the second tube must be designed to have a lower μ or otherwise must operate on a higher plate voltage than the first, so that it operates on a characteristic having a larger intercept on the axis of grid potential. (See Fig. 74, page 152.) This often necessitates heating the filament of the second tube to a higher temperature to increase the range of the characteristic. Such considerations show the important part played by the structural parameters of the tube. In some multi-stage amplifier sets it is possible to use several like tubes in series operating on the same plate voltage; that is, when the input voltage is much smaller than the limiting voltage that the first tube can handle. For example, if the amplification constant μ is 40, and the plate voltage is 120 volts, the intercept of the I_pE_g-curve is equivalent to 3 volts. If $E_g = 1.3$ volts, the input voltage can therefore have a maximum value of about 1.5 volts, since the grid can generally be allowed to become slightly positive. If this is the input voltage on the first tube, the succeeding tubes must have lower amplification constants or must operate with higher plate voltages. But if the input voltage on the first tube is, say, 5×10^{-5} volt, it can be amplified 30,000 times before it becomes too large to be handled by this type of tube operating under the conditions specified above. If each tube with $\mu = 40$ operates with an external resistance equal to four times its plate resistance, it produces a voltage amplification of 32. It would therefore take three such tubes in series to produce a voltage amplification of 30,000, and all three tubes can operate under the above conditions.

It may be remarked that if a is the voltage amplification produced by one tube, the total amplification A produced by n tubes is $A = a^n$.

Instead of using a large number of tubes in cascade to produce a high degree of amplification, use could be made of a "feed-back" arrangement, due to R. V. L. Hartley,[1] which is of special advantage when large amplification is to be produced with a few tubes in a non-inductive circuit. It will be evident that in any amplifying system the power developed in the output, which is larger than that in the input, can be greatly increased by feeding a small portion of the energy in the output back to the input, thus reamplifying that portion. This increases the output power and also the portion fed back to the input, and in this way the original input power can be amplified to almost any desired extent depending on the portion fed back and the limits of the tube characteristic.

Thus, suppose unit power be applied to the input of an amplifying system, the normal amplification of which is a-fold. The power in the output is then a. Let a fraction s of the output power be fed back to the input, so that the power returning to the input is as. The portion remaining is $a (1 - s)$. The fraction as amplified again into the output becomes $a^2 s$. Of this, a portion $a^2 s \times s = a^2 s^2$ is again fed back to the input, leaving $a^2 s - a^2 s^2 = a^2 s$ $(1 - s)$ available in the output. This process is repeated and we get in the output an amount of power given by the sum of a series of which the $(n+1)^{\text{th}}$ term is $a(1-s)a^n s^n$. Thus, if A is the total output power:

$$A = a(1 - s)(1 + as + a^2 s^2 + \ldots a^n s^n \ldots).$$

[1] R. V. L. Hartley, U. S. Patent, 1218650.

If the fraction as is less than unity the series is convergent, its sum being $\dfrac{1}{1-as}$, so that the total output power becomes:

$$A = \frac{a(1-s)}{1-as},$$

and the total amplification produced is

$$n = \frac{A}{a} = \frac{1-s}{1-as}.$$

Since $as < 1$ and a is usually large compared with unity for any good amplifier, it follows that s can be neglected in comparison with unity. The amplification n can therefore be made large by making as nearly equal to unity.

If as should be equal to or greater than unity, the above series becomes divergent and the output power increases without limit until checked by some cause determined by the nature of the circuit characteristic. The output power becomes independent of the original input; in other words, the system produces self-sustaining oscillations. If an inductive circuit is used, it may happen that as becomes greater than unity for one or more frequencies, depending upon the impedance and phase relations in the circuit, with the result that the system oscillates at these frequencies.

Hence, in order to use this scheme for amplifying currents covering a wide range of frequencies, such as telephonic currents, without producing sustained oscillations, it is advisable to use a non-inductive circuit. Such a circuit is shown in Fig. 145. The input voltage is impressed across the resistance R_1 and the grid of tube A is maintained negative in the usual way by the battery E_g. In the circuit as shown here the plate batteries E_b are connected between the plates and the output resistances R_2 and R_3. In this case, therefore, in contrast to Fig. 144, the succeeding grids tend to become too much negative due to the direct space currents in R_2 and R_3. They can become so much negative with respect to their filaments that they may choke down the space currents. It is therefore necessary to insert the grid batteries E'_g so poled as to reduce the negative potentials of the grids to the values necessary for satisfactory operation. The plate currents pass through the resistances R_2, R_3 to the common point of the circuit, which may be grounded. The alternating plate current of the third tube *C*, which may be separated from the d-c. by a condenser, passes through the recording device *S* which the system is supposed to operate and through part of the resistance R_2 to the common point.

Consider the half period during which current in R_1 flows in the direction of the arrow. This makes the grid of the tube *A* less negative with respect to its filament. The output electron current in R_2 therefore increases in the direction of the arrow. This makes the grid of *B* more negative with respect to its filament, so that the space current in *B* decreases, or increases in the direction of the arrow in R_3. The space currents in successive tubes are therefore 180° out of phase. The current in the third tube flowing

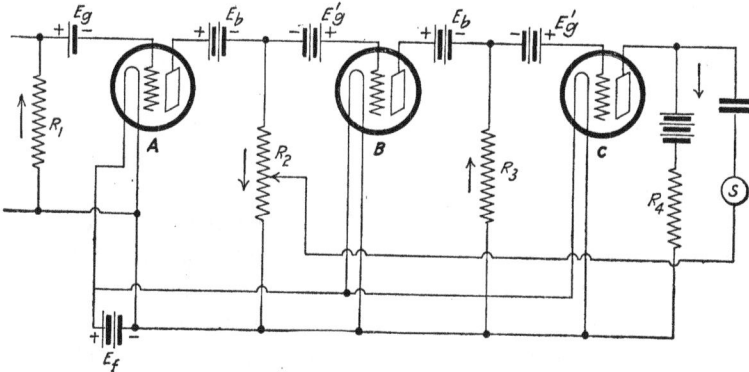

FIG. 145.

through *S* is in phase with the current in R_2, and since it must flow through part of R_2 the potential of the grid of *B* is increased over that occasioned by the space current in tube *A*. It will be apparent that in order to have the phase relations right, the output of one tube must be returned to the *input* of the *preceding tube* or one of the alternate preceding tubes. The amount of energy fed back can be controlled by varying the part of the resistance R_2 through which the feed-back current flows.

It may be remarked that the circuit needs careful adjustment to secure markedly increased amplification without making the system "sing," or produce sustained oscillations. If an inductive circuit is used the energy fed back can be controlled easily by varying the coupling between the output and input coils. In this case the phase relations can be so controlled that part of the output of a tube can be returned in proper phase to its own input. We have then simply the case of the ordinary " feed-back circuit " which is extensively used in radio reception.

Multi-stage amplifiers frequently have a tendency to sing. One way of preventing this is to feed part of the output energy back to the input in opposite phase.

The phase relations in a multi-stage amplifier set furnish a means for obtaining distortionless amplification. To illustrate this, let us consider the non-inductive two-stage amplifier shown in Fig. 144, and let us suppose that the two tubes A and B are alike. Suppose also that the resistances r and r_0 are equal. If a sinusoidal voltage e be impressed on the grid of tube A the output current flowing through r will not be sinusoidal unless r is large. It was explained in Section 58 that if the external resistance is equal to or greater than the plate resistance of the tube, the characteristic of the circuit becomes nearly linear and therefore nearly distortionless. The little distortion produced by the characteristic being not quite linear, or even the marked distortion produced when r is small, can be further reduced by using an even stage amplifier circuit. Thus, assuming that the characteristic is curved, if the voltage e_1 impressed on the input circuit is sinusoidal, then the voltage e_2 impressed on the second tube by the varying current flowing in r will be lopsided, as shown by the curve bb' of Fig. 83, page 167. That is, when the grid of A becomes less negative with respect to its filament, a greater increase in electron current is produced in r in the direction of the arrow than the decrease occasioned by the grid of A becoming more negative with respect to its filament. The potential impressed on the grid of tube B is therefore lopsided, but it is 180° out of phase with the potential on the grid of A. The increase in potential of grid B is therefore smaller than the decrease and hence if the characteristic of tube B is likewise curved, the current in its output resistance r_0 will be nearly sinusoidal. It is obvious that distortion can be reduced by this means only if the circuit contains an even number of tubes.

When the circuit is inductive, the phase relations are, of course not so simple. In such cases it is best to make the inductances in the output circuits of the tubes sufficiently large to straighten out the characteristic to the desired extent. If it is necessary to use a low impedance in the output circuit, distortion can be reduced by means of the "push-pull" circuit of E. H. Colpitts, shown in Fig. 146. It is supposed that the output circuit must be connected to a device of low impedance in which case the distortion produced when a simple circuit is used may be considerable. The way in which the "push-pull" circuit eliminates the distortion can be understood by referring to Fig. 83. The shape of the output current wave in the plate circuit of the tube A (Fig. 146) is given by the curve bb' (Fig. 83). In passing through the transformer it is resolved into the fundamental ee and the harmonic ff. The wave shape of the current in tube B is like that of

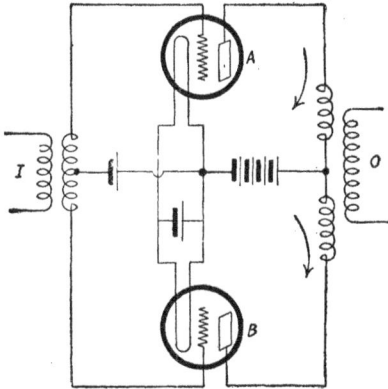

FIG. 146.

A, except that, since the potentials on the grids of the two tubes are 180° out of phase, the fundamental currents ee in the two-plate circuits will differ by 180°, while the harmonics ff will be in phase. It will therefore be seen, by referring to the current directions in Fig. 146, that the fundamentals will be additive in effect, while the harmonics will neutralize each other.

This circuit can also be used to cut out the fundamental and transmit only the harmonics, by reversing the primary coil of the output transformer of one of the tubes. This was suggested by J. R. Carson, for the purpose of modulation.

It will be apparent that a circuit like that shown in Fig. 146 requires that the two tubes be alike in their characteristics.

Fig. 147 shows an arrangement whereby the filaments and plates of a two-stage amplifier can all be operated from a single source of voltage. This circuit arrangement is of advantage when only one source of voltage is available, such as, for example, the standard city mains of 110 or 220 volts direct current. The condensers C_1 and C_2 are inserted to by-pass the alternating currents in the plate circuits in case the sections of the resistance R between the positive terminal of the voltage supply and the point connecting the plate, should become so large as to cause an undesirable waste of a-c. power in the plate circuit of the tube.

All the amplification circuits discussed thus far are unilateral; that is, they transmit and amplify currents in one direction only. For most purposes unilateral circuits are all that are needed, but for amplification of currents on telephone lines the amplifier must

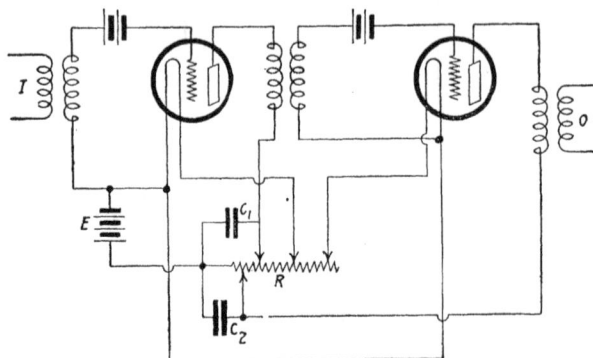

FIG. 147.

be capable of transmitting and amplifying currents in both directions, so that two-way conversations can be carried on over the line. The circuit that is commonly used in telephone practice, using thermionic tubes as amplifiers, is shown in Fig. 148.[1] It is known as the 22-type circuit (*two*-way, *two*-repeater type). The principle of this type of repeater circuit was disclosed by W. L. Richards, in 1895.[2] As applied to thermionic amplifiers, it has been in use on the lines of the Bell Telephone System since 1913. The principle consists in balancing each of the two lines against an artificial line or balancing network having an impedance equal

[1] B. GHERARDI and F. B. JEWETT, Proc. A.I.E.E., Nov., 1919, p. 1297. The reader is referred to this paper for information on the use of repeaters on long distance telephone lines.

[2] W. L. RICHARDS, U. S. Patent 542657, 1895.

to that of the line. The purpose of the balancing network can be understood from the following: The currents coming from Line W, for example, are branched off at P_1, thus furnishing the input to tube E, the amplified output of which is obtained in the output transformer T_1. Similarly the amplified currents from Line E pass through the output transformer T_2. Now, if the system into which transformer T_1, for example, feeds were not symmetrical with respect to the points P_2, the amplified current in T_1 would cause an increased input to be impressed on tube W; this in turn increases the input on tube E, and the system would produce sustained oscillations, or " sing." This is prevented by making

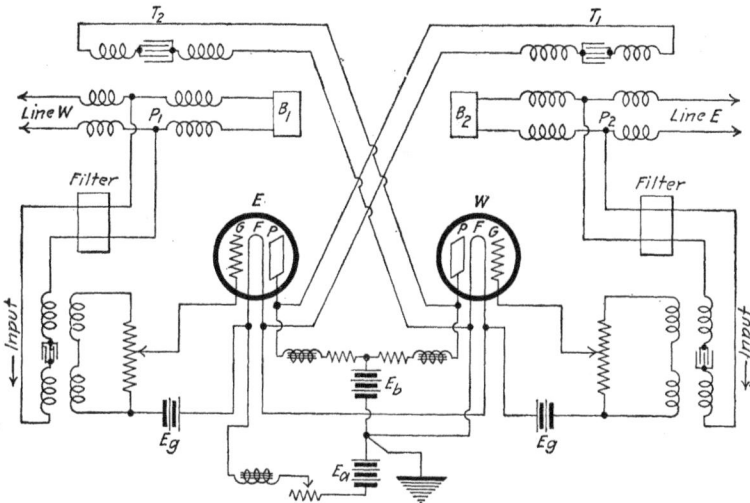

FIG. 148.

the impedance of the balancing network equal to that of the line to which it is connected, thus making the potentials of the points P_1 and P_2 independent of the currents in the corresponding output coils T_2 and T_1. This, of course, results in a division of the amplified output power, one-half becoming available and the other half being wasted in the balancing network.

In a system like this the degree of amplification obtainable without impairing the quality of transmission depends largely on the accuracy with which the lines can be balanced.

Two-way transmission can also be secured with a single repeater, by means of a circuit arrangement such as that shown in

Fig. 149. This circuit is known as the " 21-type repeater circuit " (*two*-way, *one*-repeater circuit). Instead of balancing each line with an artificial network, as in the 22-circuit, two-way transmission, with the 21-circuit, is effected by balancing the two lines directly against each other. Thus, if the impedance to the right of P is equal to that to the left of P, then the potentials of the connecting points P are independent of the amplified output current. But if the impedances of these lines are not equal, the amplified current in the output will impress increased potentials

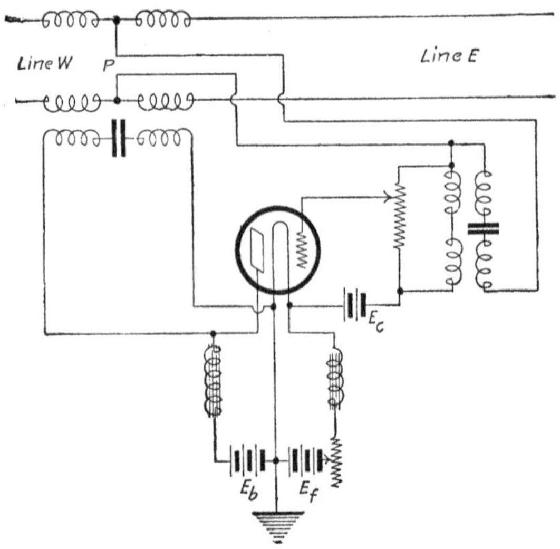

Fig. 149.

on the grid of the amplifier; these will cause further amplification, the amplification becoming cumulative, thus creating sustained oscillations or " singing."

The advantages of the 22-circuit over the 21-circuit are apparent: The latter requires that the impedances of the lines (as measured at P) leading to the two telephone substations between which the conversation is carried on, be identical—a condition which cannot always readily be realized in practice. In the 22-circuit, the two lines may have quite different impedances, the requirements being then that the balancing networks have different impedances, each balancing its own line. The repeater can then be inserted at some convenient place on the line, which need not be midway between the two stations. Furthermore, in order to create the condition of singing in the 22-circuit, it is necessary, as can readily be seen from Fig. 148, that both lines be unbalanced simultaneously. If one line and its network be perfectly balanced, an unbalance in the other will not cause singing. The 22-circuit is therefore inherently more stable than the 21-circuit.

The filters are inserted to pass only currents lying within the telephone frequency range, thus preventing the passage through the repeaters of telegraph and other signal currents that may be transmitted over the same metallic circuits. The potentiometers are inserted to adjust the amplification to the desired value.

CHAPTER VIII

THE VACUUM TUBE AS AN OSCILLATION GENERATOR

79. Introductory. Since the three-electrode tube can operate as an amplifier, the energy in the output circuit is greater than that in the input circuit. Hence, if part of the energy in the output be returned to the input, there will a further amplification of energy resulting in an increased output. If the amplified energy gets back to the input in sufficient amount, and if the phase relations of the output and input currents are right, there will be a constant reamplification and feeding back of energy from the output to the input, and the device will then produce sustained oscillations without it being necessary to supply potential variations to the grid by external means. In other words, the device will then operate as an oscillation generator. The frequency of the oscillations will be determined by the constants of the circuit, while their intensity will depend on the amount of energy fed back to the input, the shape of the characteristic curve, and the rate at which power is dissipated. There is a variety of circuit arrangements whereby this can be done. These circuits can all be divided into three main groups in which part of the energy in the output is returned to the input: (1) by resistance coupling, such as is explained, for example, in connection with Fig. 145, page 259; (2) inductive coupling; (3) capacitative coupling between output and input.

In the present chapter we shall briefly discuss the conditions that must be satisfied in order to use the three-element tube to produce oscillations of a definite frequency and amplitude. A large amount of work has been done on this phase of the subject, and no attempt will be made here to enter into a full discussion of these investigations. It is believed rather that an explanation of the fundamental principles that govern the production of sustained oscillations with the three-electrode tube, and an indication of the more important results that can be obtained, will be sufficient to enable anybody who understands them to design the circuits that may be necessary for any particular purpose.

The conditions that are necessary to make the tube act as an oscillation generator can be stated briefly as follows:

(1) The tube must be capable of amplifying. That is, it must have a unilateral impedance which is occasioned by potential variations on the grid producing a greater effect on the current in the plate circuit (output circuit) than the effect produced on the current in the grid circuit by potential variations on the plate. This property is generally expressed, as was explained in the previous chapter, by stating that an alternating potential e_g, impressed on the grid, produces an E.M.F. in the plate circuit which is equal to μe_g where μ in practice is generally greater than unity.

(2) Since the energy in the output is greater than that in the input, part of this energy can be returned to the input, but in order to insure a reamplification of this energy it is necessary to take care that the output and input currents are in phase.

(3) An oscillation circuit must be attached to the tube, having inductance, capacity and resistance of such value as to make the tube oscillate with the desired frequency. These quantities should, for best operation, also be so adjusted that the efficiency of the tube as an oscillator and the amount of power delivered to the oscillation circuit are as large as possible.

(4) The characteristic of the tube must be such that the tube constants, together with the constants in the oscillation circuit, determine the amplitude of the oscillations. In general the amplitude is limited by the factors that limit the flow of the current through the tube. These factors have been explained in Chapter IV.

80. Method of Procedure for the Solution of the Oscillation Equations. The complete solution of the oscillation equations of the vacuum tube is difficult because of the peculiar shape of the tube characteristics. It was explained in Chapter IV that the current-voltage characteristic is such that when taken over the whole range it cannot be expressed by a simple equation. For the lower part of the characteristic, where the effective applied voltage is less than the voltage drop in the filament, the current varies approximately as the $\frac{3}{2}$-power of the voltage. For high effective voltages, the exponent of the voltage decreases as the voltage increases, and finally approaches zero as the saturation current is approached. When using the tube as an amplifier in the most efficient way, we operate over such a part of the characteristic that a simple quadratic equation can be used. As was explained in Chapter VII, special precautions are taken to make the characteristic of the tube and circuit as nearly linear as possible. Now, when using the tube as an oscillation generator we seldom make use of a restricted portion of the characteristic, but, on the contrary, the plate current generally oscillates between zero and a saturation current value, and it is therefore difficult to express the current as a simple function of the applied voltage. But the conditions for oscillation can be derived without necessarily making use of a definite characteristic equation. What we shall do is to use the resistance and mutual conductance of the tube as the variable parameters in terms of which the conditions for oscillation can be expressed, and then see how these parameters depend on the characteristics of the tube.

An expression for the plate resistance of the tube was derived in the preceding chapter, for the general case in which the current varies as the nth power of the applied voltage. If the oscillations are extremely small then the resistance is given by the reciprocal of the slope of the plate current-plate potential characteristic at the point of operation. When the oscillations are finite, the resistance is approximately given by the secant joining the points of maximum and minimum current. If the characteristic is a parabola, this secant is parallel to the tangent at the point of zero alternating potential, so that the resistance is independent of the applied alternating voltage. When the characteristic is not parabolic, the resistance cannot be expressed simply by the slope of the tangent at the point of zero alternating voltage, but depends on the magnitude of the voltage. Thus, referring to Fig. 150 it will be seen that as long as the potential variations are less than AB, the resistance is practically the same as that which obtains for infinitely small oscillations around the point A. If, now, the maximum value of the potential becomes equal to AC, the line joining C' and O' is not parallel to the tangent at A', but has a smaller slope. We can take the slope of the line $O'C'$ as a measure of the resistance when the current variations extend over the whole region $O'A'C'$.

The same considerations apply to the mutual conductance which can be taken approximately to be equal to the slope of the line joining the extreme point of the characteristic over which the operation takes place.[1]

The problem of setting up the conditions for oscillation, therefore, reduces to the solution of a network involving the oscillation circuit, LCr, a fictitious generator giving a voltage equal to μe_g and a resistance r_p, as defined above. By adopting this procedure, we do not entirely ignore the curvature of the characteristic. If the resistance r_p were not dependent on the intensity

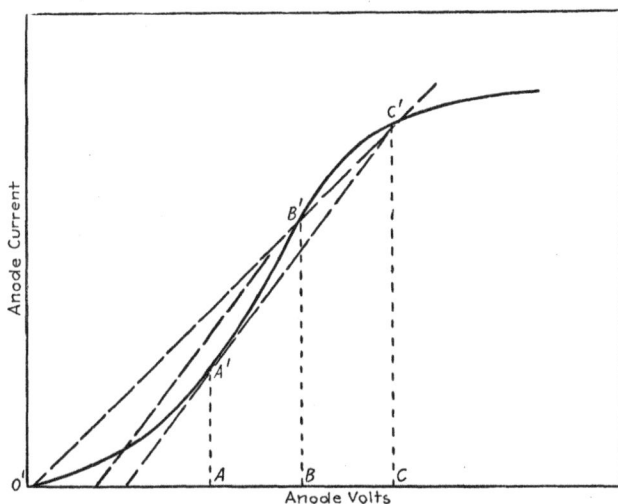

Fig. 150.

of the oscillations, the solution of the network involving the quantities enumerated above would not give an indication of the amplitude of the oscillations; however, both r_p and the mutual conductance g_m are dependent on the intensity of the oscillations. The condition for oscillation will require that r_p and g_m have values lying within certain limits, and since their values depend on the extent of the characteristic over which operation takes place, the amplitude of the oscillations will be determined by the shape of the characteristic and the constants of the external circuit.

81. Conditions for Oscillation in a Two-element Device. Let us first consider the simple case of a device containing two

[1] L. A. Hazeltine, Proc. I.R.E., Vol. 6, p. 63, 1918.

electrodes and having a resistance r, and see what are the conditions that must be satisfied in order that this device, when connected to an oscillation circuit may produce sustained oscillations. The circuit is shown diagrammatically in Fig. 151. The device is supplied with a direct current by means of the battery E, through the choke coil Ch. The oscillation circuit is represented by LC.

The condition for oscillation in such a circuit can be obtained by the simple process of setting the differential equation for the circuit and equating the damping factor to zero. In doing so we need, of course, only to consider the a-c. circuits, that is, the

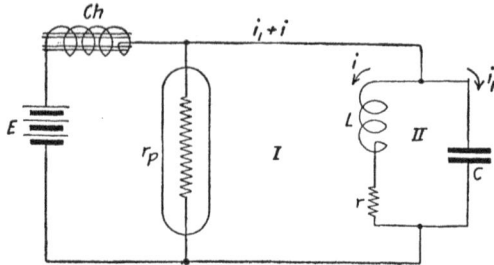

Fig. 151.

two branches I and II. Thus, summing the electromotive forces for circuit I, we get

$$r_p(i_1+i)+ri+Lpi=0, \qquad \ldots \ldots \quad (1)$$

where $p = \dfrac{d}{dt}$.

For branch II we have

$$Lp^2i+rpi=\frac{i_1}{C}. \qquad \ldots \ldots \quad (2)$$

Putting the value of i_1 given by equation (1) into (2) we get:

$$p^2i+\left(\frac{r}{L}+\frac{1}{Cr_p}\right)pi+\frac{i}{LC}\left(1+\frac{r}{r_p}\right)=0. \qquad \ldots \quad (3)$$

This is an equation of the well-known form from which it follows directly that in order that the current i shall be oscillatory, the coefficient of the linear term must be zero. That is:

$$r_p=-\frac{L}{Cr}. \qquad \ldots \ldots \ldots \quad (4)$$

From this we see that in order to obtain sustained oscillations from a device having only two electrodes, it is necessary that the device shall have a negative resistance. Examples of negative resistances have already been given in the previous pages. Thus an arc may have a negative resistance, its characteristic being of the form shown by the curve AB in Fig. 34. The resistance is given by the slope of the characteristic and this is a negative quantity for a characteristic of the kind shown. Fig. 16, page 48, shows another characteristic which over a region ABC has a negative slope and is obtained as the result of the emission of electrons from metals under the impact of electrons. Such characteristics as these are sometimes referred to as "falling characteristics."

The thermionic valve does not show a falling characteristic like the curve AB of Fig. 34, when it is sufficiently well evacuated to prevent the effects of ionization by collision from appreciably influencing the discharge. The characteristics of thermionic valves are those given and discussed in the previous chapters and it will be seen that for such devices the resistance is always positive. It is, therefore, impossible to obtain sustained oscillations from a well-evacuated thermionic valve containing only two electrodes. If such a device contains an appreciable amount of gas during the operation, the characteristic becomes unsteady and sometimes, especially at the higher voltages, exhibits regions over which the slope is negative, and when operated over that region it is, of course, possible to obtain sustained oscillations. The condition which makes this possible in a two-electrode device is unfortunately due to the cause which makes such a device unsatisfactory; namely, the presence of too much gas in the device, thus causing unsteadiness of the discharge and making reproducibility practically impossible. If a controlling electrode or grid be added to the two electrodes of a valve, the operation becomes different and we now have a device which can produce sustained oscillations with facility while at the same time satisfying all the conditions that are necessary to secure satisfactory operation in every respect, namely, freedom of gas with consequent steadiness and reproducibility of results, and comparatively long life.

82. Condition for Oscillation for Three-Electrode Tube. Let us now consider a circuit like that shown in Fig. 152. This is one of a large number of possible oscillation circuits. It is chosen here to exemplify the manner in which oscillations can be produced with an audion because this type of circuit lends itself most readily to mathematical solution. The plate current is supplied by the battery E_b through a choke coil. The oscillation circuit proper is represented by CL_2r. On account of the mutual inductance M between the coils L_1 and L_2, current variations in L_2 cause potential variations to be impressed on the grid. The oscillation circuit CrL_2 is practically non-reactive at the oscillation frequency. A current in the plate circuit establishes an E.M.F. in the circuit L_2C, 90° out of phase with the plate current (assuming that r is very small). Since the oscillation circuit is non-reactive, the oscillation current is in phase with the E.M.F. in this circuit

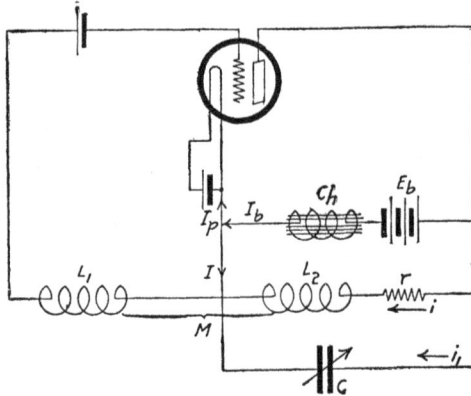

FIG. 152.

and this current induces an E.M.F. in the grid coil L_1, 90° out of phase. The potential variations impressed on the grid by the reaction of the plate circuit on the grid circuit are therefore in phase with the plate current variations.

In order to obtain an expression for the condition that must be satisfied by such a circuit to produce sustained oscillations we shall make the following assumptions which can be realized in practice, although this may not generally be the case. The approximation is, however, sufficiently good to give an indication of the quantities involved and the values that they should have in order to make it possible for such a circuit to act as an oscillation generator. We shall assume (1) that the grid is at all times maintained sufficiently negative with respect to the filament to prevent

any convection current from flowing between filament and grid.
(2) The capacities between the electrodes of the tube will be
taken to be sufficiently small to be neglected. The tube is therefore assumed to be a perfectly unilateral device; that is, there is no
reaction of the plate circuit on the grid circuit except through
inductance of the coils L_1 and L_2. (3) The oscillation circuit
will be regarded as non-reactive at the frequency of oscillation,
or at least the angle can be taken to be so small that its effect
on the phase relations of the circuit can for the present be neglected.
This is usually very nearly the case in most circuits, and circuits
can be designed for which this is always true.

A current i in the coil L_2 will induce an E.M.F. e_g in coil L_1
given by

$$e_g = M\frac{di}{dt} = Mpi.$$

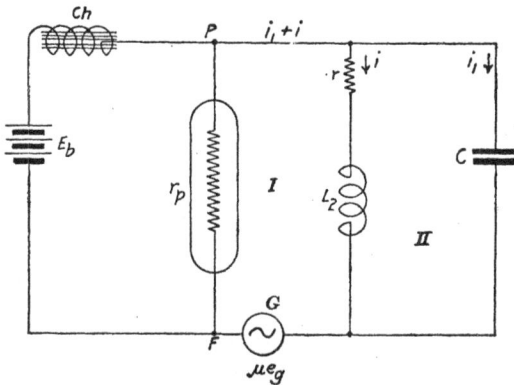

FIG. 153.

Making use of the theorem that a potential e_g impressed on the
grid introduces an E.M.F. equal to μe_g in the plate circuit, we
obtain directly for the driving E.M.F. in the plate circuit:

$$\mu e_g = \mu Mpi. \quad . \quad . \quad . \quad . \quad . \quad . \quad . \quad . \quad . \quad (5)$$

We can, therefore, simplify this circuit and give it the equivalent
form shown in Fig. 153, where r_p represents the plate resistance
of the tube, the points P and F representing the connections to
the plate and the filament. The generator G is a fictitious generator included in the circuit to represent the effect of the grid

potential variations on the plate circuit. This generator, there-fore, gives an E.M.F. equal to the value given by equation (5). The solution of the circuit now becomes extremely simple, involving only the solution of the Kirchhoff equations for the simple net-work to the right of PF. The circuit on the left of PF is a d-c. circuit and need not be considered because we are concerned now only with the a-c. values.

Summing the E.M.F.'s in branch I, we get

$$\mu M p i = r_p(i+i_1)+ri+L_2 pi,$$

which gives

$$i_1 = \frac{\mu M p i}{r_p} - \left(1+\frac{r}{r_p}\right)i - \frac{L_2}{r_p}pi. \quad \cdots \quad (6)$$

For branch II, we have

$$L_2 p^2 i + r p i = \frac{i_1}{C}. \quad \cdots \quad \cdots \quad (7)$$

Substituting the value for i_1 into equation (7) we obtain:

$$p^2 i + \left(\frac{r}{L_2} - \frac{\mu M}{Cr_pL_2} + \frac{1}{Cr_p}\right)pi + \frac{1+\dfrac{r}{r_p}}{L_2C}i = 0. \quad \cdots \quad (8)$$

This is a simple differential equation of the well-known form:

$$p^2 i + A p i + B i = 0,$$

from which we obtain directly that the condition that must be complied with to produce sustained oscillations in circuit II, is that A must be equal to zero, and putting $p=j\omega$ where j is the imaginary unit $\sqrt{-1}$, the frequency of oscillation is

$$\frac{1}{2\pi}\sqrt{B-\frac{A^2}{4}}.$$

Substituting the values of B and $A=0$ from equation (8), we find that the frequency of oscillation is given by

$$n = \frac{1}{2\pi}\sqrt{\frac{\left(1+\dfrac{r}{r_p}\right)}{L_2C}}. \quad \cdots \quad \cdots \quad (9)$$

and the condition for oscillation is

$$r_p = \frac{\mu M}{Cr} - \frac{L_2}{Cr}. \quad \cdots \quad \cdots \quad \cdots \quad (10)$$

We have seen above that the condition for oscillation in the case of a two-electrode device is that the resistance of the device must be negative and equal to $\dfrac{L}{Cr}$. This is the effective resistance of the oscillation circuit, so that when it is added to the equal and opposite resistance of the device the total resistance of the circuit, and therefore the damping, is zero. From equation (10) we see that in the case of a three-electrode tube the resistance of the tube need not be negative as long as the first term $\dfrac{\mu M}{Cr}$ is large enough. This term involves the amplification constant μ and therefore indicates directly that the ability of the audion to produce oscillations lies in its amplifying property.

In order to give an interpretation to this condition (equation (10)) let us write it in the form:

$$g_m = \begin{cases} \dfrac{\mu}{r_p} \geq \dfrac{L_2}{r_p M} + \dfrac{Cr}{M} \\[2mm] \text{or} \\[2mm] \dfrac{\mu}{r_p} \geq \dfrac{Cr}{M - \dfrac{L_2}{\mu}} \end{cases} \quad \cdot \quad \cdot \ \cdot \ \cdot \ \cdot \ \cdot \quad (11)$$

It will be recognized that $g_m = \dfrac{\mu}{r_p}$ is the mutual conductance of the tube as defined in the preceding chapter. For very small oscillations the mutual conductance is given by the slope of the plate current grid potential characteristic, while for large oscillations it can be taken to be approximately equal to the slope of the line joining the points of maximum and minimum current on the characteristic. Now, it will be recognized that as the intensity of the oscillations increases, the slope of this line becomes less and less. Equation (11), on the other hand, states that for oscillations to be sustained the mutual conductance must be greater than a quantity involving the constants of the external circuit. The right-hand side of equation (11) is also of the dimensions of a conductance and can also be represented by a line having a slope depending on the values of these constants. Suppose this line has a definite slope given by OA, Fig. 154. The oscillations will, therefore, increase in intensity, the current varying over a greater and greater range of the characteristic until the mutual conductance as given by the line BC joining the points of maximum and minimum current becomes parallel to OA.

If the mutual inductance between the plate and grid coils were decreased, the slope of the line representing the right-hand side of equation (11) would increase, say, to OA', and then the oscillations would be weaker, the plate current varying over such a range that the mutual conductance is equal to the slope of the line OA'.

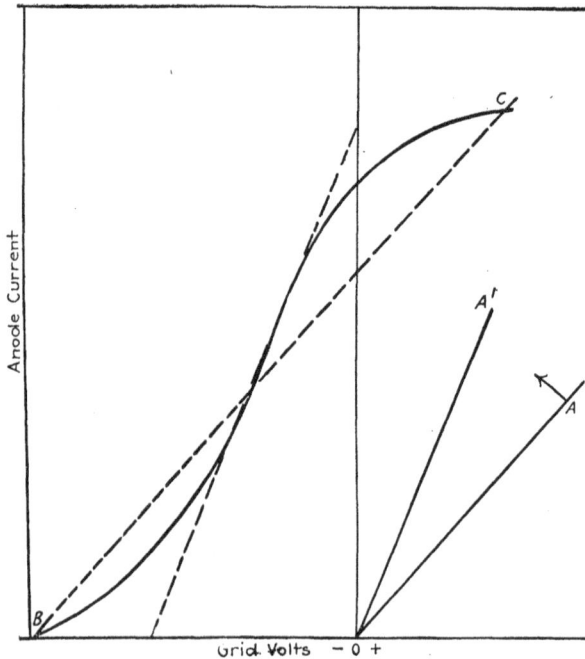

Fig. 154.

Whether or not the tube will oscillate depends not only on the coupling between the output and the input coils, but also on a number of other quantities. One of the important quantities is the amplification constant μ. Fig. 155 shows how μ influences the operation of the device as an oscillator. The line OB gives $\dfrac{\mu}{r_p}$ as a function of μ, and CD gives the expression $\dfrac{Cr}{M - \dfrac{L^2}{\mu}}$ of equation (11) as a function of μ. We shall refer to this quantity as g_0. This equation states that g_m must be at least equal to g_0; hence, for the constants of the circuit chosen in this particular case, all values of μ lying to the left of the broken line are impossible values.

The effect of the plate voltage can be shown in a similar way. It follows, for example, from the considerations given in Chapter

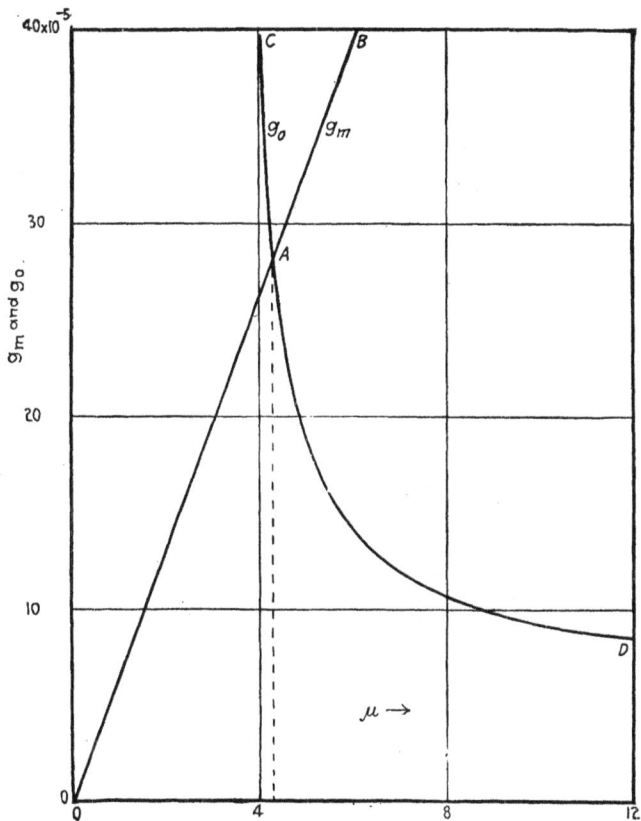

Fig. 155.

VII, that the mutual conductance is approximately proportional to the plate potential, provided the filament temperature is high enough to insure that by increasing the plate potential we do not enter into the saturation region. We can, therefore, replace $\frac{\mu}{r_p}$ and r_p by E_p and some arbitrary constants. We then obtain an expression $g_m = g_0$ where g_m is directly proportional to E_p and g_0 is a linear function of E_p. These relations when plotted as shown in Fig. 156 intersect at the point A. The condition for oscillation is that g_m must be at least as large as g_0. We see, therefore, that the tube will not oscillate until the plate voltage reaches a certain minimum value which is fixed if the other quantities, such as the coupling, etc., are fixed.

These considerations show that it is desirable to make the

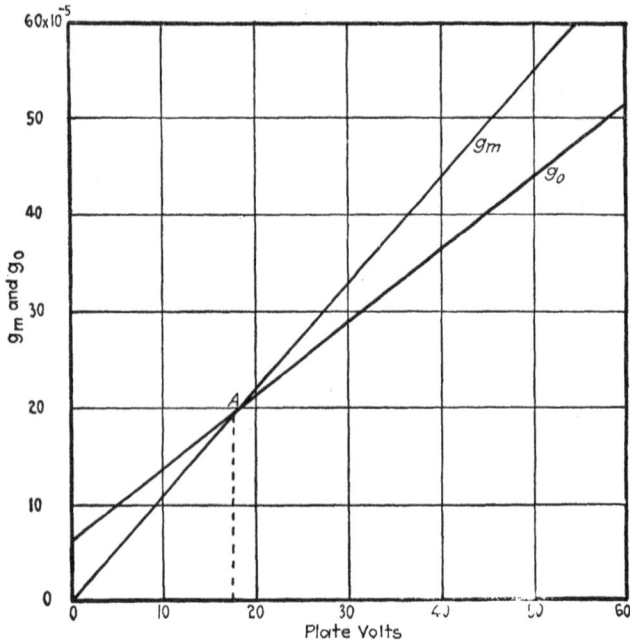

Fig. 156.

mutual conductance of the tube as large as possible. This was also found to be the case when using the tube as an amplifier. (See equation (36), Chapter VII).

The oscillation frequency as given by equation (9) is determined not entirely by the inductance and capacity in the oscillation circuit, but depends also on the plate resistance of the tube and the resistance in the oscillation circuit. Since, however, the ratio $\dfrac{r}{r_p}$ is usually very small, the frequency can generally be taken to be very closely equal to that given by the simple oscillation circuit; namely,

$$n = \frac{1}{2\pi}\sqrt{\frac{1}{L_2 C}}. \quad \cdot \quad \cdot \quad \cdot \quad \cdot \quad \cdot \quad \cdot \quad (12)$$

It will be recognized that the solution of the circuit shown in Fig. 153 does not indicate directly what the amplitude of the oscillations is. This quantity is, however, determined indirectly by the condition for oscillation. On account of the curvature of the characteristic, the mutual conductance of the tube decreases as the amplitude of the oscillations increases, in the manner explained above, until the mutual conductance reaches its minimum value. The amplitude of the oscillations in the plate circuit can then be determined from this limiting value and the characteristic of the tube. Usually, however, such a determination is not necessary.

83. Relation between Mutual Conductance of Tube and that of Plate Circuit. In the above equations, $\frac{\mu}{r_p}$ represents the mutual conductance of the tube itself. This is also the mutual conductance of the plate circuit, provided the external impedance in the plate circuit is negligibly small compared with the resistance of the tube. When this is not the case, the dynamic characteristic of the plate circuit does not coincide with the characteristic of the tube itself, but differs from it to an extent depending on the relative magnitudes of the external impedance and the plate resistance. If the external circuit is non-reactive, the dynamic characteristic of the plate circuit is the curve of noC shown in Fig. 86. If the external circuit is reactive, the dynamic characteristic of the plate circuit takes the form of the loop such as that shown in Fig. 88. In this case the quantity concerned is not a pure conductance but a complex quantity, and what we have to deal with then is the mutual admittance. In most oscillation circuits, however, the reactance is so small at the oscillation frequency in comparison with the total resistance that the angle can generally be neglected.

It can readily be seen that the mutual conductance of the circuit is less than that of the tube alone, because when the current in the external circuit is increased by an increase in the potential of the grid, the voltage drop in the external impedance causes a decrease in the plate potential, so that the resultant increase in plate current is less than would be the case if the external impedance were zero. The relation between these two mutual conductances can be obtained as follows: The alternating plate current is given by

$$i_p = \frac{\mu e_g}{r_p + Z_0}. \qquad \cdots \cdots \cdots \quad (13)$$

Putting $Z_0 = r_0 + jx_0$ we get

$$\frac{1}{Y'_m} = \frac{e_g}{i_p} = \frac{r_p}{\mu} + \frac{r_0}{\mu} + j\frac{x_0}{\mu},$$

where $Y'_m = g'_m - jb$ is the mutual admittance of the *plate circuit*. Generally the imaginary component is small in comparison with the resistance component, so that we can use the simple equation:

$$\frac{1}{g'_m} = \frac{1}{g_m} + \frac{r_0}{\mu}. \qquad \cdots \cdots \cdots \quad (14)$$

This relationship can be expressed in a somewhat different form. Since $i_p r_0 = e_p$, we get directly from equation (13): ·

$$g'_m = g_m\left(1 - \frac{1}{\mu}\frac{e_p}{e_g}\right). \qquad \cdots \cdots \cdots \quad (15)$$

The condition for oscillation can also be expressed in terms of the mutual conductance of the plate circuit instead of the mutual conductance of the tube itself. This was, for example, done by Hazeltine.[1] The quantity g in Hazeltine's equations is not the mutual conductance of the tube, but the mutual conductance of the plate circuit.

84. Phase Relations. The phase relations that exist in vacuum tube oscillator circuits have been investigated by Heising and explained with the help of vector diagrams.[2] We shall not discuss this phase of the subject beyond what is necessary for an understanding of the fundamental phenomena of such circuits. The main condition is that the plate current and grid potential must be as nearly in phase as possible. The phase relations between the various quantities are shown in Fig. 157 and can be explained with reference to Fig. 152.

I_b represents the steady direct current supplied by the battery E_b through the choke coil. We can regard this current as constant,

[1] L. A. HAZELTINE, R.I.E., Vol. 6, p. 63, 1918.

[2] R. A. HEISING, Journal of A.I.E.E., Vol. 39, p. 365, 1920.

although in actual practice it is only approximately so unless the choke coil has a very large inductance. I_p represents the instantaneous value of the plate current and I the instantaneous value of the current in the branch containing C and L_2 in parallel. This current multiplied by the instantaneous plate-filament voltage and integrated over a complete cycle, represents the a-c. power supplied by the tube. As much power is drawn from the tube as

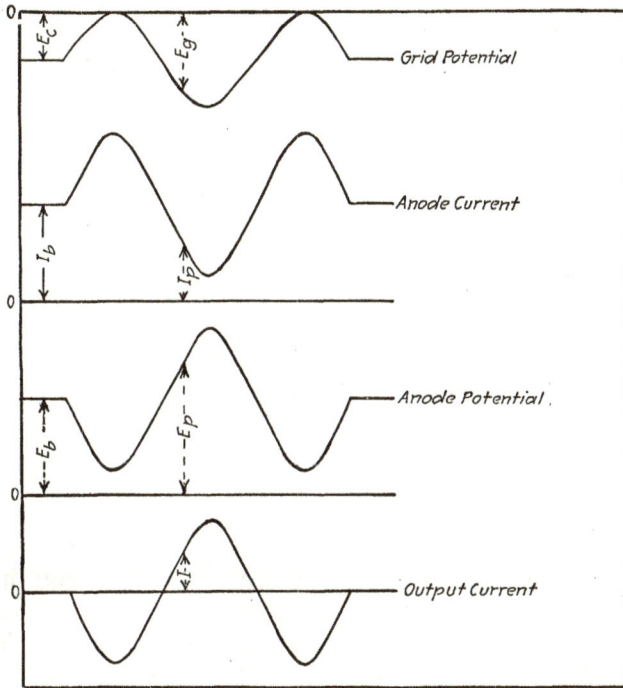

Fig. 157.

is dissipated in the oscillation circuit when the steady condition is reached.

Referring to Fig. 157, the lines marked O represent the ordinates of zero voltages and currents. The grid is maintained at a negative potential E_c. When the alternating grid potential is zero, the plate current is equal to I_b. When the grid potential oscillates, as indicated, the plate current oscillates in phase with the grid potential. The plate potential oscillates around the mean value E_b, but is 180° out of phase with the grid potential if the external circuit is non-reactive.   The current I in the branch circuit is the difference between direct current I_b drawn from the battery and the plate current I_p.   It is therefore 180° out of phase with the plate current and oscillates around zero.   We have assumed that the grid always remains negative with respect to the negative end of the filament.   If the grid becomes positive during a part of the cycle, it takes current which generally means a loss of power occasioned by heat dissipation in the grid circuit.

On account of the curvature of the characteristic, the current wave in the plate circuit, due to a sinusoidal voltage impressed on the grid circuit, is not a pure sinusoid but is distorted.   This introduces harmonics.   They can, however, be effectively tuned

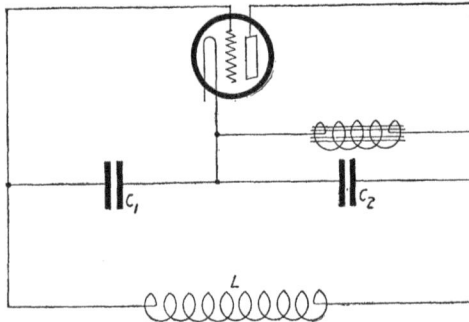

FIG. 158.

out in the oscillation circuit so that most of the energy in the oscillation circuit will be due to the fundamental.   It must be recognized that the harmonics cause a waste of power.   These considerations apply in general to the fundamental, the effect of harmonics being neglected.

85. Colpitts and Hartley Circuits.   The circuit shown in Fig. 152 is only one of a large number that can be used with a vacuum tube oscillator.   It was chosen there for its simplicity, although it is not the most commonly used type of circuit.   Two circuits that are frequently used are those shown in Figs. 158 and 159, known as the Colpitts and Hartley circuits, respectively.   The main difference between these circuits is that in the one the coupling between output and input circuits is capacitive and in the other it is mainly inductive.   If we neglect the effect of the electrostatic capacities between the electrodes of the tube, the oscillation circuits are $C_1 C_2 L$, for the Colpitts circuit, and $L_1 L_2 C$, for the Hartley circuit. The effect of the inter-electrode capacities will be considered below.

The conditions for oscillation for these circuits have been given by Hazeltine, Heising [1] and others. Taking, for example, the case of the Hartley circuit, the condition for oscillation can be expressed by:

$$g_m = \frac{\mu C r (L_1 + L_2 + 2M)}{(L_2 + M)[\mu(L_1 + M) - (L_2 + M)]}, \quad \cdot \quad \cdot \quad \cdot \quad (16)$$

where g_m = mutual conductance of the tube,

M = mutual inductance between L_1 and L_2,

$r = r_1 + r_2$.

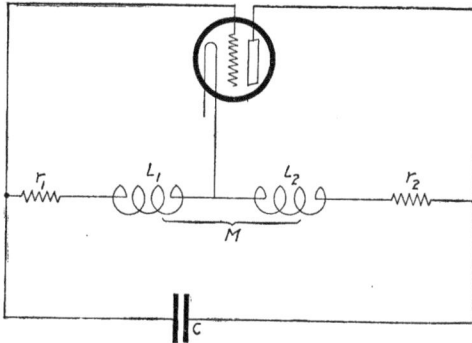

FIG. 159.

From this equation it follows that there is a certain relation between the voltages established in the plate and grid coils, which makes the tube oscillate most readily. Since the conditions for oscillation state that the right-hand side of equation (16) must not be greater than g_m, it follows that the tube will oscillate most readily when this expression is a minimum.

Putting

$$c_g = \frac{k}{L}(L_1 + M);$$

$$e_p = \frac{k}{L}(L_2 + M);$$

$$e_g + e_p = \frac{k}{L}(L_1 + L_2 + 2M) = kL$$

$$\frac{e_p}{e_g} = n,$$

[1] Loc. cit.

where k is a constant, **we find:**

$$g_m = \frac{\mu C r}{L} \frac{(1+n)^2}{n(\mu-n)}. \quad \cdots \cdots \quad (17)$$

This is a minimum for

$$n = \frac{\mu}{\mu+2}. \quad \cdots \cdots \cdots \quad (18)$$

For tubes having a high value of μ, therefore, L_1 should be approximately equal to L_2. If μ is low, on the other hand, the best condition can necessitate making L_2 considerably smaller than L_1.

86. Tuned Grid-circuit Oscillator. This type of circuit which is commonly used in the reception of radio signals, is shown in Fig. 160. If it is assumed, as before, that the grid is maintained

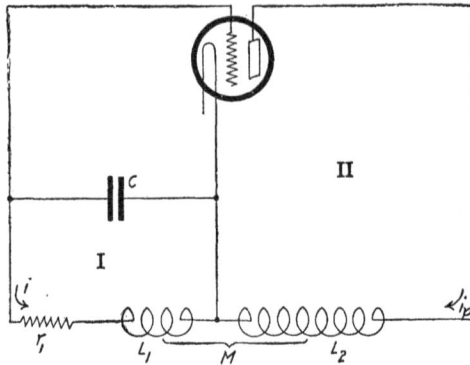

FIG. 160.

at a sufficiently high negative potential to insure that there is no appreciable convection between filament and grid, the condition for oscillation for this circuit can also be easily obtained. The potential e_g applied to the grid is given by

$$e_g = \frac{1}{C} \int i \, dt, \quad \cdots \cdots \cdots \quad (19)$$

and the electromotive force induced in the plate circuit through the tube, on account of the effect of the grid potential on the current is μe_g. There is another electromotive force induced in the plate circuit, namely, Mpi, and is due to the mutual reactance of the grid circuit on the plate circuit through the coils L_1 and L_2. The electromotive force induced on the oscillation circuit, due to the current i_p in the plate circuit, is Mpi_p. Equating these E.M.F.'s in the circuits, we get for circuit I:

$$Mp^2i_p = L_1p^2i + r_1pi + \frac{i}{C}, \quad \ldots \ldots \quad (20)$$

and for circuit II:

$$Mp^2i + \frac{\mu i}{C} = r_ppi_p + L_2p^2i_p. \quad \ldots \ldots \ldots \quad (21)$$

Eliminating i_p from these two equations, the equation for i becomes:

$$(L_1L_2 - M^2)p^3i + (r_pL_1 + rL_2)p^2i + (rr_pC + L_2 - \mu M)pi$$
$$+ r_pi = 0, \quad . \quad (22)$$

which is of the form

$$p^3i + Ap^2i + Bpi + Di = 0.$$

This is a cubic equation and has one real and two complex roots. The condition which makes the damping zero is $D = AB$. That is:

$$\frac{r_p}{r_prC + L_2 - \mu M} = \frac{r_pL_1 + rL_2}{L_1L_2 - M^2}. \quad \ldots \ldots \quad (23)$$

In most circuits rL_2 can be neglected in comparison with r_pL_1. With this approximation the condition for oscillation becomes:

$$g_m \geq \frac{\mu}{r_p} = \frac{rC}{M} - \frac{M}{L_1r_p}. \quad \ldots \ldots \quad (24)$$

The right-hand side of this equation contains two terms, one of which is directly proportional to M, and the other inversely proportional to M. There appears, therefore, to be an optimum value for the mutual inductance between the input and output which makes g_m a minimum.[1]

87. Effect of Inter-electrode Capacities—Parasitic Circuits. We have assumed in the above that there is no reaction of the plate circuit on the grid circuit through the tube itself. In some types of circuits the capacities between the electrodes cause the circuits to behave differently from what is to be expected. The simple circuit shown in Fig. 160 can, for example, be drawn in the manner shown in Fig. 161, where the capacities between the electrodes of the tube are indicated C_1, C_2 and C_3. Such a circuit, therefore, has more than one degree of freedom, a number of oscillation cir-

[1] S. BALLANTINE, Proc. I.R.E., Vol. 7, p. 159.

cuits being added to the main oscillation circuit CL_1. Of these parasitic circuits, the most important one in the diagram shown

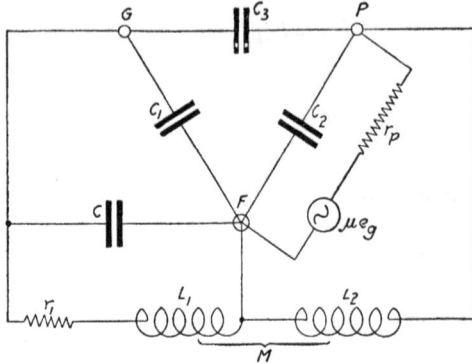

Fig. 161.

is the circuit formed by the capacity C_3 between grid and plate, and the inductance L_1 and L_2 in series, the total inductance being

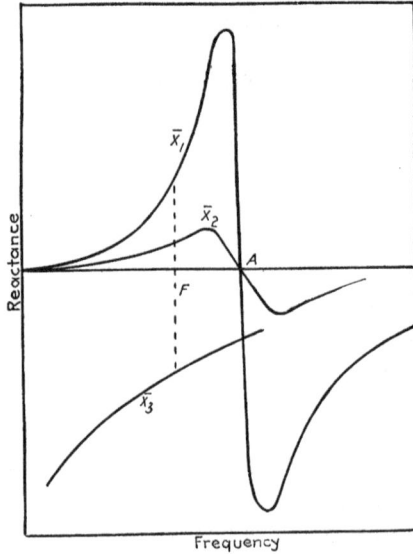

Fig. 162.

L_1+L_2+2M. The effect of the capacity C_3 is to make the frequency of oscillation different from that which would be obtained from a simple circuit CL_1. The reactance-frequency curve of the circuit CL_1 is shown by the curve marked X_1 (Fig. 162). For frequencies lower than that corresponding to the point A, this circuit has a positive or inductive reactance. The effective reactance, due to the coil L_2 and its coupling with the oscillation circuit L_1C, is given by the curve X_2. At frequencies below A, the total reactance is inductive, and oscillations will occur at such a frequency that the inductive reactance is equal to the capacitive reactance due to the capacity C_3 between grid and plate. The oscillation frequency is, therefore, that corresponding to the point F instead of the point A, as would be the case if the effect of the grid-plate capacity were negligible. This circuit, therefore, behaves somewhat like a Hartley circuit in that the plate coil L_2, and the oscillation circuit L_1C, together, act like an inductance in

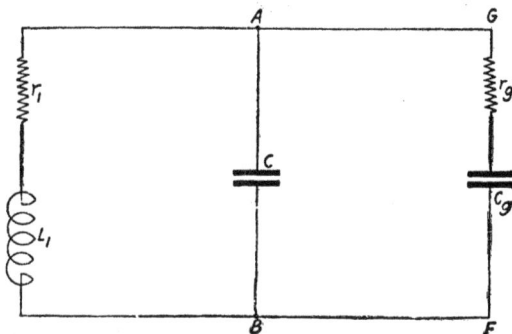

FIG. 163.

parallel with a capacity. In the Hartley circuit, the capacity between grid and plate is simply in parallel with the oscillation circuit capacity C. This circuit is, therefore, more suitable for use at high frequencies.

88. Regeneration. The effect of the inter-electrode capacities can cause a tube to produce oscillations even when there is no mutual inductance M between the output and input coils. It was explained in Chapter VII, Sections 69–71, that on account of these capacities there is an effective impedance between filament and grid, which depends not only on the capacities between the electrodes but also on the constants of the output circuit. This impedance can generally be represented by a resistance r_g and a reactance due to the effective input capacity C_g. The input circuit can, therefore, be drawn as shown in Fig. 163. The impedance of the circuit formed by C in parallel with r_g and C_g in series is:

$$Z = \frac{1+j\omega r_g C_g}{-\omega^2 r_g C_g C + j\omega(C+C_g)}. \quad \cdot \quad \circ \quad \circ \quad \bullet \quad \bullet \quad \bullet \quad (25)$$

The real component r is

$$r = \frac{r_g}{r_g^2 \omega^2 C^2 + \left(\dfrac{C+C_g}{C_g}\right)^2}. \quad \cdot \quad \cdot \quad \cdot \quad \cdot \quad \cdot \quad (26)$$

The first term of the denominator in this equation is usually negligibly small compared with the second, so that the total resistance of the circuit is:

$$r_1 + \left(\frac{C_g}{C+C_g}\right)^2 r_g. \quad \cdot \quad \cdot \quad \cdot \quad \cdot \quad \cdot \quad (27)$$

If oscillations are impressed on this circuit, the rate at which they would die out depends, of course, on the value of the total resistance. If r_g is negative, the total resistance will be reduced; that is, there will be a smaller consumption of power in the input circuit and the tube will give a greater amount of amplification. If r_g is negative and the effective resistance to the right of AB is equal to the resistance r_1, i.e., if the total resistance of the circuit is zero, the circuit will produce sustained oscillations. We can take expression (27) as a measure of the damping, δ, due to the resistance in the circuit. The increase in amplification, due to a reduction in this resistance results in the effect that is sometimes referred to as "regeneration." A measure of the regenerating effect is given by $\frac{1}{\delta}$. Now, it was shown in Chapter VII that r_g is positive when the external plate circuit is non-reactive or contains only capacitive reactance. If, on the other hand, the reactance in the plate circuit is inductive and the angle of the impedance in the plate circuit is large enough, then r_g is negative. In Fig. 164 are plotted curves showing the relation between the regenerative effect and the ratio of the external impedance in the plate circuit to the plate resistance. The values of r_g and C_g, used in computing these curves, were obtained from equations (54) and (56) of Chapter VII. The quantity $\frac{1}{\delta}$ was computed with the values so found and with arbitrarily assumed values of r_1, as indicated in the curves. The curve for $r_1 = 5.4$ ohms stretches to infinity, indicating that over the range of the ratio $\dfrac{Z_0}{r_p}$ from about .8 to about 1.2, the tube produces sustained oscillations, due to the reaction of the plate circuit on the grid circuit through the electrostatic capacities of the tube.

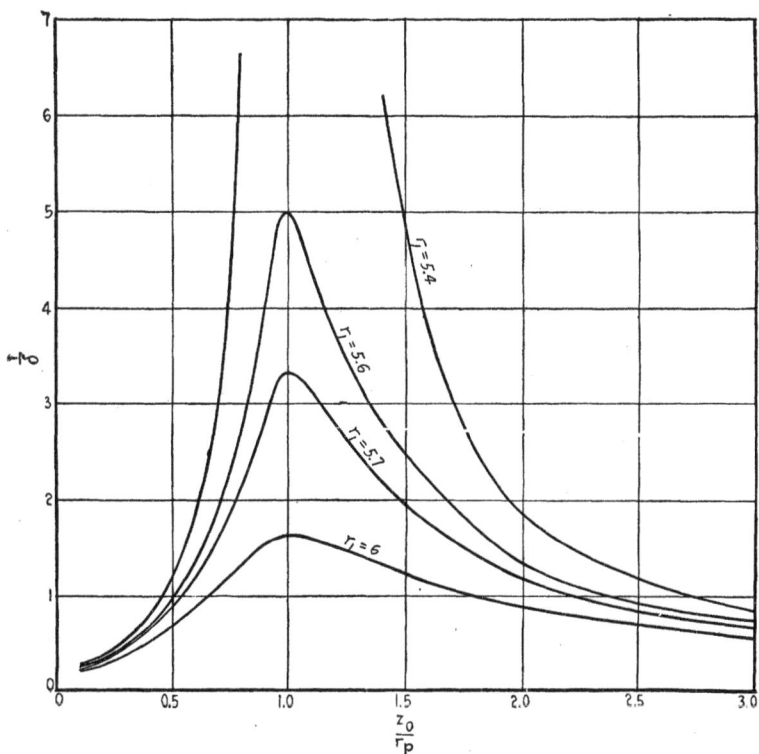

FIG. 164.

An interesting result shown by these curves is that the maximum regenerating effect is obtained when the external impedance in the plate circuit is equal to the plate resistance. This, it will be remembered, is also the condition for maximum power amplification derived in Chapter VII, for the case of the simple amplifier.

According to expression (27) the regenerating effect becomes greater the smaller the capacity C in the oscillating circuit.

J. M. Miller [1] has computed curves giving the signal strength as a function of the *inductance* in the plate circuit. These curves are similar to those shown in Fig. 164, except that they are more symmetrical. They have the same general form as a curve giving experimental results published by Armstrong.[2]

89. Complex and Coupled Circuits—Meissner Circuit. Complex circuits can be reduced to simple circuits by the addition of the reactances of the separate branches. Thus, the circuit shown in Fig. 160 constitutes a complex circuit if the capacity between grid and plate becomes effective in determining the frequency of the oscillations. The reactances of the branches I and II are indicated by the curves X_1 and X_2 of Fig. 162, while the reactance

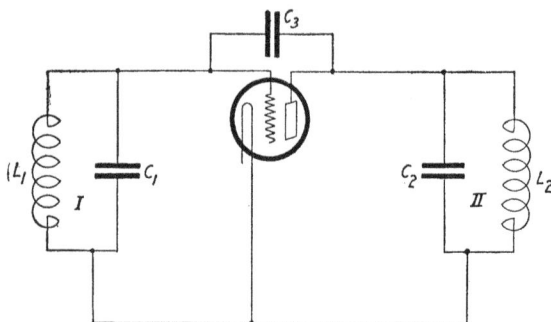

FIG. 165.

due to the capacity C_3 between grid and plate is indicated by X_3. The frequency of the oscillations is determined by the value of these reactances which makes the total reactance of the circuit zero.

In general, a complex circuit such as that shown in Fig. 165 can be regarded as a simple circuit in which the oscillation circuits L_1C_1 and L_2C_2 act as inductances or capacities, according as the reactance between grid and plate is capacitive or inductive. The reactance-frequency curve of a parallel oscillation circuit like L_1C_1 has a shape such as the curve X_1 in Fig. 162. Now, if the react- ·nce between grid and plate is capacitive or negative, the total

---

[1] J. M. MILLER, Bureau of Standards Bulletin 351.

[2] E. H. ARMSTRONG, Proc. I.R.E., Vol. 3, p. 220, 1915.

reactance can only become zero at a frequency F, which is lower than the natural oscillation frequency A of the simple parallel circuit L_1C_1 or L_2C_2. From the curve it is seen that at frequencies below A the reactance of the simple circuit L_1C_1 is inductive (positive), so that the complex circuit shown in Fig. 165 can be represented by the simple circuit of Fig. 166. This is a Hartley circuit. If, on the other hand, the reactance between plate and grid were inductive, the frequency of oscillation of the circuit will be such as to make the sum of this inductive reactance and the reactances of the branch circuits I and II equal to zero. This would require that the reactances of these branch circuits be capacitive and therefore the frequency of oscillation will be higher than the natural frequency of the circuits I and II separately.

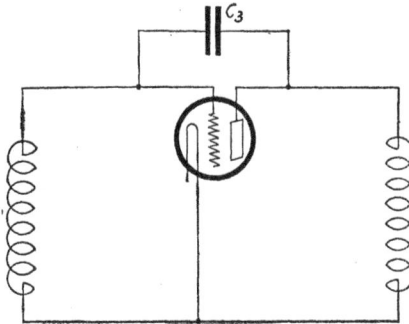

FIG. 166.

In this case the total circuit reduces to one like that shown in Fig. 166, except that instead of the two inductances, we have two capacities, and instead of the capacity C_3, we have an inductance. In other words, the complex circuit is reduced to a simple Colpitts circuit.

Coupled circuits can be treated in much the same manner. The Meissner circuit shown in Fig. 167 is an example of a coupled circuit.[1] The two oscillation circuits are $L_1L_2C_3$ and LC. The effects that can be obtained with such a circuit by varying the constants of the oscillation circuits have been discussed by Heising in the paper cited above. The reactance-frequency curves of such a circuit are shown in Fig. 168, where X_1 represents the reactance of the circuit $L_1L_2C_3$, and X_2 the reactance due to

[1] A. MEISSNER, " Electrician," Vol. 73, p. 702, 1914.

coupling with the oscillation circuit LC. The sum of the two reactances is indicated by the curve X. It is seen that there are three frequencies for which the total reactance is zero. The tube will not oscillate at the frequency F because this represents an unstable condition, but it can oscillate at the two frequencies F_1 and F_2. Usually, however, it oscillates at only one frequency. By suitably adjusting the coupling between the oscillation circuit LC, or choosing the constants of the circuits, the reactance-frequency curve of the combination can take such a form that the three frequencies practically merge into one. This can be done by making the coupling loose, or by making the total inductance

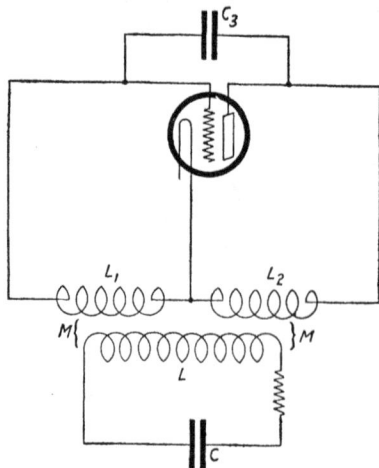

Fig. 167.

L_1+L_2 large compared to C_3. This is usually the case with most tubes when the desired frequency is not very high.

90. Circuits Comprising a-c. and d-c. Branches. The circuits shown above indicate only the a-c. branches. These are the only branches that need to be considered in determining the conditions for oscillation and the frequency. In practice we also need d-c. sources of power supply, and it is often necessary to separate the a-c. and d-c. circuits. This can be done readily by applying the simple and well-known rule to separate the d-c. from the a-c. branches by means of inductances and capacities. In doing so, however, it is necessary to adjust these inductances and capacities to such values that they do not appreciably influence the behavior

of the oscillation circuit proper, or introduce parasitic circuits
that would result in a loss of power.

Fig. 169 shows, as an example, the Hartley circuit as it is commonly used in practice. The plate battery E_b is inserted directly
in the circuit connecting the plate to the inductance L_2. It is
usually not necessary to separate the direct and alternating current in this branch of the circuit. The capacity C_s and resistance
R_s are used here instead of a battery to maintain the grid at an

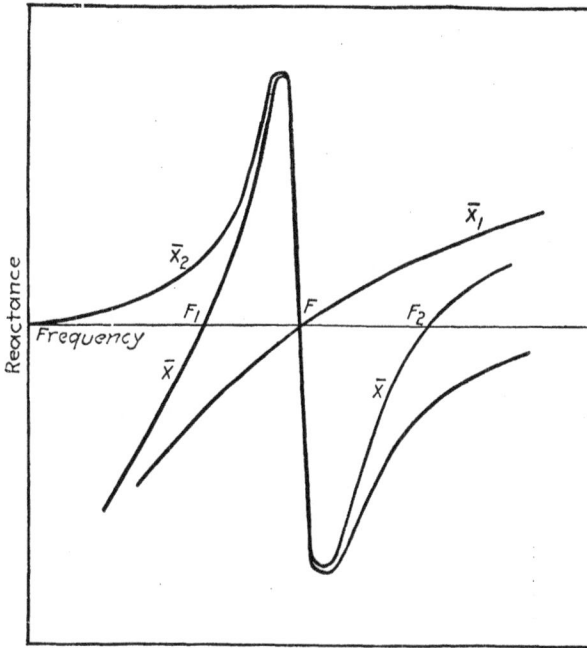

Fig. 168.

appropriate negative potential with respect to the filament. This
means of maintaining the grid negative operates only when there
is a convection current between filament and grid; that is, when
the grid becomes positive during part of the time that the a-c.
potential on the grid is positive. When the grid becomes positive
it attracts electrons, and current flows through the resistance R_s.
During the rest of the cycle there is no flow of electrons from filament to grid. There is, therefore, established a rectified current

through the resistance R_s, and this lowers the potential of the grid with respect to the filament.

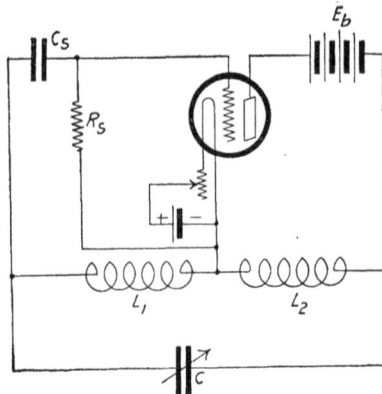

Fig. 169.

Fig. 170 shows a Colpitts circuit as it can be used in practice. In this case the resistance R_s is replaced by a choke coil Ch_1. The alternating and direct current in the plate circuit are separated by means of the choke coil Ch_2 and the capacity C_b. The

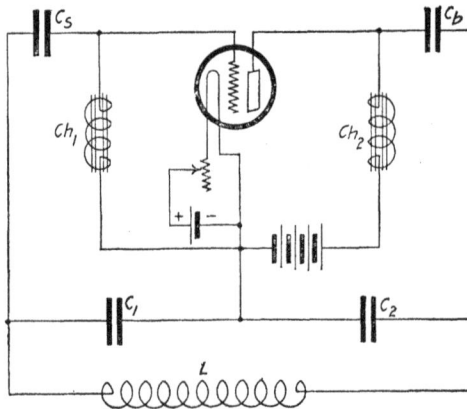

Fig. 170.

inductance of this choke coil is usually chosen as high as possible, or at least so high that its impedance is several times the plate resistance of the tube. The capacities C_s and C_b are chosen sufficiently large so that they do not appreciably affect the operation of the oscillation circuit LC_1C_2.

91. Effect of Grid Current. In deriving the conditions for oscillation above, it was assumed that the grid is at all times maintained sufficiently negative with respect to the filament to prevent any convection current from flowing between filament and grid. In practice this is usually not the case. The grid usually becomes positive to an extent depending on the adjustments of the circuit constants. Thus, when using the condenser and resistance to maintain the grid negative, as shown in Fig. 169, the grid must become positive during part of the cycle. The rectified current through R_s maintains the grid at a steady negative potential, and the value of this potential will be greater the greater the grid current becomes. Hence, the fraction of a period during which grid current flows, that is, the amount of grid current, will be determined by the rate at which it leaks off through the resistance R_s. Now, the grid and plate potentials are approximately 180° out of phase. This is shown, for example, in Fig. 157, which was drawn for the case in which the plate is connected to a non-reactive circuit. The amount of current flowing to the grid depends not only on the grid potential but also on the potential of the plate. The higher the plate potential the more readily will the electrons be drawn through the openings of the grid, and the smaller will be the grid current. But if the plate becomes less positive at the same time that the grid becomes more positive (as shown in Fig. 157), there is a tendency for the grid current to become much greater, and if the potential variations impressed on the grid become so great that the maximum positive grid potential becomes equal to or perhaps greater than the simultaneous minimum plate potential, then the grid can rob the plate of so much current that the characteristic representing the plate current as a function of the grid potential becomes apparently saturated at a current value which is lower than the actual saturation current at the temperature of the filament. This effect is shown in Fig. 171 which represents the static characteristics of the tube for various fixed values of the plate potential. When the tube operates in an oscillation circuit, which is so adjusted that the reactance in the plate circuit is practically zero, the dynamic characteristic of the plate circuit is represented by the curve AOB. The normal grid potential, when the tube does not oscillate, is represented by the value E_c. When oscillating, the plate potential decreases when the grid potential increases and vice versa, so that the characteristic is given by AOB instead of the static characteristic $A'O'B'$. At B the plate and grid potentials become comparable, and the plate current begins to decrease. This plate current is less than the emission current; that is, the current represented by the total number of electrons leaving the filament. The point B represents the maximum potential that the grid can acquire without causing too much waste of power.

The mutual conductance of the plate circuit can be represented by the slope of the straight line joining A and B. The instantaneous value of this quantity is zero at B and A, and has a

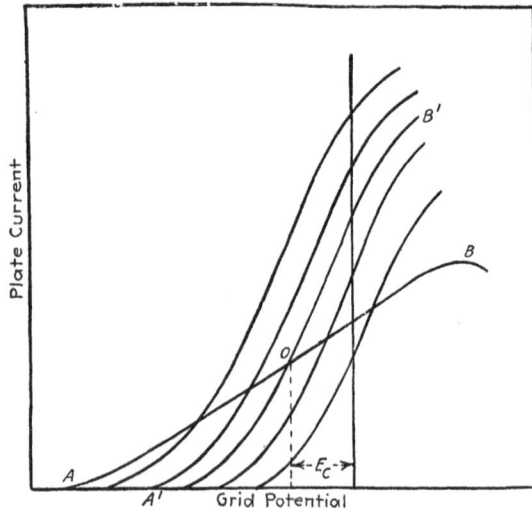

FIG. 171.

maximum value at O, but the mean value which is determined by the integrated slope of the curve is finite.

92. Output Power. The value of the grid potential at which the bend B in Fig. 171 occurs, depends on the resistance of the external plate circuit. The current in this resistance r_0 causes a potential drop which reduces the potential on the plate, since the voltage of the plate battery remains constant. The larger this resistance the greater will be the decrease in the plate potential when the current in the plate circuit increases, and the sooner will the bend in the characteristic occur. Also, the greater the external resistance, the smaller will be the slope of the dynamic characteristic. Fig. 172 shows the dynamic characteristics for a number of different values of the external plate resistance r_0. The characteristic *OB* is obtained for the largest, and *OE* for the smallest external resistance. If the tube operates, for example, in a circuit like that shown in Fig. 152, the impedance of the parallel circuit L_2r and C is given by

$$Z = \frac{r(1-\omega^2CL_2)+\omega^2L_2Cr}{(1-\omega^2CL_2)^2+\omega^2C^2r^2} + j\frac{\omega L_2(1-\omega^2CL_2)-\omega Cr^2}{(1-\omega^2CL_2)-\omega^2C^2r}. \tag{28}$$

The resistance or real component of this impedance at the resonance frequency is given by $r_0 = \dfrac{L_2}{Cr}$. The characteristics

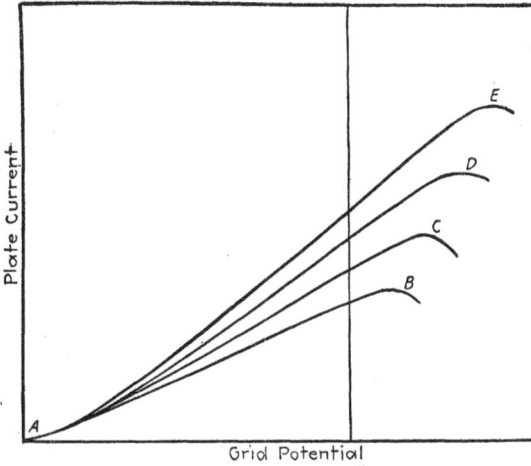

Fig. 172.

OB, *OC*, etc., correspond to different values of this resistance. The maximum power output in each case is obtained when the grid potential rises up to the value indicated at the bend. If, now, the output for each of these resistances r_0 be plotted as a function of this resistance, we obtain a curve such as that shown in Fig. 173, which shows a maximum for a particular value of the external resistance r_0. This resistance is equal to the plate resistance of the tube. From the maximum value given in Fig. 173, the ratio of the inductance to the capacity of the oscillation circuit can be determined, which gives the maximum output power. In general, maximum output power does not necessarily mean maximum efficiency. This will become evident from considerations in the next section.

93. Efficiency.—The efficiency of the oscillator can be expressed by the ratio of the power supplied to the oscillation circuit to the power drawn from the source of plate voltage (battery, generator, etc.). Strictly speaking, the overall efficiency should take account also of the power expended in heating the filament. In high power oscillators this power can usually be neglected, as will become evident from the following considerations. Of

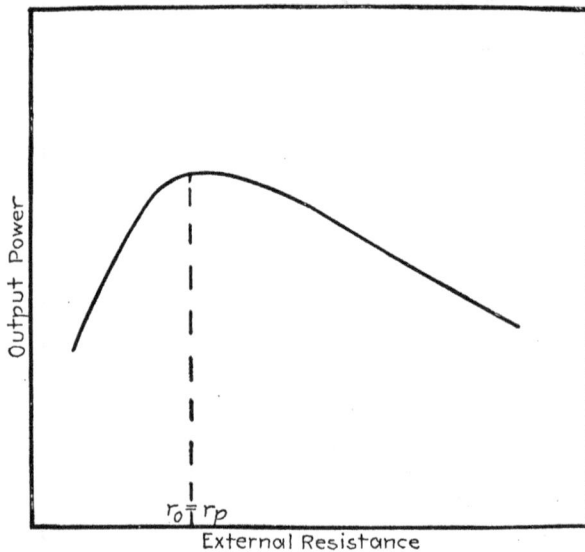

FIG. 173.

the power drawn from the source of plate voltage E_b, part is dissipated at the plate and serves no useful purpose. This we shall refer to as P_p. The remainder, P_0, of the power drawn from the plate battery, is delivered to the oscillation circuit. If we neglect the power expended in heating the filament, the efficiency can be expressed as:

$$\text{Efficiency} = \frac{P_0}{P_b} = \frac{P_0}{P_p + P_0},$$

where P_b represents the power supplied by the plate battery and is equal to the product of the direct current I_b and direct voltage E_b in the plate circuit.

For a fixed value of the efficiency, the power drawn from the plate battery is a measure of the power supplied to the oscillation circuit. This can be increased by increasing either the plate current or the plate voltage. In the former case, the filament area or the filament temperature, or both, must be increased and then the power expended in heating the filament may become comparable with, or even greater than, the power in the plate circuit. If, on the other hand, the power in the plate circuit be increased by increasing the plate battery voltage, instead of the current, the power expended in heating the filament becomes relatively smaller and smaller. It will be evident that even if the saturation current obtained from the filament remains relatively small, the power can be increased to almost any desired value by sufficiently raising the voltage of the plate current supply. Of course, in doing so the operating point of the characteristic rises to higher current values and may even fall on the saturation part of the characteristic, but this can be prevented by increasing the negative potential on the grid or increasing the value of the amplification constant μ. As was explained in Chapter VII, the plate potential necessary to give any chosen value of plate current increases as μ is increased. The limitation to increase in power by this means is not inherent in the tube, but is determined almost entirely by the available source of high voltage direct current. The plate voltage is usually supplied by one of three means: Battery, d-c. generator, or vacuum tube rectifying system, such as those explained in Chapter VI. Batteries are costly and are seldom used for high-power tubes, while high-voltage d-c. generators are at present inefficient. Vacuum tube rectifying systems have been used successfully and are capable of giving very high voltages but care must be taken to smooth out the rectified current wave, and the extent to which it is smoothed out by means of filters, for example, depends on the resistance of the load in its output. If satisfactory d-c. generators could be made to give from 10,000 to 20,000 volts, it would be possible to get several kilowatts output power from tubes that are relatively inexpensive and simple to make.

In considering the efficiency, we shall neglect the power dissipated in heating the filament.

Let E_v, E_g, I_p be the instantaneous plate and grid potentials and plate current, and E_b, E_c, I_b, the corresponding d-c. values when the alternating components are zero. Let e_p, etc., be the corresponding R.M.S. values, and e'_p the maximum a-c. values. From the curves shown in Fig. 157, it follows that

$$E_p = E_b + e'_p \sin pt, \quad \ldots \ldots \ldots \quad (29)$$

and referring to Fig. 152, we see that

$$I_p = I_b - I = I_b - i' \sin pt. \quad \ldots \ldots \quad (30)$$

The power dissipated at the plate is given by

$$P_p = \frac{1}{2\pi} \int_0^{2\pi} E_p I_p dt \quad \ldots \ldots \ldots \quad (31)$$

or, putting in the values from the above two equations and integrating:

$$P_p = E_b I_b - \frac{e'_p i'}{2} = P_b - P_0. \quad \ldots \ldots \quad (32)$$

The power dissipated at the plate is therefore equal to the power supplied by the plate battery minus the power supplied to the oscillation circuit, and the efficiency is

$$\frac{e'_p i'}{2E_b I_b}. \quad \ldots \ldots \ldots \quad (33)$$

This is never greater than 50 per cent and becomes equal to 50 per cent if the plate current oscillates over the whole range of the characteristic and the maximum value of the alternating plate potential is such as to reduce the plate potential to zero at the moment when the grid has its maximum positive potential. Under these conditions $e'_p = E_b$, and $i' = I_b$.

This expression was derived on the assumption that the values of E_p and I_p are always within the limits of the characteristic. There is, however, another way in which the tube can be operated, which gives higher efficiency. This can be done by so proportioning the plate and grid potentials that the plate current flows only during a small part of the cycle. Taking, for example, the case in which the plate and grid potentials are so adjusted that the operating point does not lie on the characteristic, but is situated beyond the intersection of the characteristic with the axis of the grid potential as indicated at A (Fig. 174), it will be seen that the plate current flows only during a part of the half period during which the a-c. component of the grid potential is positive. During the time that the plate current is zero, the power dissipated at the plate is, of course, zero. When current flows to the plate the potential of the plate decreases on account of the voltage established in the external resistance. If the current could become so large that the potential of the plate is reduced to

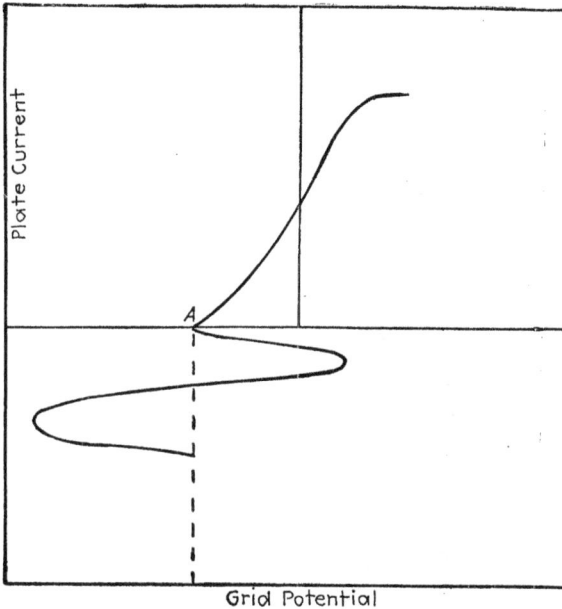

FIG. 174.

practically zero, the power dissipated at the plate is again nearly zero, so that the total power dissipated at the plate becomes very small. In the extreme case in which the current at a given grid potential rises suddenly to such a high value that the plate potential is almost immediately reduced to zero, the total power dissipated at the plate would become zero and then the efficiency would be 100 per cent. This is, of course, a theoretical limit which never obtains in practice. The plate potential would hardly ever drop down to zero because when it drops so low as to become about equal to the simultaneously occurring instantaneous value of the grid potential, the electrons coming from the filament would be diverted to the grid and the plate current would decrease, so that in general the plate potential would not at any time become lower than the grid potential.

The increase in efficiency, when operating the tube somewhat

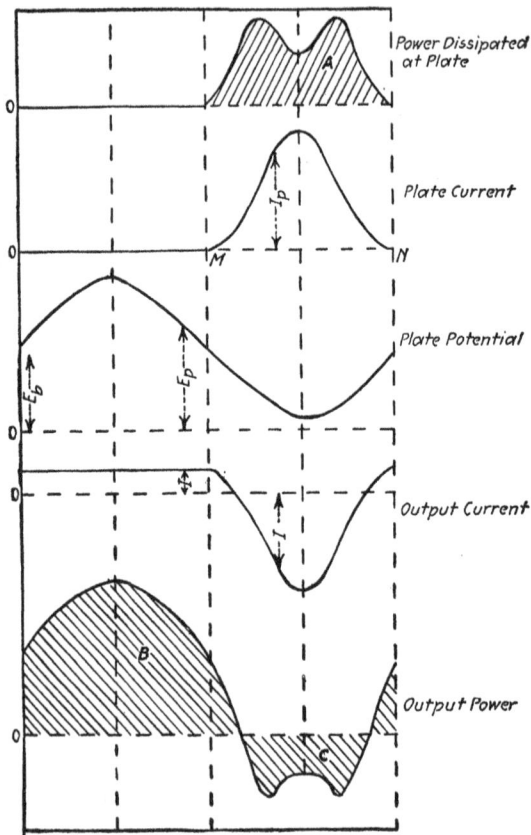

FIG. 175.

in the manner described above, can be explained with reference to the curves in Fig. 175. These curves are drawn for the conditions that current to the plate flows only during part of a cycle, and that the plate potential when the current is a maximum is reduced to a small value. For such an irregular set of curves the simple analysis given above, and which led to equation (33), cannot readily be applied. But the efficiency can still be expressed by the equation

$$\frac{P_0}{P_0+P_p} = \frac{\int E_p I dt}{\int E_p I dt + \int E_p I_p dt}. \quad \cdots \quad (34)$$

In Fig. 175, the horizontal dotted lines represent the axes of zero values. The plate current is represented by the curve just above the plate potential curve. The output current is represented by I, and indicated by the curve below the plate potential curve. This current has a mean value indicated by the dotted line. The plate current I_p flows only during the period indicated by MN. During this period the plate potential drops and the power dissipated at the plate, which is given by $\int E_p I_p dt$, is represented by the area A. The output power, on the other hand, is given by $\int E_p I dt$, and is represented by the difference between the shaded areas B and C. This power, for the conditions chosen, is greater than the power dissipated at the plate, the efficiency for the values chosen here being about 70 per cent.

Fig. 176 shows oscillograms of the currents and voltages taken under such conditions that the plate current remains zero during about half of a complete cycle. For this I am indebted to Mr. J. C. Schelling. The photograph shows two sets of curves taken on the same film. I_0 represents the current in the oscillation circuit and is 90° out of phase with the plate current I_p. The grid current I_g is in phase with the plate current and the grid potential, and these quantities are nearly 180° out of phase with the plate potential E_p.

J. H. Morecroft [1] has computed the efficiency for a number of assumed shapes of the plate current wave. These curves are shown in Fig. 177. Below the figure are indicated the power dissipated at the plate, the output power and the efficiency for each of the assumed shapes of the plate current waves. In the last case, where the current wave is assumed to be rectangular, the efficiency rises to a value of 82 per cent. This represents the best condition as far as output power and efficiency are concerned, but in general the output power decreases as the efficiency increases because this assumed shape of wave is not obtained in

[1] Transactions of A.I.E.E., 1919.

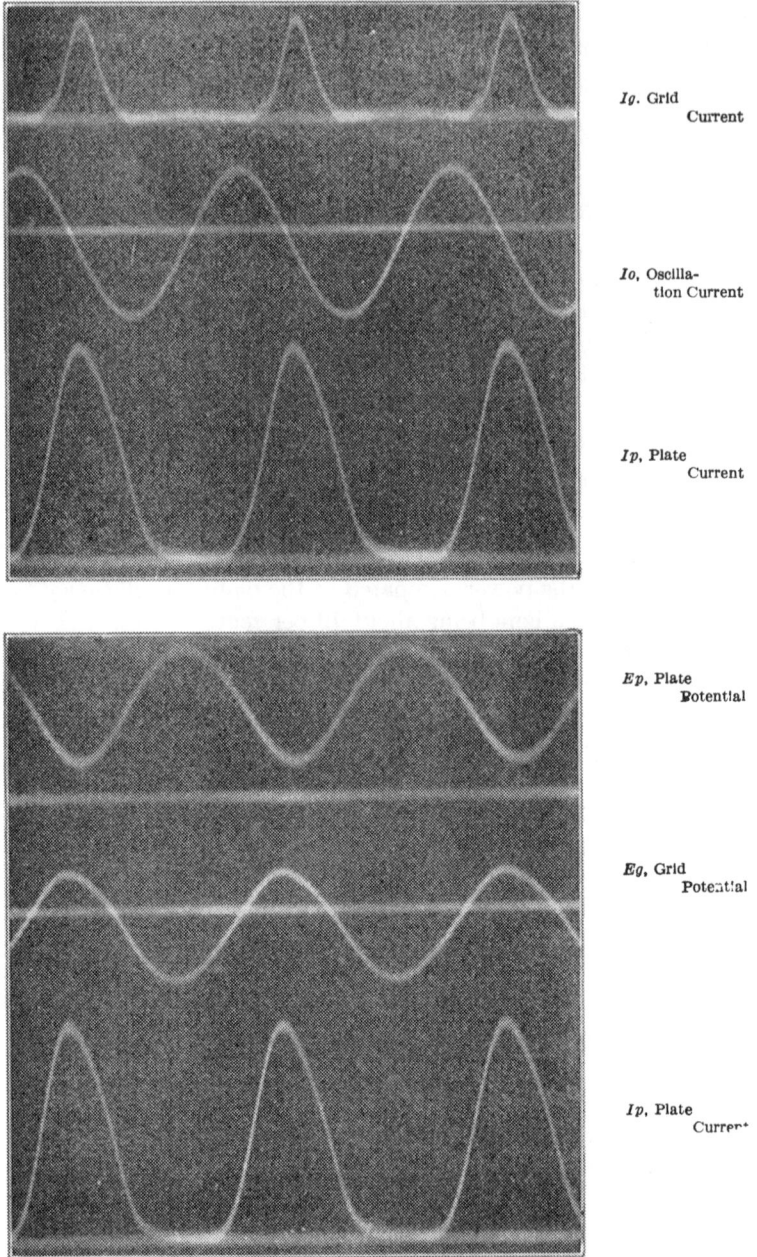

I_g. Grid
Current

I_o, Oscilla-
tion Current

I_p, Plate
Current

E_p, Plate
Potential

E_g, Grid
Potential

I_p, Plate
Current

FIG. 176.

practice. In practice it is therefore necessary to compromise between efficiency and output power.

The importance of high efficiency becomes apparent when considering that the limiting factor in a tube is the amount of power that can be dissipated at the plate. This is limited by the factors discussed in Section 31. If we write η for the efficiency, we see from equation (34) that the output power is given by

$$P_0 = \frac{\eta P_p}{1-\eta}.$$

Since P_p is a fixed quantity for a given tube, the output power could be made very large by making η large. For example, if

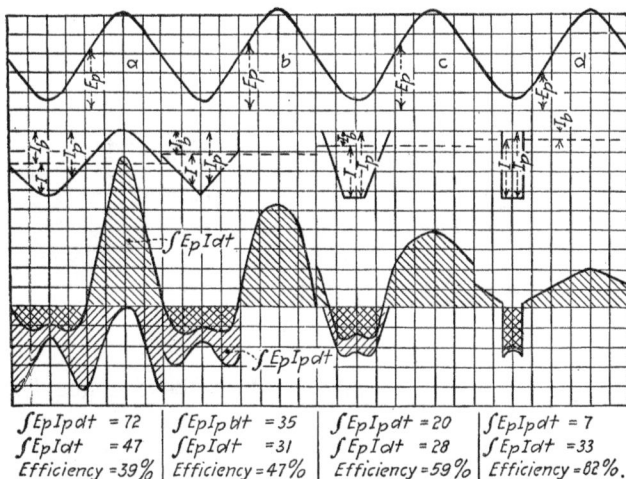

$\int E_p I_p dt = 72$	$\int E_p I_p dt = 35$	$\int E_p I_p dt = 20$	$\int E_p I_p dt = 7$
$\int E_p I dt = 47$	$\int E_p I dt = 31$	$\int E_p I dt = 28$	$\int E_p I dt = 33$
Efficiency = 39%	Efficiency = 47%	Efficiency = 59%	Efficiency = 82%.

Fig. 177.

the efficiency could be as high as 90 per cent, and if the plate is capable of dissipating, say, 500 watts, then the output power would be about 4.5 kw. This would not require a very large tube. The total area of the plates that would be necessary for a dissipation of about 500 watts ranges from about 50 to 100 sq. cm., depending on the material used for the plates. When using a tube in this manner, it is necessary to remember that when the tube stops oscillating, for example, when the oscillation circuit is detuned, the total power supplied by the plate battery will be dissipated at the plate and may cause the liberation of too much gas, or even melt the plates. It is therefore necessary to insure that whenever the oscillations should stop, the plate battery be immediately cut out, or its voltage be sufficiently reduced.

94. Method of Adjusting Coupling between Output and Input. In order to obtain the best operation with a tube, it is necessary to adjust properly the coupling between the output and input circuits. In most circuits this is readily done by making use of any of well-known means of changing the mutual reactance. In some circuits, however, changing the coupling also changes the oscillation frequency. Fig. 178 shows, for example, a Col-

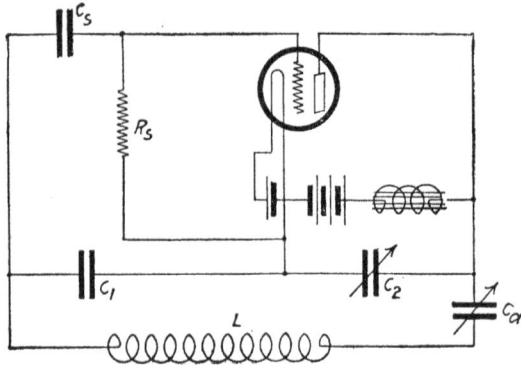

Fig. 178.

pitts circuit; that is, a circuit in which the coupling between input and output is capacitive. The coupling is changed by changing the condenser C_2, but it will be seen that this at the same time changes the frequency of the oscillation circuit LC_1C_2. This is usually taken care of by inserting another condenser C_a, the capacity of which is then so adjusted as to bring the frequency back to its original value. Such a circuit requires two adjustments when it is necessary to change the coupling while keeping the frequency constant.

A circuit which avoids this double adjustment has been described by R. A. Heising.[1] This circuit is shown in Fig. 179. The oscillation circuit is given by LC_1C_2 and the mutual reactance between the output and input is varied by varying the contact Q. This adds an inductive reactance to the capacitive

[1] Loc. cit.

reactance, thereby changing the mutual reactance between the output and input circuits. It will be recognized that if the plate is connected to the point Q', the circuit is the same as that shown in Fig. 178, with the capacity C_a omitted. This means of adjusting the coupling does not appreciably change the oscillation frequency.

95. Influence of the Operating Parameters on the Behavior of the Oscillator. It will be realized that there are a large number of factors that determine the operation of a vacuum tube oscillator. The most important of these factors are the filament current, d-c. plate and grid potentials, plate and grid coupling and oscillation circuit resistance. When it is a mere matter of obtaining an

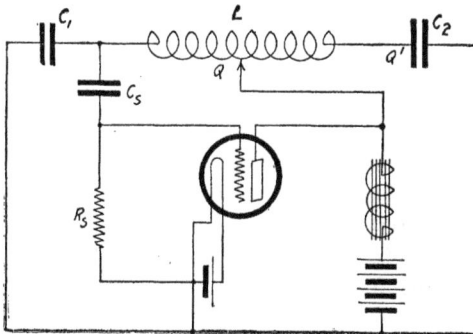

Fig. 179.

alternating current by means of the vacuum tube, very few adjustments will serve the purpose. It will usually be found that the tube starts oscillating immediately on closing the plate and filament circuits. If it fails to oscillate a slight increase in filament current or plate voltage, or both, will set the tube oscillating. If, on the other hand, it is desired to obtain maximum power output at the maximum efficiency consistent with it, the adjustments have to be made carefully, but with a little practice the whole operation reduces to a simple one. Some of the operating parameters are fixed by the limits of the tube and circuit. For example, the tube may be designed to operate on a certain range of filament current and plate battery voltage. This automatically fixes two parameters. The manner in which the behavior of the oscillator is influenced by these various parameters can be explained with reference to the following diagrams. These represent in a general way, what can be expected with commonly used types of tubes. The nature of these curves could be expected to vary somewhat with different types of tubes.[1] One of the most important variables is the filament current. The influence of the filament current on the operation of the tube can be understood by referring to Figs. 17 and 18, that were discussed in the beginning of Chapter IV. Fig. 17 gives the relation between the output current and the plate or anode voltage. When using the tube as oscillator, we operate over the sloping part OA of the characteristic. The three sets of curves shown are for different values of the filament current. Fig. 18, on the other hand, gives the relation between the anode current and the temperature of the filament or the filament current. But the sloping part of this characteristic represents a temperature of the filament which is so low that the plate potential is sufficiently high to draw all the electrons away to the plate as fast as they are emitted from the filament. The condition which may be characterized as temperature saturation is represented by the horizontal portion CD of the curve, and obtains when the number of electrons drawn to the plate is less than the total number emitted. The part CD of Fig. 18, corresponds to the sloping part of OA of Fig. 17; hence, for a given d-c. plate potential it is necessary that the temperature of the filament be so high that we operate on the horizontal part of the plate current, filament current characteristic. If this is not the case, the variation in output current with the variation in the grid potential is too small to produce oscillations. The dependence of the oscillation current and the plate current upon the filament current is indicated in Fig. 180. If the filament current is below a certain value given by A, the tube does not produce sustained oscillations. Filament currents below this value correspond to the saturation part of the curve giving the plate current as a function of the plate potential. If the filament current is raised beyond the value indicated by A, the tube starts oscillating and the oscillation current increases until, when temperature saturation is obtained, it shows no further increase with increase in filament current. In order to secure the best operation, therefore, the filament current should not be less than the value indicated by B. On the

[1] A variety of experimental curves have been obtained by Heising with a standard VT-2 type of tube and published in the Journal of the A.I.E.E., May, 1920.

other hand, the filament current should not be increased much beyond this value because that would shorten the life of the tube.

If the tube is operated with a resistance R_s in the grid circuit, as indicated in Fig. 169, for example, the value of the oscillation current obtained depends on this resistance, in the manner shown in Fig. 181, where the lowest curve represents the highest leak resistance R_s in the grid circuit. The oscillation current is less for the higher resistance, but the horizontal part of the curve

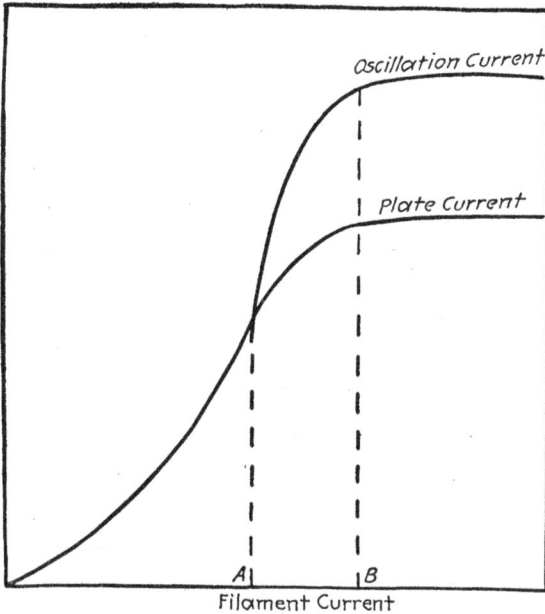

Fig. 180.

starts at a lower filament current. On the other hand, if the oscillation current be plotted as a function of the filament current for various values of the plate potential, a set of curves is obtained similar to that shown in Fig. 181, except that the lowest curve would represent the case for the lowest plate potential, so that although the output can be increased by increasing the plate battery voltage, the horizontal part of the curve is reached at a higher filament current. The filament current at which the bend in the curve occurs can be taken to represent the safe temperature of the filament. It will be seen then that the safe temperature increases with increase in plate potential and decreases with increase in the grid leak resistance. By making use of these two variables, plate potential and grid leak resistance, a compromise can be effected to give the best output for the longest life of the filament.

The relation between the oscillation current and plate potential is shown in Fig. 182.[1] The tube starts oscillating at a plate potential depending on the adjustments of the circuit. If the plate voltage be raised, the oscillation current increases almost

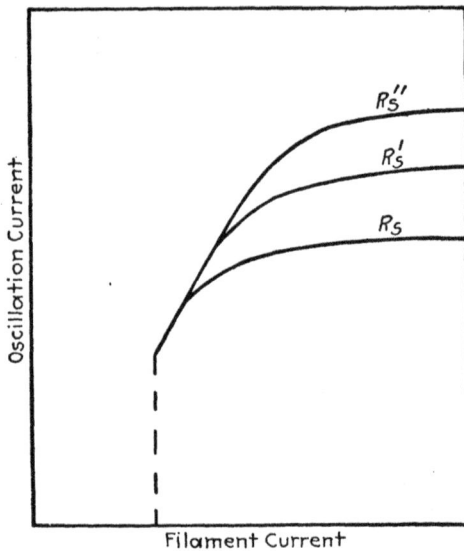

Fig. 181.

linearly with it. As the grid leak resistance is increased, the slope of this line becomes less and the oscillation current for given plate potential becomes less. The value of leak resistance R_s that gives satisfactory operation usually lies in the neighborhood of 5000 to 10,000 ohms.

When a grid battery is used to maintain the grid at an appropriate negative potential, the tube behaves differently from the manner explained above, where the negative grid potential was maintained by means of the grid leak resistance R_s. For example, with a battery in the grid circuit the oscillations will usually not

[1] R. A. HEISING, loc. cit.

start until the plate voltage is raised to a higher value than that necessary when using the grid leak resistance instead of a battery.

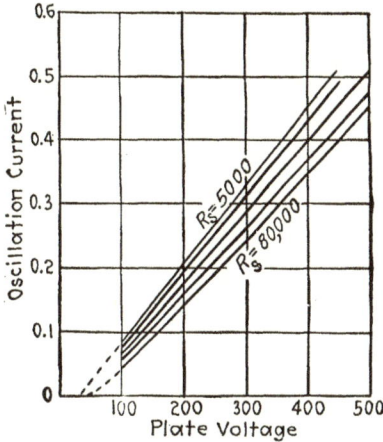

Fig. 182

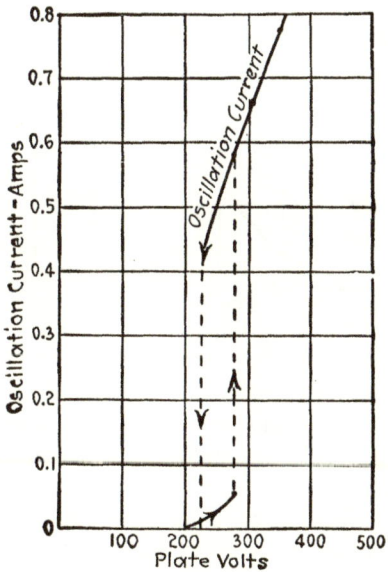

Fig. 183.

If the plate voltage be further increased, the oscillation current increases almost linearly as indicated in Fig. 183. But if the plate voltage be again reduced, the oscillations will persist until this voltage reaches a value which is quite appreciably lower than that necessary to start the oscillations.

The output power as a function of the plate battery voltage can be represented by a curve like that shown in Fig. 184, which shows a rather rapid increase as the plate voltage is raised. To obtain increase in output power by increasing the plate voltage, it is, of course, necessary to insure that the filament current is

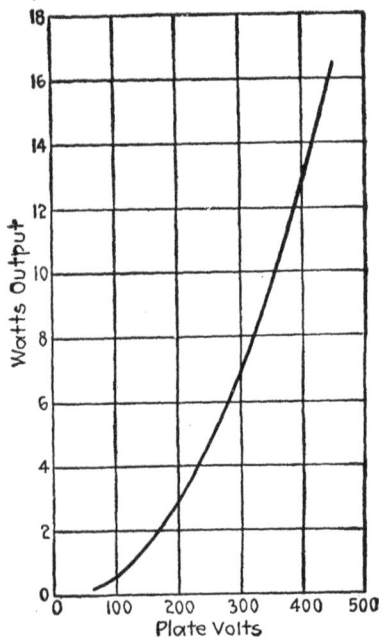

Fig. 184.

high enough to prevent the space current from becoming saturated. Also, the power delivered to the oscillation circuit depends on the resistance of this circuit and the resistance of the tube. The latter depends on the d-c. plate potential so that in general an increase in the plate battery voltage would necessitate a readjustment of the capacity and inductance in the oscillation circuit to give the maximum output power.

96. Range of Frequency Obtainable with the Vacuum Tube Oscillator. Circuits for Extreme Frequencies. The vacuum tube has been used to give oscillations having a frequency ranging from a fraction of a cycle per second to many millions of cycles per second. For low frequencies, the frequency is determined almost entirely by the inductance and capacity in the oscillation circuit, and the only limitation to this end of the scale is the size of the inductances and capacities. For very high frequencies, the frequency of the oscillation is determined mainly by the electrostatic capacity between the electrodes of the tube and by the inductances and capacities of the wires connecting the electrodes. The upper limit to the frequency obtainable depends mainly on the inter-electrode capacities.

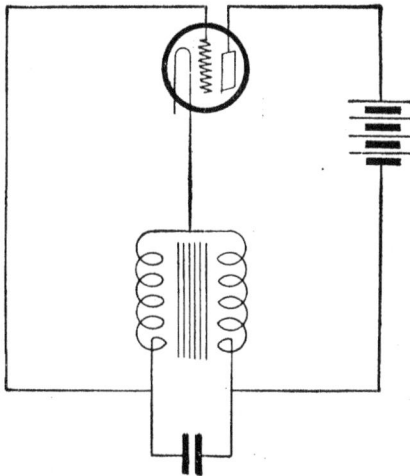

FIG. 185

When very low frequencies are desired, it is best to use a Hartley circuit, in which the two coils L_1 and L_2 of Fig. 159 take a form of an iron core transformer such as is shown in Fig. 185. By means of such a circuit it has been possible to obtain frequencies as small as a fraction of a cycle per second.

When it is desired to obtain exceptionally high frequencies, the inductances in the oscillation circuit can be reduced to the utmost extent, until they take the form of short straight wires connecting the electrodes. The capacity between grid and plate forms the capacity of the oscillation circuit. A circuit which has been used, for example, by W. C. White,[1] to obtain a fre-

[1] General Electric Review, Vol. 19, 771, 1916.

quency of fifty million cycles per second is shown in Fig. 186. The grid inductance is furnished by the connecting wire GAF and the plate inductance by the connecting wire PDB. The plate current is supplied by the battery E_b through the choke coil L_1. C_1 represents a by-pass for the high frequency and is so large that it does not affect the oscillation frequency. W represents a long pair of parallel wires connected to the system through the small capacities C_2 and C_3. By suitably adjusting the bridge H, stand-

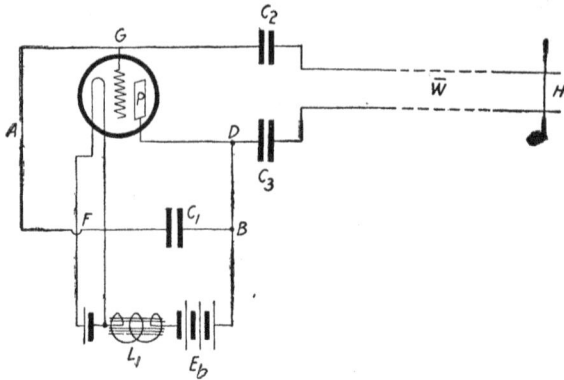

Fig. 186.

ing waves can be obtained. In White's experiments these waves were about 6 meters long. This circuit represents a very simple means of demonstrating standing waves. The vacuum tube is much superior to the induction coil frequently used in laboratories for this demonstration experiment. By using tubes that are specially designed to have low electrostatic capacities between its electrodes, it is possible to obtain waves of a few feet in length.

CHAPTER IX

MODULATION AND DETECTION OF CURRENTS WITH THE VACUUM TUBE

97. Elementary Theory of Modulation and Detection. In the applications of the vacuum tube considered so far, it is desirable that the characteristic of the plate circuit be as straight as possible. For example, when using the device as an amplifier, it was explained in Chapter VII that the external impedance in the plate circuit is usually made so large that the current voltage characteristic of the plate circuit is sufficiently straightened out to enable us to neglect quantities of higher order than the first. When using the tube as an oscillation generator, it is also desirable to have a linear characteristic because the curvature introduces harmonics which result in a waste of power.

In the following we shall consider those applications of the vacuum tube which depend directly on the curvature of the characteristic. The two most important of these applications are the use of the tube to modulate high frequency currents for purposes of signaling and the detection of high frequency currents.

When considering the second order quantities that enter into the characteristic of the device, it is generally not possible to express the characteristic by a simple equation, but we can still apply the equation derived in Section 22, Chapter III, which holds generally for three electrode devices. Neglecting the small quantity ϵ we can write this equation in the form

$$I_p = f\left(\frac{E_p}{\mu} + E_q\right). \quad \ldots \ldots \ldots \quad (1)$$

In general the function f is not linear and, therefore, if a sinusoidal voltage be impressed upon the input of the tube, the output wave will be distorted in the manner explained in Section 57. For such a condition we can express the varying current in the output

315

as a function of the sinusoidal voltage impressed upon the input
by a simple power series,

$$J = a_1e + a_2e^2 + \ldots, \quad \ldots \ldots \quad (2)$$

where J represents the varying current and will in general have
the form of a lopsided wave, and will, therefore, comprise currents
of different frequencies and a direct-current component. This
series, it has been found, usually converges so rapidly that we can
neglect all quantities of higher order than the second. Experimental proof of this will be given later on. The first term of
equation (2) represents a current having the same frequency
as that of the input voltage e. The second term is the one which
gives rise to modulation and detection effects.

To evaluate the coefficients a_1 and a_2, we can proceed in the
manner given by J. R. Carson.[1] Carson has considered two
cases, namely, when the output circuit contains a pure non-
inductive resistance and secondly when the output circuit contains
a general impedance. In order to derive an expression for the
coefficient a_2 in terms of the parameters of the tube, we shall
discuss only the first case, namely, in which the plate circuit
contains only a pure resistance.

The quantities to be considered in the circuit can be expressed
as follows:

$$\left. \begin{array}{l} I_p = I_b + J \\ E_p = E_b + v \\ E_g = E_c + e \end{array} \right\}, \quad \ldots \ldots \ldots \quad (3)$$

where I_b, E_b and E_c represent the d-c. values of plate current and
potential, and grid potential, and I_p, E_p and E_g are the quantities
that obtain as a result of the variations J, v and e superimposed on
the d-c. values. Substitution of equation (3) into (2) gives:

$$J = P_1(\mu e + v) + P_2(\mu e + v)^2 + \ldots, \quad \ldots \quad (4)$$

where P_1, P_2, etc., are given by

$$P_n = \frac{1}{n!}\left(\frac{\partial^n I_p}{\partial E^n{}_p}\right)_{E_p = E_b}. \quad \ldots \ldots \quad (5)$$

[1] J. R. Carson, Proceedings I.R.E., Vol. 7, p. 187, 1919.

The physical significance of the differential parameters P_n become apparent when they are evaluated with the help of the characteristic equation (1). Thus,

$$
\left.
\begin{aligned}
P_1 &= \frac{\partial I_p}{\partial E_b} = \frac{1}{r_p} \\
P_2 &= \frac{1}{2}\frac{\partial^2 I_p}{\partial E_p{}^2} = -\frac{1}{2}\frac{r'_p}{r_p{}^2}
\end{aligned}
\right\} , \quad \ldots \ldots \quad (6)
$$

where r_p is the plate resistance of the tube and r'_p is the variation in the plate resistance due to the curvature. These equations now enable us to evaluate the coefficients a_1 and a_2 of equation (2). To do this it should be noted that the variation v in plate voltage is equal and opposite to the voltage drop established in the external resistance r_0 due to the varying current J in the output. Hence, substituting $v = -r_0 J$ into equation (4), we get:

$$
J = P_1(\mu e - r_0 J) + P_2(\mu e - r_0 J)^2 + \ldots , \quad \ldots \quad (7)
$$

This equation now gives the varying current J in terms of the input voltage e and the parameters of the tube and circuit. To express J as an explicit function of the input, we can substitute the series for J given by equation (2) into equation (7) and equate coefficients of like powers of e. When this is done the expression for the varying current J becomes,

$$
J = \frac{\mu e}{r_p + r_0} - \frac{1}{2}\frac{\mu^2 r'_p r_p e^2}{(r_p + r_0)^3} + \ldots , \quad \ldots \quad (8)
$$

If the characteristic is linear, the plate resistance r_p is constant and, therefore, its derivative r'_p is zero. This makes the second term of the above equation zero. Hence, replacing the varying values J and e by the R.M.S. values i_p and e_g, equation (8) reduces to equation (22) given in Chapter VII.

The second term of equation (8) represents the property of the tube that enables it to act as a modulator and detector. The value of the coefficient given by the second term in equation (8) will be helpful in the interpretation of the equations that follow. For the present we shall use equation (8) in the simple form

$$
J = a_1 e + a_2 e^2, \quad \ldots \ldots \ldots \quad (9)
$$

to explain how the second term is instrumental in producing modulation and detection.

98. Modulation. Suppose that a tube be inserted in a circuit such as that shown in Fig. 187. Let high frequency currents be impressed at H. F. and low frequency currents, lying within the audible range, at L. F. The total input voltage on the tube is then,

$$e = e_1 \sin pt + e_2 \sin qt, \quad \cdots \quad (10)$$

where $\dfrac{p}{2\pi}$ and $\dfrac{q}{2\pi}$ represent the high and the low frequencies respectively. In order to obtain the output current, we have to substitute this expression for e in equation (9). When using the

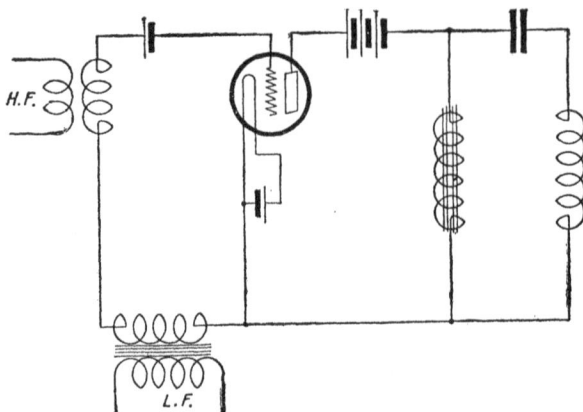

Fig. 187.

tube as a modulator, we are interested only in currents having frequencies lying within the range $\dfrac{p+q}{2\pi}$. Hence, substituting (10) into (9), evaluating and dropping all terms having frequencies lying outside of this range, we obtain,

$$J = a_1 e_1 \sin pt + 2a_2 e_1 e_2 \sin pt \sin qt. \quad \cdots \quad (11)$$

This expression represents a wave of varying amplitude as shown in Fig. 188, the amplitude of the high frequency carrier[1] wave

[1] The word "carrier" is here used as a general term to indicate the high frequency wave, which is modulated by the signaling wave. It has also a more specific meaning in which it refers to the transmission of high frequency currents over wires.

varying in accordance with the audio frequency wave impressed on the input of the tube.

We can, for purposes of explanation, write equation (11) in the form,

$$J = A \sin pt(1 + B \sin qt). \quad \ldots \quad (12)$$

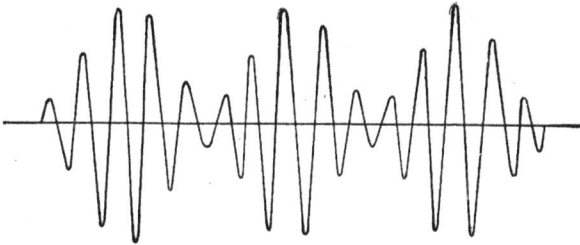

FIG. 188.

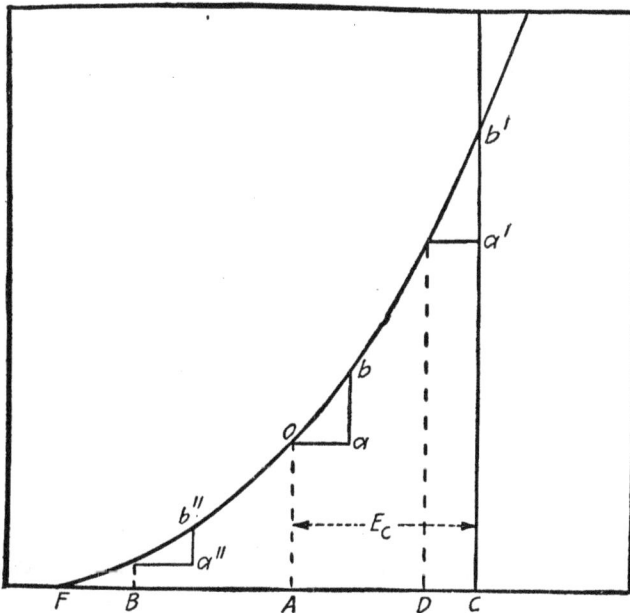

FIG. 189.

The way in which a wave of the type shown in Fig. 188 is produced by the vacuum tube, becomes apparent when we consider the characteristic. For example, Fig. 189 shows the plate current grid potential characteristic. Suppose a constant potential

E_c be applied to the grid so that the normal plate current is represented by the ordinate AO. Now let a high frequency voltage of amplitude Oa be superimposed on this constant grid potential. The output current wave will then have the amplitude given by ab. If the grid potential be increased to the value BC and a high frequency voltage of the same amplitude as before be impressed on the input, the output current wave will have an amplitude $a''b''$, and this is smaller than before. If, on the other hand, the grid potential be reduced to the value DC, the amplitude of the

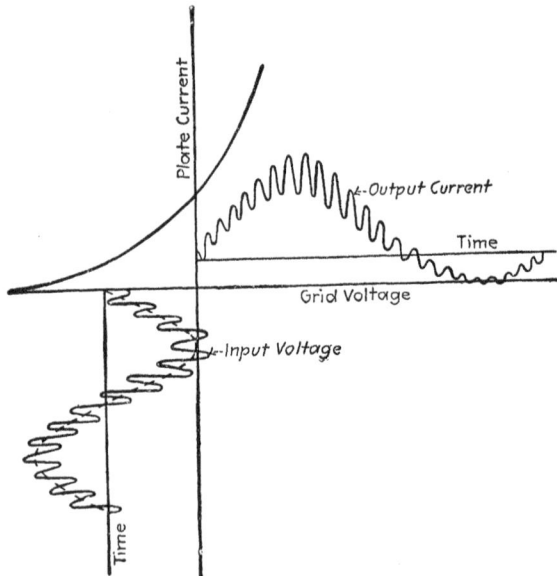

FIG. 190.

output current wave for the same amplitude of input becomes greater and is represented by $a'b'$. If now, we impress on the input not only a high frequency voltage of constant amplitude Oa, but also at the same time a low frequency having an amplitude equal to say AB, then the amplitude of the output current wave will alternately increase and decrease at a frequency equal to that of the low frequency wave impressed on the input and the result is an output current wave of the shape shown in Fig. 188. Fig. 190 represents the input and output waves. The input wave is a high frequency of constant amplitude superimposed on a low frequency, while the output wave is a high frequency of varying amplitude superimposed on a low frequency. If the output circuit (Fig. 187) be tuned to the high frequency, the low frequency current variations are filtered out, thus resulting in the wave shown in Fig. 188.

If the low frequency voltage impressed on the input circuit has such a value that the maximum negative potential of the grid becomes equal to CF (Fig. 189), the current is reduced to zero and the modulated output wave then takes the form shown in Fig. 191. The wave can then be said to be completely modulated. When this happens the coefficient B in equation (12) is unity and the maximum amplitude of the high frequency wave when the grid has its minimum negative potential is then $2A$. In some measurements it is very important to insure that the wave is completely

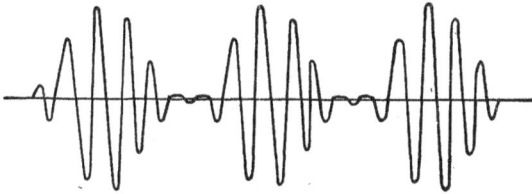

Fig. 191.

modulated as will become evident later on when we come to consider measurements on the detecting efficiency of tubes.

The second term in equation (11) gives a measure of the extent to which a wave is modulated. The coefficient a_2 is given by equation (8), namely,

$$a_2 = -\frac{1}{2}\frac{\mu^2 r_p r_p'}{(r_p + r_0)^3}. \quad \cdots \cdots \cdots (13)$$

The amplitude of the modulated wave is, therefore, proportional to

$$e_1 e_2 \cdot \frac{\mu^2 r_p r_p'}{(r_p + r_0)^3}. \quad \cdots \cdots \cdots (14)$$

It is, therefore, proportional to the product of the amplitudes of the audio and the radio input voltages and to the curvature r_p' of the characteristic. The modulated output power is also proportional to $\dfrac{r_0}{(r_0 + r_n)^6}$, and this is a maximum when r_0 is equal to

$\frac{1}{5}r_p$, a result which was stated by Carson.[1] If we put $r_0 = nr_p$ expression (14) may be written:

$$\frac{e_1 e_2 r'_p}{(1+n)^3} \cdot \left(\frac{\mu}{r_p}\right)^2, \quad \cdot \quad \cdot \quad \cdot \quad \cdot \quad \cdot \quad (15)$$

which shows that the value of the device as a modulator depends on the ratio of μ to r_p. This quantity which is the mutual conductance of the tube has also been found to be a measure of the figure of merit of the tube as amplifier and as oscillation generator.

99. Modulation Systems. The results derived above can be interpreted by stating that a device will operate as a modulator if it has a varying resistance characteristic; the resistance to the radio frequency currents is varied in accordance with audio frequency currents. There are, therefore, two main systems whereby modulated currents can be transmitted over a line or from an antenna. The first is exemplified in Fig. 187. Radio frequency and audio frequency voltages are impressed on the grid and the resulting modulated current in the output of the tube is transmitted over a line or antenna of constant impedance. The antenna must then be tuned to a frequency range $\frac{p \pm q}{2\pi}$, where $\frac{p}{2\pi}$ is the carrier or radio frequency and $\frac{q}{2\pi}$ the audio frequency.

In telephony $\frac{q}{2\pi}$ covers a range of from about 100 to 2000 or 3000 cycles per second. The antenna must, therefore, be tuned so that it has approximately the same impedance for frequencies covering a range of about 2000 cycles. This is also a condition for ordinary wire telephony which requires that the telephone line should be capable of transmitting this whole range of frequencies with about equal facility. The only difference is that in ordinary wire telephony the frequencies cover a range up to about 2000, whereas in carrier or radio telephony the frequencies cover the same range but their actual values are in the neighborhood of the carrier frequency.

The other system consists in impressing a high frequency directly on the antenna or line and then varying the resistance of the antenna in accordance with audio frequency. Such a system is shown schematically in Fig. 192, which shows the modu-

[1] Loc. cit.

lator M in shunt with the antenna inductance, that is, the antenna inductance is shunted by the plate resistance of the modulator tube. The amount of current in the antenna, which is supplied by the high frequency oscillation generator, will, therefore, depend on the resistance of the tube M. This resistance is varied in accordance with the speech voltages impressed on its input in the manner explained with reference to Fig. 189.

A modification given by R. A. Heising [1] is shown in Fig. 193. The oscillation circuit shown here is the same as that given in Fig. 179, the capacity of the antenna forming the capacity C_2 of Fig.

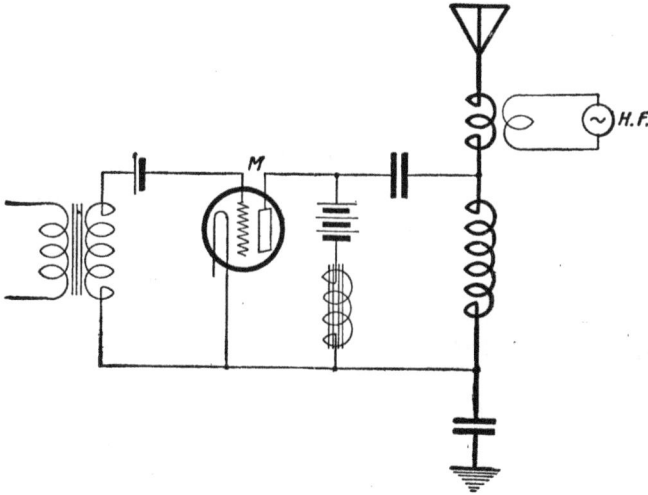

Fig. 192.

179. The oscillator and modulator are both supplied by a battery through a low frequency choke coil which insures that they are both supplied with constant direct current. The speech or audio frequency voltage is impressed on the grid of the modulator by means of the transmitter through the transformer as indicated. Between the plates of the modulator and oscillator is a high frequency choke. If the telephone transmitter is not actuated, the oscillator tube supplies high frequency currents of constant amplitude to the antenna. If now an audio frequency voltage be impressed on the grid of the modulator, audio frequency currents are established in the output circuit of this tube and consequently

[1] See CRAFT and COLPITTS, Proceedings, A.I.E.E., Vol. 38, p. 360, 1919.

the potential of the plate of the oscillator varies in accordance with the low frequency, thus producing low frequency variations in the amplitude of the high frequency oscillations obtained from the oscillator and impressed on the antenna. The coupling is adjusted by sliding the contact Q as explained in connection with Fig. 179 in Chapter VIII.

A number of modulating and transmitting circuits have been suggested. The circuit shown in Fig. 194 [1] is another illustration of the application of the principles given in the foregoing. This circuit is so arranged that the high frequency is impressed at H. F. in such a way that the grids of both tubes are in phase. The high

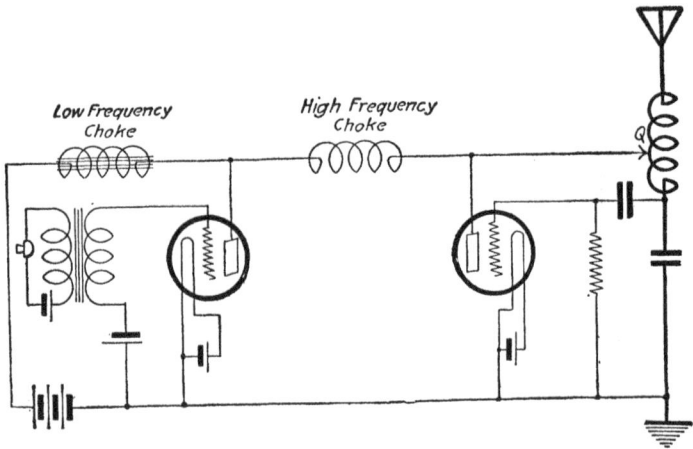

Fig. 193.

frequency currents in the output coils, therefore, flow in opposite directions and the output in the secondary of the transformer T_0 is zero. But if the audio frequency voltage is impressed as indicated at L. F., the grid of the one tube becomes positive when the other becomes negative so that the resistance of the one tube is reduced while that of the other is increased. This causes an increase in the amplitude of the high frequency currents flowing through the one tube and a decrease in the amplitude of those flowing through the other tube. In this way, therefore, energy is radiated only during the time that the tube is actuated by the

[1] British Patent 130219, 1918.

speech voltage. What is transmitted then is only the waves
given by the second term of equation (11).

100. Detection. The mechanism of detection is identical
with that of modulation and is due to the same cause, namely,
the curvature of the characteristic. In general, therefore, the
equations derived above are applicable also to the problem of
radio detection with the thermionic tube. The only difference
is that in this case we are concerned with a different range of
frequencies. While in the case of the modulator, the output is
tuned to radio frequencies, in the detector the output is tuned
to audio frequencies because the problem of detection involves

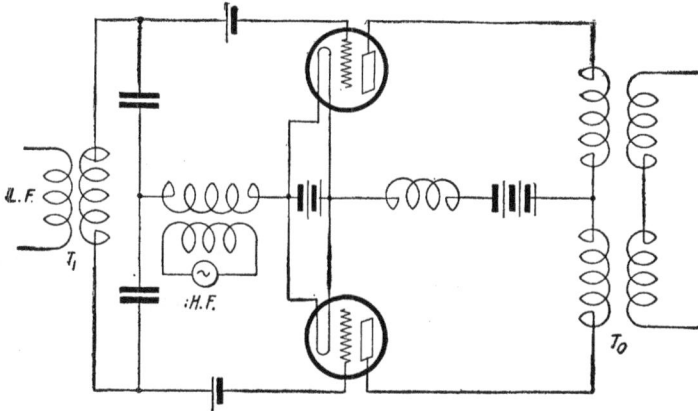

FIG. 194.

transforming high frequency into low frequency currents so that
they can become audible. We can, therefore, use equation (9)
to determine the low frequency output of a detector. In this
case, however, we are not concerned with the first term at all.
For example, if a radio frequency $e \sin pt$ be impressed upon the
input of a detector, the output current is given by:

$$J = a_1 e \sin pt + \frac{a_2 e^2}{2} - \frac{a_2 e^2}{2} \cos 2\,pt. \quad . \quad . \quad . \quad (16)$$

The first term is simply an inaudible high frequency. It need,
therefore, not be considered and we can write instead of equation
(9) the equation for the instantaneous detecting current I_d as

$$I_d = ae^2 \sin^2 pt, \quad . \quad . \quad . \quad . \quad . \quad . \quad (17)$$

where a is written for a_2. We shall refer to a as the *detection coefficient*. Its value in terms of the parameters of the circuit is given by equation (13).

Equation (16) contains only high frequency components and a d-c. component. The d-c. component of equation (16) makes possible the detection of high frequency incoming currents impressed on the input of the detector, if the output of the detector contains a d-c. current measuring instrument which is sensitive enough to indicate a change in the plate current given by the second term of equation (16). When a telephone receiver is used in the output continuous incoming waves of constant amplitude cannot be detected, because equation (16) does not contain an audio frequency term. The incoming waves must either be modulated high frequency waves or if they are continuous waves of constant amplitude, the heterodyne method must be used to detect them (see Section 109). If a modulated high frequency wave such as that given by equation (12) be impressed on the detector, the instantaneous value of the detecting current is given by

$$I_d = a[A \sin pt \ (1 + B \sin qt)]^2. \quad . \quad . \quad . \quad (18)$$

In evaluating this expression, all terms containing frequencies that lie outside the audible range can be neglected. This gives:

$$I_d = aA^2 B \sin qt - \frac{aA^2B^2}{4} \cos 2 \ qt. \quad . \quad . \quad . \quad (19)$$

Now q in equation (12) represents the low frequency component of the modulated wave. It is seen, therefore, that in view of the curvature of the characteristic of the tube the output current contains a term of the same frequency as the audio frequency with which the carrier wave was modulated. It also contains a term having twice the audio frequency. This term is, however, usually so small as not to cause any appreciable distortion of the wave in the output of the detector.

In deriving these expressions it is assumed that the grid does not take appreciable current. The circuit in which the detector can be used to comply with the above equations is shown in Fig. 195. The input circuit LC is tuned to the frequency of the incoming oscillations and the grid is kept negative with respect to the filament by means of the battery E_g. The condenser C_1 serves as a by-pass to the high frequency currents in the output circuit, the audio frequency component of the output passing through the telephone receiver.

A receiving circuit that is commonly used and in which the battery E_g is replaced by a condenser, will be discussed later on. (Section 103.)

It will be apparent that the reason why equation (19) contains a term having the same frequency $\frac{q}{2\pi}$ that is used to modulate the wave at the transmitting station, is because the incoming wave, which is given by equation (12), contains both the terms $A \sin pt$ and $AB \sin pt \sin qt$. If the incoming wave were of the

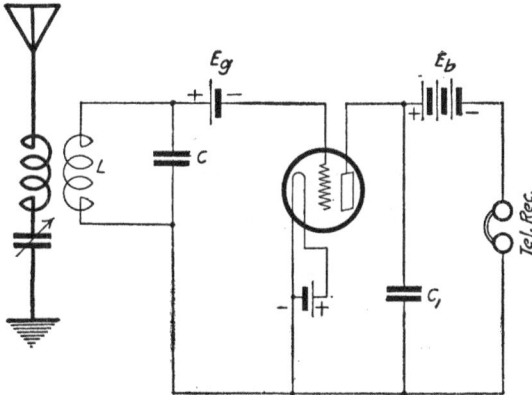

Fig. 195.

form $C \sin pt \sin qt$, simple trigonometry will show that the only audio component of the current in the output of the detector is one which has double the modulating frequency, the audio detecting current being given by:

$$I_d = C^2 \sin^2 pt \sin^2 qt, \quad \ldots \ldots (20)$$

which on evaluating and dropping inaudible terms becomes:

$$I_d = C^2 \cos 2qt. \quad \ldots \ldots \ldots (21)$$

It follows therefore that in order to obtain the modulating frequency $\frac{q}{2\pi}$, the waves impressed on the input of the detector

must be made to include a wave of the desired strength having
the frequency $\frac{p}{2\pi}$.

**101. Root Mean Square Values of Detecting and Modulated
Currents.** The above equations give the instantaneous values
of the currents or voltages considered. The R.M.S. values can
readily be obtained. Thus, the R.M.S. value i_d of the detecting
current, the instantaneous value of which is given by equation
(19) is:

$$i_d = a\sqrt{\frac{A^4 B^2}{2}\left(1+\frac{B^2}{16}\right)}. \quad \cdot \quad \cdot \quad \cdot \quad \cdot \quad \cdot \quad (22)$$

If we neglect the small double frequency quantity given by the
second term in the parenthesis, i_d reduces to the common form

$$i_d = \frac{aA^2 B}{\sqrt{2}}. \quad \cdot \quad \cdot \quad \cdot \quad \cdot \quad \cdot \quad \cdot \quad \cdot \quad \cdot \quad (23)$$

The R.M.S. value of the modulated input voltage as given by
equation (12) can be obtained by putting $p = nq$, since p is large
compared with q. ($\frac{q}{2\pi}$ covers frequencies ranging to 2000 or
3000 cycles per second, while $\frac{p}{2\pi}$ is generally of the order of several hundred thousand cycles per second). The R.M.S. of the
modulated wave which can be taken as the effective input voltage e_g on the grid of the detector, then becomes:

$$e_g = \sqrt{\frac{A^2}{2}\left(1+\frac{B^2}{2}\right)}, \quad \cdot \quad \cdot \quad \cdot \quad \cdot \quad \cdot \quad (24)$$

and involves B which is a measure of the extent to which the
wave is modulated. If the wave is completely modulated $B = 1$,
as was explained in Section 98. In this case, remembering that
the peak value of the high frequency is $2A$ we find that the ratio
of the R.M.S. to the peak value is $\frac{4}{\sqrt{3}}$ instead of $\sqrt{2}$ as in un-
damped waves.

**102. Relation between Detection Coefficient and the Operating
Plate and Grid Voltages.** The detection coefficient a depends
on the values of the d-c. plate and grid voltages so that in determining the value of a tube as a detector, this relationship must be

taken into account. If the detecting current i_d be measured as a function of the effective voltage $E_\gamma = \left(\dfrac{E_p}{\mu} + E_g + \epsilon\right)$ it will be found that as this voltage is increased by increasing either E_p or E_g, the detecting current at first increases, reaches a maximum, and then decreases. It is assumed that the grid is at all times negative with respect to the negative end of the filament. Now the detection coefficient a is given by the second derivative of the characteristic, and is a measure of the detecting current, that is, the audio frequency component in the output. The maximum of

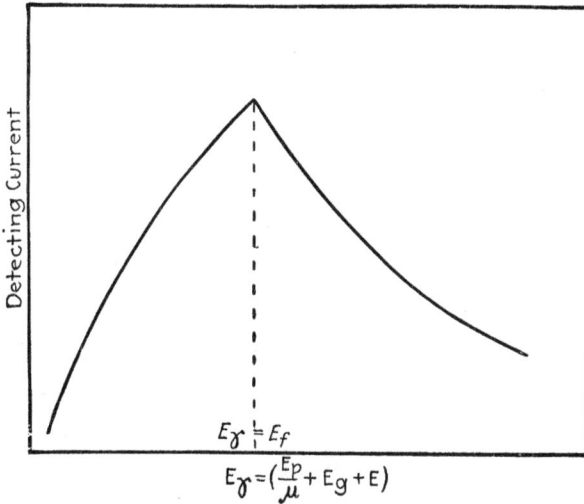

FIG. 196.

detecting current such as shown in Fig. 196 is due to the potential drop in the filament due to the heating current. It can be accounted for if we take regard of the voltage drop in the filament in giving an expression for the current as a function of the plate or grid voltage. It was shown in Section 28 that if this be considered, the characteristic of the tube can be expressed by means of two equations, one which holds for values of the applied plate potential less than the potential drop in the filament and the other for larger values of the plate potential. These two equations are given as equations (17) and (19) of Chapter IV. They were derived for the case of a simple valve containing only anode and cathode. But we can, to a first approximation, apply the considerations given there to the three electrode device if we replace the plate potential by the expression $E_\gamma = \left(\dfrac{E_p}{\mu} + E_g + \epsilon\right)$ so that we can write the characteristic equations in the form

$$I_p = KE_\gamma^{5/2} \text{ for } E_\gamma \leqq E_f, \quad \ldots \quad \ldots \quad \ldots \quad (25)$$

$$I_p = K[E_\gamma^{5/2} - (E_\gamma - E_f)^{5/2}] \text{ for } E_\gamma \gtrless E_f. \quad \ldots \quad (26)$$

where E_f is the voltage drop in the filament.

These two equations can be represented by a continuous curve closely approximating a parabola. The detecting current, or the second derivative of equations (25) and (26) when plotted as a function of the effective voltage on the other hand, shows a distinct maximum, which occurs at a value of the effective voltage E_γ equal to the voltage drop in the filament. The simple rule, therefore, to obtain the best results when using the tube in the circuit shown in Fig. 195 is to make

$$\frac{E_p}{\mu} + E_g + \epsilon = E_f. \quad \ldots \quad \ldots \quad \ldots \quad (27)$$

Fig. 197 shows an experimental curve in which the detecting current is plotted as a function of the plate potential E_p, the grid potential remaining constant. For $\mu = 12$, $\epsilon = -0.5$, $E_f = 2.5$ volts and $E_g = 0$, the maximum according to equation (27) occurs at a plate potential of about 36 volts.

The condition given by equation (27) states that the potential difference between a plane coincident with that of the grid and the positive end of the filament is zero. This condition holds generally even when the grid is connected to the positive end of the filament instead of to the negative. If it is connected to the positive end the condition for maximum detecting current is $E_\gamma = 0$. This has also been verified experimentally. The condition $E_\gamma = 0$ when the grid is connected to the positive end of the filament does, of course, not mean that the space current is zero because since E_γ is positive when reckoned from all points on the filament other than the extreme positive end.

When using the tube in the simple circuit shown in Fig. 195 it is necessary to make sure that electrons do not flow to the grid. This is usually secured by putting in the negative grid battery E_g.

In practice, especially when receiving weak signals, it is usually not necessary to insert this battery because the potential variations impressed on the grid seldom exceed a small fraction of a volt, and, under these conditions, the current flowing to the grid is usually negligibly small. There is, however, a factor which must be considered, namely the contact potential difference between

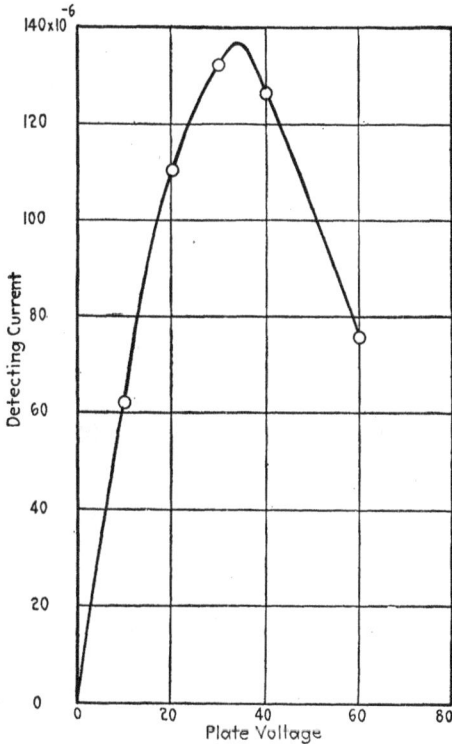

Fig. 197.

the filament and the system constituting grid and plate. The quantity ϵ in the characteristic equation gives a measure of this effect. If the filament, grid and plate are of the same material ϵ will usually be practically zero, but if the filament is, for example, of a different material ϵ may be either positive or negative, but it seldom exceeds the value of about 1 volt. If it is positive it means that the grid is intrinsically positive with respect to the filament and, therefore, to secure best operation it is necessary to insert a grid battery to maintain the resultant potential of the grid negative with respect to the negative end of the filament. In the case of tubes containing oxide coated filaments, ϵ is usually negative. In such a case, therefore, the grid battery can be dispensed with altogether. The quantity ϵ will differ from zero whenever the electron affinity of the filament is different from that of the grid, the contact potential difference between the two being equal to the difference between their electron affinities expressed in volts (see Chapter III).

103. Detection with Blocking Condenser in Grid Circuit. The method of detection discussed above and which can be carried out in practice with a circuit like that shown in Fig. 195 is perhaps not used as commonly as another type of circuit which is shown

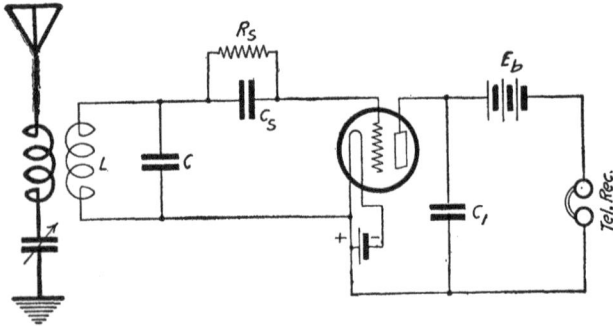

FIG. 198.

in Fig. 198. The difference between these two circuits is that Fig. 198 contains in the grid circuit a condenser C_s shunted by a high resistance leak R_s. The mechanism of detection with such a circuit is different from that in which the blocking condenser is omitted. In the latter case the best results are obtained when the grid is maintained at a sufficiently high negative potential to prevent any convection current from flowing between filament and grid, the detection depending only on the curvature of the *plate-current characteristic*. When the blocking condenser is used the detection depends on the curvature of the *grid-current characteristic*, the potentials of the elements being so proportioned that convection current does flow from filament to grid.

In order to explain how the tube detects with a condenser in the grid circuit, let us first indicate briefly how the tube operates without the blocking condenser. Fig. 199 shows the processes involved in this case. Modulated high frequency potential variations are impressed on the grid. On account of the curvature of the characteristic the high frequency current variations in the plate circuit can be represented by a lopsided wave curve. This effect was explained in Section 57. Such a lopsided wave gives rise to

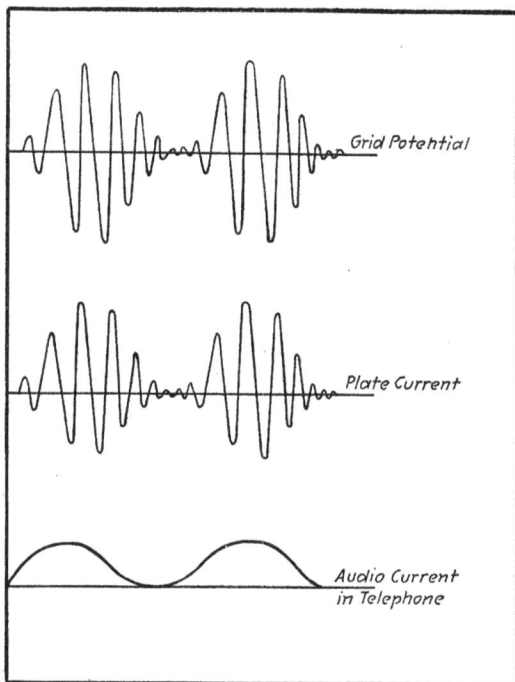

FIG. 199.

the audio frequency component as shown in the bottom curve of Fig. 199.

When the blocking condenser is used in the grid circuit the operation of the tube as a detector is as follows: Suppose the incoming oscillations are again high frequency currents modulated by a low frequency as shown by the uppermost curve of Fig. 200. Suppose for the present that the resistance R_s is omitted. When the grid potential becomes positive with respect to that of the filament, electrons are attracted to the grid. During the next half cycle when the grid potential becomes negative the electrons cannot escape from the grid because they are trapped on the insulated part of the circuit comprising the grid and the one plate of the condenser C_s. During the next positive loop of the incoming wave the grid attracts more electrons, which are also trapped so that they cannot escape from the grid during the succeeding negative loop. In this way the grid builds up a negative potential and the high frequency potential variations on the grid vary around

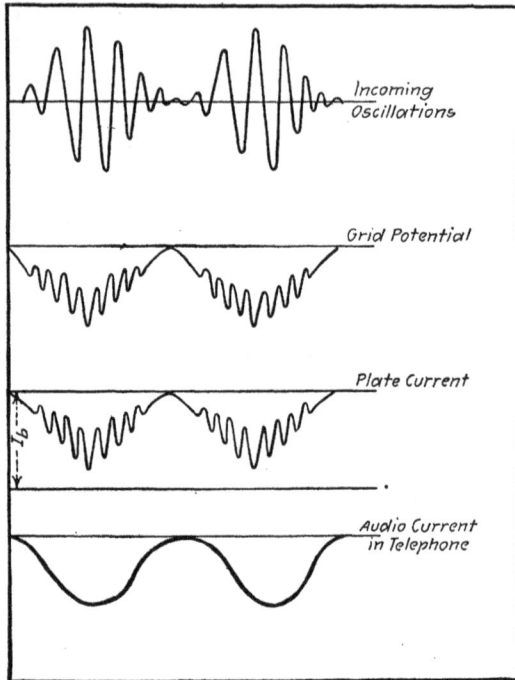

Fig. 200.

the mean value of grid potential which becomes more and more negative as the strength of the incoming oscillations increases. This reduces the plate current, and if the condenser C_s and the insulation of the part of the circuit comprising C_s and the grid were perfect the plate current would be permanently reduced, and this would make the tube inoperative. To prevent this a high resistance leak R_s is shunted across the condenser C_s, its value being so proportioned that the electrons cannot leak off through this resistance to any appreciable extent in a time comparable with the period of the high frequency oscillations but do leak off in a time which is of the order of magnitude of the low frequency variations of the amplitude of the high frequency oscillations. The result is that the potential of the grid takes such values as represented by the second curve of Fig. 200. This causes the plate current wave to assume the shape shown in the diagram. The high frequency variations in the plate circuit pass through the condenser C_1 (Fig. 198) inserted in the output circuit and the current in the telephone receiver takes the shape shown by the bottom curve of Fig. 200.[1]

In order to secure the best results with this type of circuit it is necessary to operate on that part of the grid voltage, grid current characteristic which shows the greatest curvature, and simultaneously adjust the plate potential to such a value that the operating point on the plate current, grid potential characteristic lies in the region where this characteristic is steepest. This usually requires that the grid be maintained at a positive potential with respect to the negative end of the filament. The simplest way to secure this is to connect the grid circuit to the positive end of the filament as shown in Fig. 198 instead of to the negative end as is commonly done in other circuits. This makes the filament negative with respect to the grid, the average potential difference between them being in the neighborhood of the value where the grid current characteristic has its greatest curvature. The best value for the capacity C_s usually lies between about 150–500 micro-microfarads while the leak resistance R_s should be of the order of two megohms.

If the detecting current be measured as a function of the effective voltage $E_\gamma = \left(\dfrac{E_p}{\mu} + E_g + \epsilon\right)$ a curve is obtained like that shown in Fig. 201. When the blocking condenser is not used we have seen the relation between detecting current and effective voltage gives a maximum as shown in Fig. 197.

104. Method of Measuring the Detecting Current. The measurement of the detecting current under conditions approximating those met with in practice has always been a difficult matter because it involves the measurement of very small alternating currents. Their values under practical conditions range from about 10^{-6} ampere down to 10^{-8} ampere and sometimes less. This makes it

[1] E. H. ARMSTRONG, El. World, Vol. 64, p. 1149, 1914.

entirely impossible to use hot wire instruments. The telephone receiver is a very sensitive device for indicating small alternating currents, but does not directly give a measure of the value of the currents in the receiver. The audibility method, which will be discussed later on, has been suggested to measure detecting currents with a telephone receiver. It consists in shunting the telephone receiver with a variable resistance and adjusting this resistance until the current in the telephone receiver is just large enough to make it possible to discriminate between the dots and dashes of the incoming signals. The ratio of the total current in the receiver and shunt resistance, that is, the detecting current

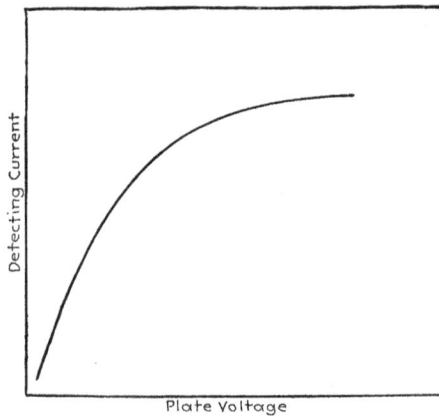

FIG. 201.

to the current in the receiver alone, measures what is known as the "audibility." This method is not very reliable, and its accuracy depends to a large extent on the conditions under which the measurements are made (see Section 108).

The following method requires only that two notes of the same pitch be adjusted to equal intensities.[1] It is, comparatively speaking, very accurate, and does not depend nearly so much on the conditions under which the measurements are made. The principle of this method can be explained with reference to Fig. 202. The incoming high frequency oscillations are impressed on the grid in the usual way. In order to measure the small

[1] H. J. VAN DER BIJL, Phys. Rev., Vol. 13, p. 311, 1919; Proc. Inst. Radio Engineers, Vol. 7, p. 603, 1919.

detecting current in the output of the detector we use a generator
U, giving a note of the same pitch as that of the detecting current,
and then attenuate the current from the generator by means of a
receiver shunt S until the current i_d has the same value as the
detecting current delivered by the tube. W is a switch whereby
the telephone receiver can be connected either to the output of
the tube or to the output of the generator. The shunt and series
resistances of the receiver shunt are adjusted until the tone heard
in the receiver is of the same intensity for both positions of the
switch W. The receiver shunt has been described in Section 72.

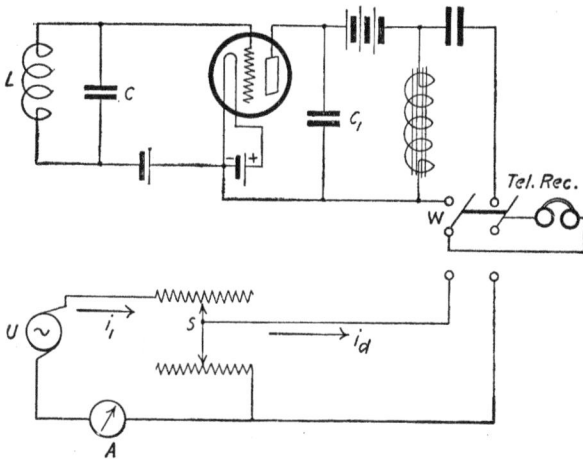

FIG. 202.

The shunt and series resistances are varied in definite steps by the
simple operation of turning a dial, these steps being so proportioned that the impedance in the output of the generator U
remains constant for all adjustments of the shunt. The current
i_1 delivered by the generator into this impedance is so large that
it can easily be measured with a hot wire instrument A such as a
thermo-couple. It was shown in Section 72 that the relation
between the current i_1 and the branch current i_d flowing through
the receiver is

$$\frac{i_1}{i_d} = e^{\alpha d}, \quad \cdot \quad \cdot \quad \cdot \quad \cdot \quad \cdot \quad \cdot \quad \cdot \quad (28)$$

where α is the constant of the receiver shunt and d expresses the
current attenuation produced by the shunt in terms of length of

the cable or line having an attenuation constant equal to α per unit length. For the standard cable of reference commonly used in telephony $\alpha = 0.109$ per mile, d being expressed in miles (see Section 72). Expressing the above equation in common logarithms we get

$$\log i_a = \log i_1 - \frac{d}{K}, \quad \ldots \ldots \quad (29)$$

where

$$K = \frac{2.303}{\alpha} = 21.13. \quad \ldots \ldots \quad (30)$$

Now i_1 is measured by means of the instrument A, and d is a known value depending on the adjustment of the receiver shunt in the manner explained in Section 72; hence, if the shunt be so adjusted that the tone in the receiver is of the same intensity for both positions of the switch W we can, from the above equation, obtain the detecting current i_a.

The impedance of the telephone receiver should, of course, have such a value that the best operation is obtained. If necessary, we can, to secure this, insert a transformer between the telephone receiver and the output of the tube. Furthermore, the detecting current depends on the value of the voltage variation impressed on the grid and upon the extent to which the incoming wave is modulated. The R.M.S. value of the voltage can be measured by means of a resistance r and an a-c. galvanometer G as shown, for example, in Fig. 208. When comparing tubes for their operation as detectors the input need not be measured, nor is it necessary to insure that the incoming wave is completely modulated as long as these quantities remain the same throughout the measurements. When measuring the detecting efficiency of the tube, however, it is necessary, as will be explained later on, to measure these quantities.

This method of measuring the detecting current has been found to be very useful when studying the influence of the operating parameters such as the d-c. plate and grid potentials on the detecting current. When making such measurements it is customary to express the detecting current simply in terms of the adjustment d of the receiver shunt instead of computing the actual value of the detecting current from equation (29). It will be noticed, however, that d increases when the detecting current decreases. It is therefore advantageous to calibrate the receiver shunt in terms of D–d when D is an arbitrary constant.

105. Measurement of the Detection Coefficient. The method described above makes it possible to measure the detection coefficient if the relation between the detecting current and the voltage impressed on the input is known. If the tube is used without a blocking condenser in the grid circuit the detecting current can be given by the equation

$$i_d = ae_g^2. \quad \ldots \ldots \quad (31)$$

When the tube is used with a blocking condenser this relation also holds fairly accurately provided the input voltage is small, generally not greater than about half a volt. If we put this value of i_d into equation (29) we get

$$d = 2K \log_{10} e_g + C, \quad \ldots \ldots \quad (32)$$

where

$$C = K (\log i_1 - \log a). \quad \ldots \ldots \quad (33)$$

Hence if we measure the relation between the input voltage e_g and the setting d of the receiver shunt we obtain a straight line from the intercept of which the detection coefficient a can be determined. The intercept C is obtained when $e_g = 1$. This gives

$$\log a = \log i_1 - \frac{C}{K}. \quad \ldots \ldots \quad (34)$$

The detection coefficient can therefore be obtained to any desired degree of accuracy by taking a sufficiently large number of observations of e_g and d.

The circuit whereby such measurements can conveniently be made is shown in Fig. 203. In this circuit the source of audio frequency current used to modulate the high frequency current also supplies the current with which the detecting current in an output of the detector is compared. U is the generator of the audio frequency currents. This can be a vacuum tube oscillator or a microphone generator such as that described in Section 72. (See Figs. 114 and 115.) Its output passes through a filter F, which transmits only frequencies of about 800 cycles. This current is sufficiently large to be measured with a thermo-couple A. but after passing through the receiver shunts S and S' it is attenuated until the intensity of the tone heard in the receiver T is equal to the detecting current coming from the detector tube D. By means of the switch W either the detecting current or the current from the generator U can be passed through the telephone receiver. If the switch is thrown to the left the current from U passes directly through the receiver after being attenuated by the receiver shunt. When W is thrown to the right the output of U is impressed on the input circuit of the modulator M. This low fre-

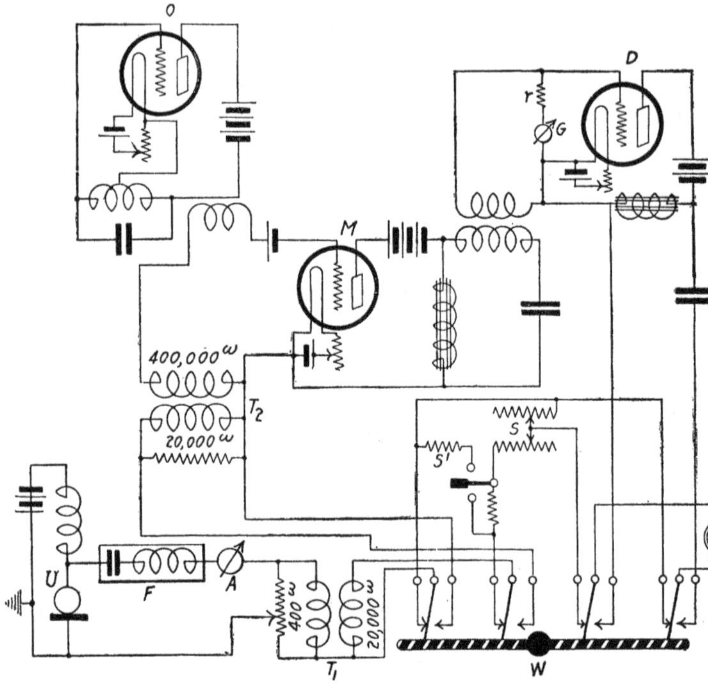

Fig. 203.

quency current is therefore used to modulate the high frequency current also impressed on the input of M and obtained from the vacuum tube oscillator O. The output of the modulator is impressed on the detector D, the voltage between filament and grid of D being measured by means of the resistance r and thermogalvanometer G. It follows then from the equations developed above that the audio frequency output of the detector is of the same pitch as the current supplied by the generator U, thus making the adjustment of the receiver shunt S for equal intensities of these two notes a comparatively simple matter. It is true that the output of the detector also contains a note of double frequency, as shown, for example, by equation (19), but this double frequency note is usually so weak as not to cause any trouble.

The circuit shown in Fig. 203 requires that certain precautions be taken to obtain reliable results; for example, it is necessary to make sure that the output impedance of the generator U remains constant for both positions of the switch W. Thus, supposing that the impedance of the telephone receiver T is 20,000 ohms, it is necessary to make the input impedance of the transformer T_2 which is placed in the input circuit of the modulator M also 20,000 ohms. This transformer is usually wound to have a high output impedance in order to impress the highest input voltage on the grid of the tube to which it is connected for the lowest amount of power expended in the input. The transformer T_1 is inserted when the impedance of the generator U is different from that of the telephone receiver T. In order to adjust the current from U to the desired value the primary of transformer T_1 is shunted with a resistance and the connection to the output of the generator is made by means of a sliding contact as indicated in the diagram. The receiver shunt S' has a fixed value, giving an attenuation equal to the maximum attenuation given by the variable shunt S, and can be inserted when the detecting currents to be measured cover a greater range than can be taken care of by one receiver shunt. Receiver shunts are seldom made to cover a greater range of attenuation than 30 miles of standard cable $\left(\dfrac{i_1}{i_d} = 26.3\right)$.

In making measurements of this kind it is necessary to insure that the modulated wave impressed on the input of the detector is completely modulated. The necessity for this can readily be seen by referring to equations (23) and (24), which give the R.M.S. values of the detecting current and the modulated voltage on the input of the detector. From these equations it will be seen that for a constant modulated input voltage e_g the detecting current depends on B and this, we have seen, is a measure of the extent to which the wave is modulated. This can also be seen by referring to Figs. 188 and 191. Two waves such as those shown in these figures may have the same heating effect as measured, for example, by means of a resistance and thermo-galvanometer, but they will not produce the same detecting effect when they are impressed on the detector. In the limiting case in which the wave is not modulated at all ($B' = 0$) the R.M.S. of the input voltage will have a finite value $\dfrac{A}{\sqrt{2}}$, but the detecting current will be zero (equation 23). In order to insure that measurement of the detection coefficient shall have any meaning the extent to which the wave impressed on the input of the detector is modulated must be kept constant, and the simplest way to do this is to completely modulate the wave, thus making $B = 1$. This can readily be done in practice in the following way: Referring to Fig. 204, which represents

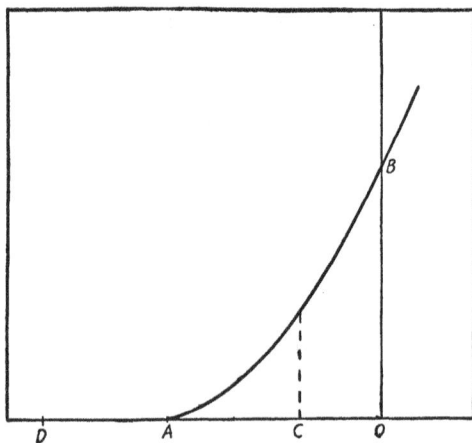

FIG. 204.

the relation between plate current and grid potential, it is evident from the explanations given in Section 53 that the intercept OA which represents the negative grid potential necessary to reduce the plate current to zero is $\dfrac{E_p}{\mu}$. If we now apply a constant grid potential $E_g = \dfrac{E_p}{2\mu}$ and make the peak value of the low frequency input voltage equal to this quantity, then the amplitude of the high frequency oscillations is reduced to zero every time that the grid acquires its maximum negative potential CA and then the wave will be completely modulated. The simplest way to secure this in practice is first to adjust the negative grid battery in the input circuit of the modulator to a value $\dfrac{E_p}{\mu}+\dfrac{E_p}{2\mu}$; that is, to a value given by OD (Fig. 204), and then gradually increase the strength of the low frequency input voltage until a d-c. meter placed in the output of the modulator just indicates a current flow in the output of the modulator. The peak value of the input potential is then equal to DA or $\dfrac{E_p}{2\mu}$. The voltage of the high

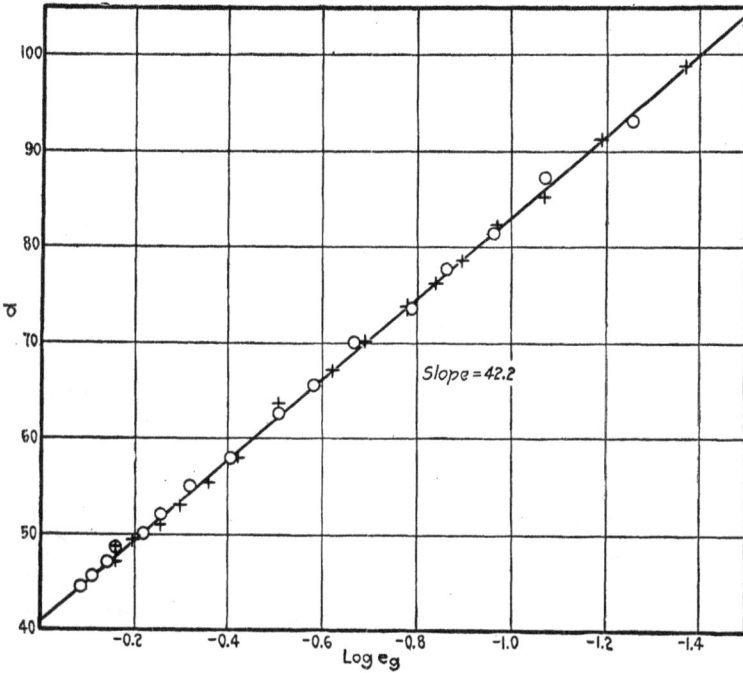

Fig. 205.

frequency impressed on the modulator can also be measured in the same way and should in general be somewhat smaller than the low frequency voltage. Finally the grid battery in the input circuit of the modulator is adjusted to the value OC, before the measurements on the detecting current are undertaken.

Fig. 205 shows some experimental results that were obtained with the circuit shown in Fig. 203. The ordinates indicate the setting of the receiver shunt for different values of the input voltage e_g, the logarithms of which are plotted as abscissæ. According to equation (32) these points should lie on a straight line. Furthermore, if equation (31) holds the slope of this line should be 2K; that is, 42.26, since K for the receiver shunt used is 21.13. The crosses and circles represent observations made by two different observers on different days. The slope of the line drawn through them is 42.2. These measurements were made without a blocking condenser in the grid circuit and prove that in deriving the equations in the previous pages we were justified in assuming that the power series given by equation (2) converges so rapidly that we can neglect quantities of higher order than the second, and that therefore the detecting current is given by a simple equation (31).

From the intercept C of this line (log $e_g = O$), and the value of the current i_1 the detection coefficient can be obtained directly with the help of equation (34). In the case to which the experimental results given in Fig. 205 apply the current i_1 as measured by a meter inserted in the 20,000 ohm line was 3.10×10^{-3} ampere and the intercept for $e_g = 1$ is 40.8. From this we obtain for the detection coefficient $a = 36.2 \times 10^{-6}$ amp./(volts)^2.

106. Detecting Efficiency. The detection coefficient a gives a measure of the audible component of the current in the output of the detector and depends on the impedance of the telephone receiver. It is therefore not suitable for expressing the figure of merit of the tube as a detector. The impedance of the telephone receiver should be chosen to give maximum response. The audio frequency output power is the quantity which gives a better indication of the behavior of the tube, and is given by the product of the square of the detecting current and the resistance of the telephone receiver. The power developed in the receiver depends, of course, also on the power developed in the input circuit, that is, on the strength of the incoming oscillations, the figure of merit of the tube as a detector being given by the ratio of audio frequency output power to radio frequency input power. This is a difficult quantity to measure. It was shown, for example, in Section 71 that the power expended in the grid circuit depends on the constants of the output circuit. The reaction of the output on the input circuit through the electrostatic capacities between the electrodes of the tube causes the tube to behave as if it had an effective input impedance. If the output circuit contains only a pure resistance, the resistance component of this effective impedance between filament and grid has a positive value, which, however, is usually small compared with the reactance component. If, on the other hand, the external circuit contains an inductive impedance, the grid resistance may be negative, thus giving rise to a generation of power in the input. At very high frequencies, it was shown in Section 71, the resistance component of the input impedance is practically zero, but the input voltage may be considerably reduced on account of the input circuit being shunted by the capacity between filament and grid. When the tube is used with a blocking condenser in the grid circuit there is a convection current between filament and grid, thus giving rise to an input resistance which must be added to that caused by the inter-electrode capacities. If, on the other hand, the tube is used without a blocking condenser in the input circuit, in which case the grid should be kept negative with respect to the filament, the input resistance is due entirely to the reaction of the output circuit to the input through the capacities of the tube, and can be made as small as we please by properly adjusting the constants of the circuit. Most of the input power is then dissipated in the input coil and condenser. The input power can therefore be dissipated at the grid, in the external input circuit and in a fictitious input resistance occasioned by the reaction of the output circuit on the input. The relative amounts of power dissipated in these parts depends on the adjustments of the circuit. It is for this reason usually better to express the figure of merit of the device as a detector in terms of the audio frequency output power for a given high frequency *voltage* impressed on the input because there is a definite relation between these quantities. The equations developed in the previous sections express the quantities considered in terms of the input voltage, and therefore hold whatever may be the effect of the circuit and the inter-electrode capacities on the input power. Expressing the detecting efficiency δ in terms of the relation of output audio frequency power to input radio frequency voltage we have:

$$\delta = a^2 r_0, \quad \cdot \quad \cdot \quad \cdot \quad \cdot \quad \cdot \quad \cdot \quad \cdot \quad (35)$$

where r_0 is the resistance in the output of the detector and a is the detection coefficient. The curve shown in Fig. 205 was obtained with a circuit in which the telephone receiver used had an impedance of 20,000 ohms and a resistance of 6400 ohms at about 800 cycles per second. The detecting efficiency of the tube on which these measurements were made is therefore 8.1×10^{-6} watt/(volt)^4. The smallest amount of power dissipated in this receiver which could still give a signal that is barely audible is about 3×10^{-12} watt. The high frequency input voltage necessary to give the least audible signal with this particular tube and telephone receiver is therefore about 0.025 volt.

These measurements were made on a Western Electric type $VT-1$ tube (Fig. 127). This tube operates on a plate voltage of about 20 volts. The power consumed in heating its filament ranges from 2.2 to 3.5 watts. On account of the small amount of power involved when using the tube as a detector it is desirable to make the filament as small as possible in order to reduce the power expended in heating it. The limitation to the decrease in the size of the filament is due mostly to mechanical difficulties, but smaller types of tubes have been developed, of which the one shown in Fig. 131 (page 244) is a sample. This tube only requires a small fraction of a watt to heat its filament, the filament operating on a voltage ranging from 1.0 to 1.5 volts so that it can be used with a dry cell. The detecting efficiency of this little tube was found to be about 4.3×10^{-6} watts/(volts)^4. A ratio of two in the output power corresponds to a difference of about three standard cable miles, which is not a big difference. A difference of one standard cable mile is hardly noticeable unless comparison be made directly.

107. Comparison of Detectors. The circuit shown in Fig. 203 is not always suitable for use where a large number of tubes are to be tested, because it requires accurate calibration and careful adjustment of the operating parameters such as the radio frequency voltage impressed on the input of the detector, the audio frequency current delivered by the generator U, etc. The constancy with which vacuum tubes can be made, however, makes it possible to test tubes by comparison methods that are simple to operate. The tubes to be tested can then be compared with a standard tube that was carefully calibrated by means of such a circuit as shown in Fig. 203. If a tube is well evacuated it will retain constant operation over a considerable length of time. The writer has, for example, used a " standard " Western Electric tube whose detecting efficiency did not change to any noticeable extent in the course of about a year, during which time it was in frequent use.

A simple circuit whereby detectors can be compared is shown in Fig. 206. The input voltage can be adjusted to the desired value by adjusting the resistance r and need not be known accurately, it being sufficient to know that it lies within the range of voltages used in practice. By means of the switch W either one of the detectors can be inserted in the circuit and the receiver shunt S adjusted until the note in the receiver T has the same intensity for both positions of the switch W. The key K serves to throw

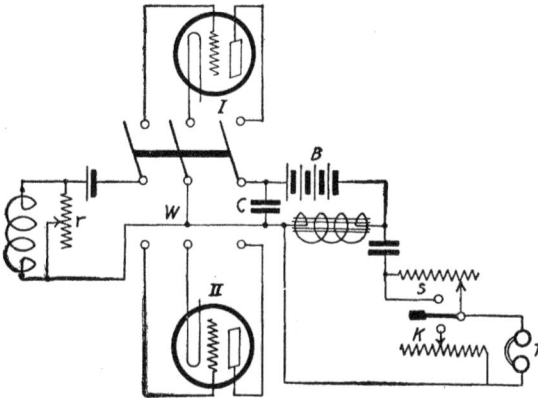

FIG. 206.

the shunt into or out of the circuit according as W connects the tube of higher or lower efficiency. The capacity C is a radio frequency leak and the output circuit is connected across the choke coil, which insures that the d-c. potential on the plate remains constant for all adjustments of the resistances of the receiver shunt S.

If i_d and i'_d be the detecting currents obtained from the tubes I and II, and a and a', their detection coefficients, then since the input voltage is the same for both, we have

$$\log \frac{i_d}{i'_d} = \log \frac{a}{a'} = \frac{d - d'}{K}, \quad \cdots \quad \cdots \quad (36)$$

where d and d' are the adjustments of the shunt in units depending on the units of K. The detecting efficiency of the tube under test

is then given by

$$\delta' = \left(\frac{a'}{a}\right)^2 \delta. \quad \ldots \ldots \ldots \quad (37)$$

Fig. 207 shows the complete circuit as it can be used for comparing detectors in practice. This circuit contains an amplifier tube connected to the output of the detectors. When the detector is to be used in practice with an amplifier such a circuit is desirable to insure that the detector is tested under conditions approximating as closely as possible to practical conditions. If the detector is not to be used with an amplifier the receiver shunt and telephone

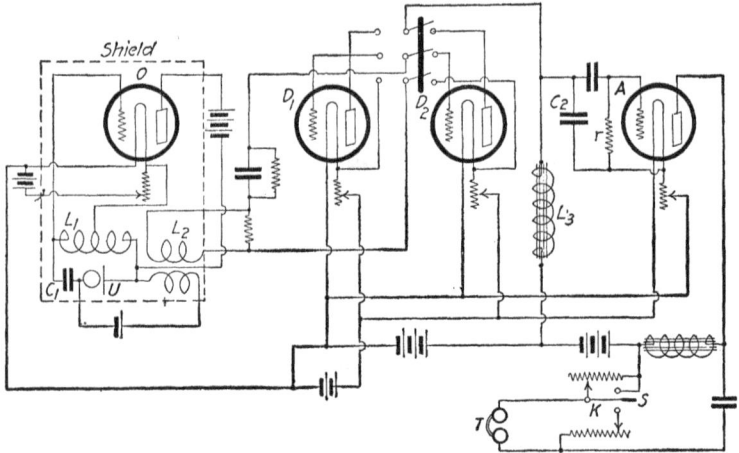

Fig. 207.

receiver can be connected directly to the output of the detector as shown in Fig. 206. To obtain a modulated high frequency test wave would ordinarily require a radio frequency oscillator, an audio frequency oscillator and a modulator, but in *comparing* detectors it is not necessary that the wave be completely modulated, and under such conditions a modulated high frequency wave can be obtained very easily by means of a vacuum tube oscillator and a microphone generator such as the one described in Section 72. In Fig. 207 U is the microphone generator, the carbon button of which is inserted directly in the oscillation circuit C_1L_1. When this generator is in operation the resistance of the carbon button varies periodically at an audio frequency, thus causing audio frequency variations in the amplitude of the radio frequency oscillations produced by the oscillator tube. Modulation produced in this way is not complete, but in this case complete modulation is not necessary because the input is the same for both tubes. The condenser C_2 in the output of the detector serves as a high frequency leak and the resistance r is inserted to prevent the grid from acquiring a negative charge. Its value is in the neighborhood of 2 megohms.

108. Audibility Method of Measuring the Detecting Current. The audibility or "shunted telephone" method has been frequently applied to the measurement of the strength of received signals in long distance radio communication, and has also been used to obtain an idea of the sensitiveness of detectors. In this method the telephone receiver is shunted by means of a resistance r_s which is adjusted until the signal heard in the receiver is just barely audible. If i_d is the detecting current and i_0 the least audible current in the receiver, then the "audibility" is given by

$$\frac{i_d}{i_0} = \frac{Z_0 + r_s}{r_s}, \quad \ldots \ldots \ldots \quad (38)$$

where Z_0 is the impedance of the receiver. In using a method like this it must, of course, be remembered that Z_0 cannot be replaced simply by the resistance of the receiver as is sometimes done, but the reactance and the motional impedance of the receiver must be taken into consideration.

This method is open to other serious objections. In the first place, it is liable to considerable error because the measurement of least audible signals is made difficult by the influence of extraneous noises such as room noises and static. The least audible current depends furthermore to an appreciable extent upon the condition of the observer, so that the current necessary to give the least audible signals will vary from time to time even with the same observer. These disadvantages make the method unreliable for purposes of determining the detecting efficiency of a tube with any degree of accuracy. Secondly, the way in which the audibility method is ordinarily used does not make provision for the change in the effective impedance of the output circuit to the detecting current when the shunt resistance r_s is varied. This would give misleading results, since the detecting current depends upon the relative values of the

internal output impedance of the tube and the external impedance
in the output circuit. It is therefore necessary in all measurements of this kind to adjust these impedances properly and keep
them constant throughout the measurements. If the audibility
method is to be used the "audibility box" should be so designed
that any variation in the shunt resistance is accompanied by an
addition or subtraction of an equivalent resistance so as to keep
the total impedance of the circuit constant. This can be done
with the scheme that will now be described. This scheme was
used by the writer [1] to determine to what extent the audibility
method may give reliable results if precautions are taken to eliminate sources of error other than those which depend only on the
psychological and physiological influences on the observer. The
fact that the current necessary to give the least audible signal
has different values for different observers and is therefore incapa-
ble of objective determination does not of itself rule out the
audibility method for the measurement of signal strength, since
the detector set could first be calibrated by determining the
audibility for known input signals and then used by the same
observer to make the final measurements. Hence assuming that
extraneous noises could effectively be cut out, the possibility of
adapting this method to such measurements would depend upon
the extent to which the observer's conception of least audible
signal remains constant during the time that elapses between his
calibration of the set and the making of the final measurements.
It is hardly necessary to say that the whole set must remain
unchanged, especially the tube and the telephone receiver.

The circuit whereby the least audible signal can be studied
under constant circuit conditions is shown in Fig. 208. To cut
down the current in the telephone receiver a receiver shunt is used
such as that described in Section 72. This shunt contains a
series and shunt resistance, both of which are variable, instead
of simply a shunt resistance. The receiver shunt is so calibrated
that the series and shunt resistances are changed simultaneously
in such a way that the total impedance to the detecting current
in the output circuit of the tube remains constant. A choke coil
L by-passes the direct current in the plate circuit and insures that
the d-c. potential of the plate remains constant for all adjustments
of the receiver shunt. This is necessary because the shunt is so

[1] H. J. van der Bijl, Proc. I.R.E., Vol. 7, p. 624, 1919.

designed that for all its adjustments the *impedance* of the output circuit remains constant, but the *resistance* does not remain constant, and therefore if the choke coil and capacity were omitted, thus making it necessary for the d-c. plate current to pass through the shunt, the potential of the plate would be different for different adjustments of the receiver shunt. It is very important to keep the relation between the impedances constant because the detecting current depends very markedly on the value of the external impedance in the plate circuit.

The wave impressed on the input of the detector can be a spark signal wave or a modulated wave which is interrupted. For test purposes such a wave can easily be obtained with an arrangement such as that shown in Fig. 207, where the oscillator tube O and the microphone generator U together form a simple system for pro-

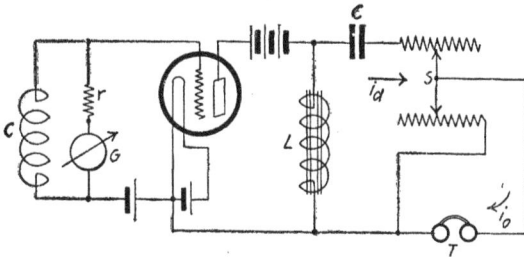

FIG. 208.

ducing modulated waves. The output of this oscillator system can then be passed through an omnigraph to produce the signals. The R.M.S. of the input can, as before, be measured by means of a resistance r and galvanometer G (Fig. 208).

The use of the receiver shunt makes it possible to express the audibility in a simple way. If i_d be the detecting current and i_0 the least audible current in the receiver the audibility $\dfrac{i_d}{i_0}$ can be expressed in miles of standard cable by making use of the equations developed in Sections 104 and 105. Thus, taking the case in which the detecting current is proportional to the square of the input voltage, we obtain

$$d = 2K \log e_g + K(\log a - \log i_0), \quad \ldots \quad (39)$$

which thus gives a linear relation between the logarithm of the input voltage and the audibility when expressed in terms of miles

d instead of current ratio, the relation between d and the current ratio being given by equation (28). The intercept of this line, obtained by putting $e_g = 0$, is

$$K (\log a - \log i_0), \quad \ldots \ldots \quad (40)$$

and gives a measure of the audibility efficiency of the tube expressed in miles/(volt)^2.

The simple linear relation (39) makes it possible to obtain the audible efficiency as an average of a large number of observa-

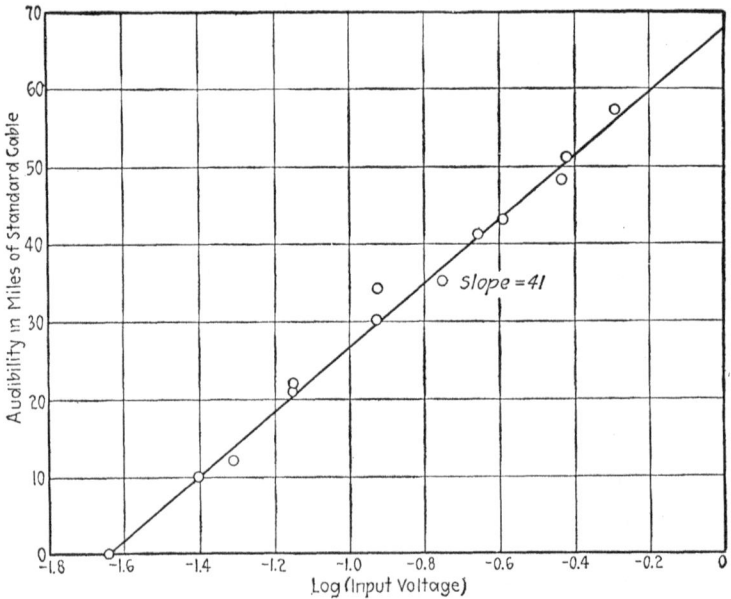

Fig. 209.

tions. A number of such observations plotted against the logarithm of the corresponding input voltages are shown in Fig. 209. The individual points vary considerably compared with the observations shown in Fig. 205, which were obtained with the other method discussed above, but these points are evenly grouped about a straight line the slope of which is about 41. The theoretical value of the slope is 42.26. The slope of the line depends upon the attenuation constant of the shunt and the simple quadratic relation (17) between the detecting current and the input voltage. The intercept of this line, on the other hand, is influenced also by the detection coefficient a of the tube and by what constitutes the least audible signal i_0 for the particular telephone receiver used and for the observer at the particular time of making the measurements. The attenuation factor K of the receiver shunt is determined merely by resistances of wires, and does, of course, not vary to any noticeable extent. By using a good tube for which the detection coefficient remains constant such measurements can therefore be used to give an idea of the extent to which the determination of least audible signal depends on the observer and how the "least audible current" varies from one observer to another. Experiments conducted by the writer [1] along this line and in which observations were made by four different observers over a period of eight days, the total number of observations being something like 350, showed that the maximum variation in the audibility expressed as a current ratio for the four observers was almost 600 per cent, but the variation in the measurements of each observer over a period of about a week averaged about 100 per cent. It must, however, be remembered that a variation of say 50 per cent in the audibility expressed as a current ratio is not a serious matter. In fact, such a variation would hardly be noticeable unless, of course, the comparison be made directly. The more satisfactory way of expressing the audibility is to use the logarithmic scale, that is, to express it in terms of length d of cable or line. When the audibility is expressed in this way the maximum variation observed was 28 per cent.

These measurements have also shown that for a tube like the average $VT-1$ the smallest input voltage (R.M.S.) that can just give the least audible signal is of the order of 0.03 volt. If the incoming signals are weaker an amplifier must be attached to the output of the detector. The minimum input voltage depends, of course, also on the sensitiveness of the telephone receiver used in making the measurements.

109. Heterodyne Reception with the Audion. The heterodyne method of reception consists in supplementing the incoming high frequency currents with a locally generated current of a frequency which differs from that of the received current by an amount which lies within the audible range. This method makes it possible to detect continuous waves of constant amplitude, and furthermore greatly increases the strength of the audible current in the receiver

[1] Loc. cit., p. 623.

placed in the output of the detector. The manner in which the heterodyne method increases the detecting current can readily be seen. Thus taking the case in which the detecting current is proportional to the square of the input voltage as given by equation (17), if we impress on the input circuit of the detector two high frequency voltages $e_1 \sin pt$ and $e_2 \sin qt$, the detecting current is given by

$$I_d = a(e_1 \sin pt + e_2 \sin qt)^2. \quad \ldots \quad (41)$$

On evaluating this expression and dropping all terms representing frequencies that lie outside of the audible range we get

$$I_d = a e_1 e_2 \cos (p - q)t. \quad \ldots \ldots \ldots (42)$$

If $\frac{p}{2\pi}$ is 100,000 cycles, for example, and $\frac{q}{2\pi}$ 99,000 cycles, then the above expression represents a current having a frequency of a thousand cycles per second and is therefore audible. Furthermore, the locally generated voltage e_2 can be made much larger than the voltage e_1 of the incoming waves, and therefore the detecting current can be very much increased.

Equation (42) holds, strictly speaking, only when the grid is maintained at a sufficiently high negative potential with respect to the filament to prevent any electrons from flowing from the filament to the grid. Under these conditions the detecting current has been found to be proportional to the product $e_1 e_2$ of the input voltages. These measurements can be made with a circuit like that shown in Fig. 202. Since the detecting current in this case can easily be made so large that it can be measured with a thermo-couple without increasing the strength of incoming signal beyond a practical range, it is also possible to use a circuit like that shown in Fig. 210. A high frequency voltage of small amplitude is impressed at A and represents the incoming signal wave received by the antenna. A locally generated high frequency is impressed at B, and is adjustable. By means of the switch W a thermo-galvanometer G can be used to measure the detecting current i_d and the effective voltage impressed on the grid. In using such a method it is, however, necessary to insure that when the switch connects the galvanometer to the output of the tube the impedance of the input circuit is not changed. This can be taken care of by inserting a resistance r_1+r_2 on the input side having a value equal to the resistance of the galvanometer plus the resistance r_2, which may be added in the galvanometer circuit to measure the input voltage. The key K should therefore be opened when the galvanometer is connected to the input and closed when it is connected to the output of the tube.

Results obtained by my associate, Mr. R. H. Wilson, are given in Fig. 211, which shows the relation between the detecting current measured in this way and the product e_1e_2 of the input voltages for three different values of the voltage of the battery in the grid circuit. In general, the curves do not coincide on the

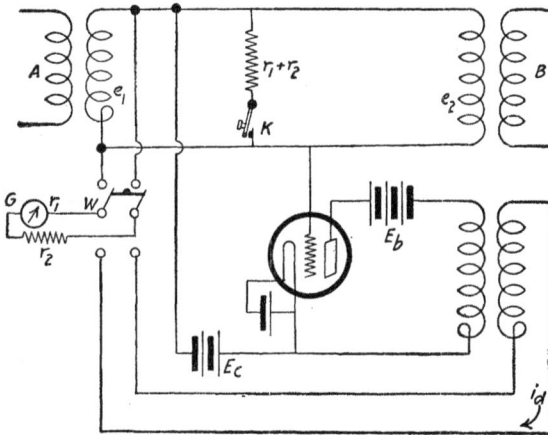

Fig. 210.

straight part. The values shown in this figure were obtained by reducing the observations to a common value to superimpose the curves on one another. For each of these grid voltages the voltage of the plate battery was adjusted to maintain the space current constant. In other words, the operating points on the characteristics for the three cases are shown at A, B and C, of Fig 212, where the three curves represent the characteristics obtained with three different plate voltages. Fig. 211 shows a linear relation between the detecting current and the product e_1e_2 up to a value of this product depending on the value of the d-c. grid potential chosen. The point at which the observed detecting current deviates from the straight line is obtained when the sum of the peak values of the voltages impressed on the grid becomes greater than the negative d-c. grid potential so that current begins to flow in the grid circuit, thus causing a waste of power. It is seen therefore that the strength of the signal can be increased by increasing the negative grid potential and at the same time

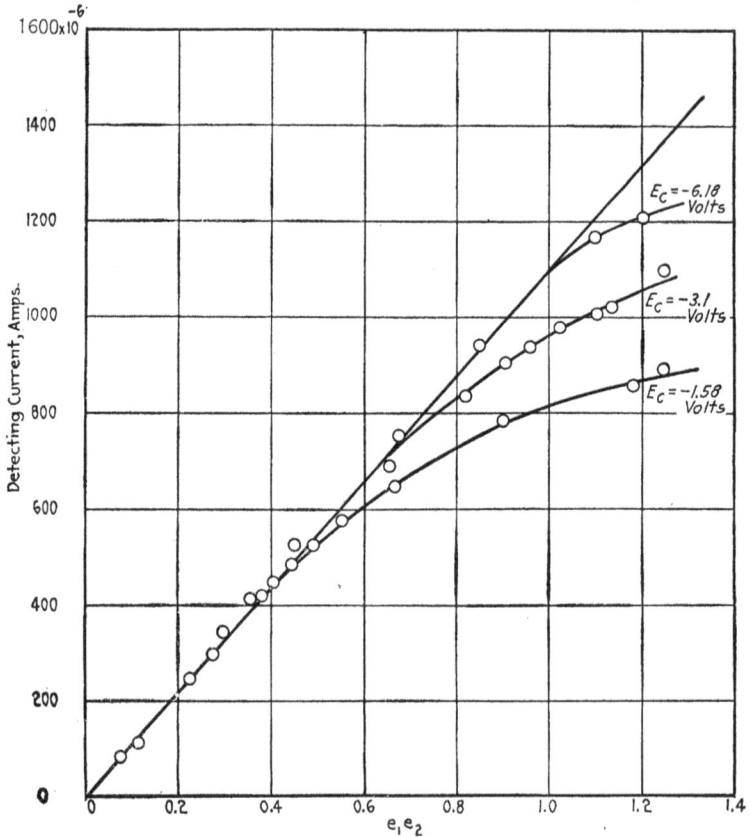

FIG. 211.

increasing the plate potential. In the case of the lower characteristic the input voltage cannot much exceed the value OA, but in the case of the characteristic with the high plate potential the input can be made as large as OC, without allowing electrons to flow from filament to grid. Increasing the sum of the input voltages superimposed on the negative grid potential beyond the value necessary to reduce the current to zero does not seem to cause a deviation from the linear relation between the detecting current and the product of the input voltages. The observations shown in Fig. 211 represent in a general way what is to be expected, although it is possible that with different tubes and different circuits peculiarities of behavior causing a deviation from such a simple relation can manifest themselves.

When the heterodyne method of reception is used with a blocking condenser in the grid circuit the relation between the detecting current and the input voltage is not as simple as when the blocking condenser is omitted. In general, the detecting

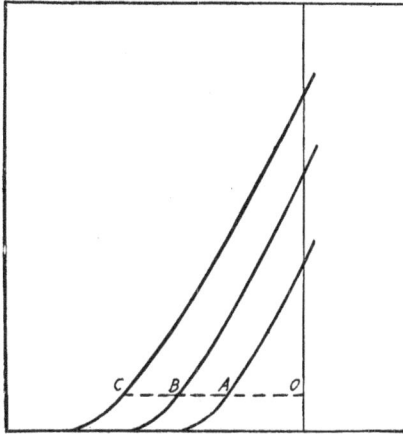

FIG. 212.

current has a maximum value at a value of the product e_1e_2 depending on the operating parameters of the tube, and the maximum detecting current is generally larger than when the blocking condenser is omitted. Fig. 213 shows the general relation between the detecting current and the product of the input voltages. The three curves are obtained with three different values of the d-c. grid potential, the plate battery voltage having a constant value in this case of 22 volts. These measurements were made with a standard $VT-1$ tube. It will be noticed that the best reults are obtained when the grid is kept positive with respect to the filament. This is in accordance with the explanation given in Section 103, where it was shown that detection with a blocking condenser in the grid circuit depends on the curvature of the grid current characteristic. The detecting current increases with increase in the curvature of this characteristic and also becomes greater the steeper the plate current grid potential characteristic is. If the relation between the maximum detecting current for a constant voltage of the grid battery be plotted as a function of the plate battery voltage a curve is obtained such as that obtained in Fig. 214. This curve shows a rapid increase in detecting current with increase in the plate voltage until a plate battery voltage of about 34 volts is reached. The bend in the curve at higher voltages is due to the

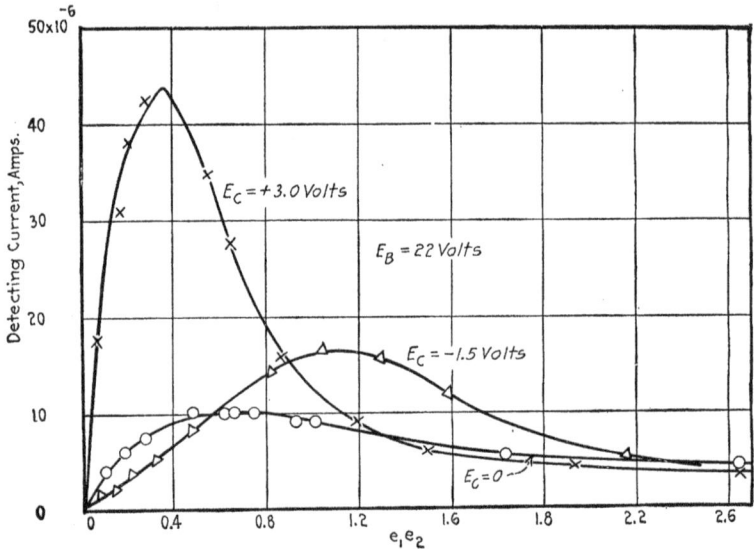

Fig. 213.

plate current characteristic becoming less steep when voltage saturation is approached.

110. Zero Beat or Homodyne Method of Receiving Modulated Waves. The heterodyne method of reception can be used when receiving continuous unmodulated waves, as is done in radio telegraphy, but cannot be applied to the reception of modulated waves as in radio telephony. In this case we can make use of the simple systems which consist in first detecting the modulated high frequency waves and then amplifying the low frequency component with the amplifier tube attached to the output of the detector. Or the received high frequency currents can first be amplified and then detected. This latter scheme has the advantage of greater sensitivity because the voltage impressed on the input of the detector which is connected to the output of the amplifier is amplified and the detecting current we have seen above

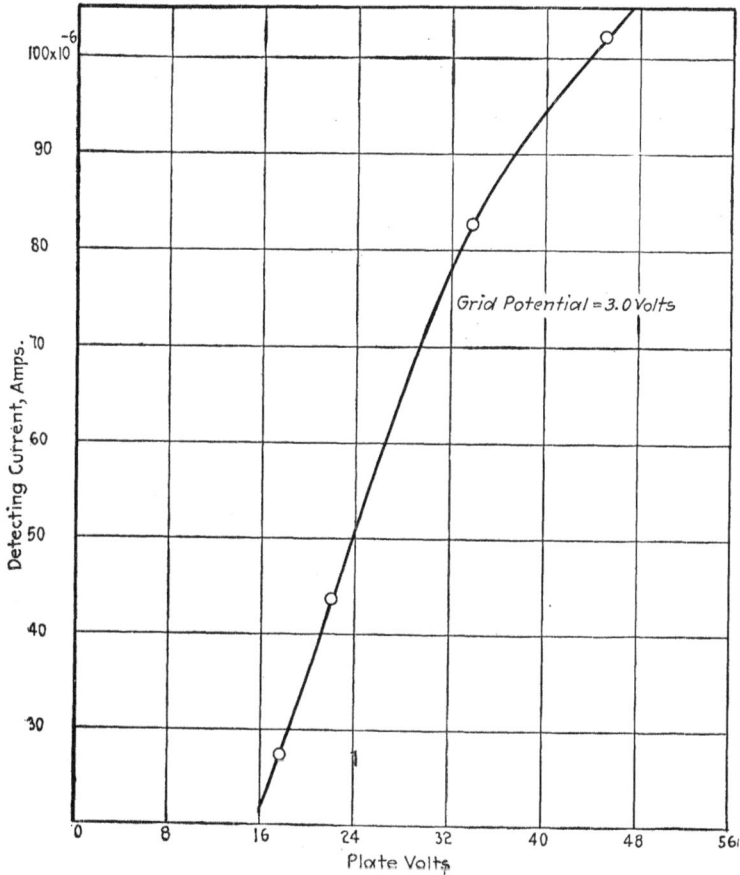

Fig. 214.

is practically proportional to the square of the input voltage. Such a circuit, however, has a tendency to sing. This can usually be prevented by feeding back part of the output energy to the input circuit so that the currents in the feed-back and input coils cause potential variations on the grid that are 180° out of phase.

There is, however, another method, whereby modulated high frequency currents can be received. This method is analogous to the heterodyne method except that the local source of high frequency currents has a frequency equal to that of the received carrier wave instead of differing from it by an amount which lies within the audible range. Thus, if we impress on the input of the detector a modulated high frequency voltage $A \sin pt$ $(1+B \sin qt)$ and also a continuous high frequency voltage $e \sin pt$ having the same frequency $\dfrac{p}{2\pi}$ as the carrier component of the received wave, then it follows, similarly to the cases considered above, that the output of the detector contains a current of the same frequency as the current with which the carrier wave was modulated at the transmitting station. This means of receiving has been termed the "homodyne method." It is hardly necessary to say that when using this method of reception care must be taken to keep the two high frequencies very closely in tune. For example, $\dfrac{q}{2\pi}$ in the above expression represents speech frequencies which only cover a range of about 2000 cycles per second. Now, if the carrier wave has a frequency of say 200,000 cycles per second, it follows that the locally generated source of high frequency must be kept at 200,000 cycles within a very small fraction of 1 per cent. If, for example, it differs from this frequency by $\frac{1}{4}$ per cent the locally generated and the received currents form a beat note of about 500 cycles, resulting in speech which would be unintelligible. The way to tune in the locally generated high frequency is to listen to the beat note, the pitch of which changes rapidly as the two frequencies approach each other, until when they are exactly alike the beat note ceases and intelligible speech is heard in the telephone receiver.

111. The Feed-back Receiving Circuit. It will now be apparent from considerations given in this and the last previous chapter, that a very simple and effective means of receiving can be obtained by using the heterodyne principle together with the amplifying property of the tube. In other words, the detector tube is used to generate its own high frequency by feeding back part of the output energy into the input. The circuit arrangement is shown in Fig. 215. The coil L_1, which forms part of the output circuit, is coupled to the input so that part of the output energy is returned to the input. The condenser C_2 serves as a by-pass for the high frequency currents. This set is operated very simply by tuning the input and varying the coupling between L_1 and the input until a note of the desired pitch and intensity is heard in the receiver. Measurements that have been made on this kind of circuit show that when the input voltage impressed on the grid reaches a certain small value the telephone receiver in the output suddenly gives a relatively loud note, and generally a coupling which is slighly greater than that necessary to do this is the best. This circuit is sometimes referred to as the regenerative circuit. It is also possible to make the output react on the input without direct electromagnetic coupling, in which case the coupling is

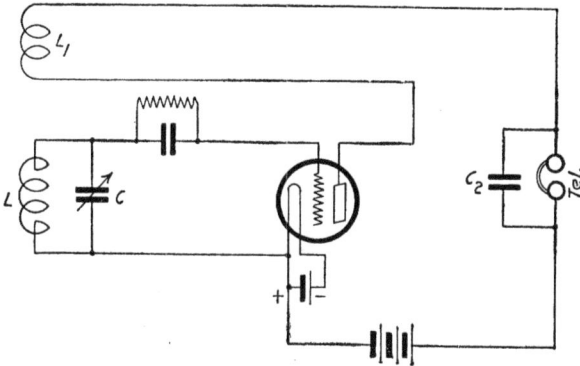

Fig. 215.

supplied by the electrostatic capacities between the electrodes of the tube. The regenerative action of the tube has been discussed in Section 88.

112. Radio Transmitting and Receiving Systems. A large number of radio systems are now in use and no attempt will be made to describe them here. They can be understood or even designed with ease once the principles of operation of the tube are understood. A circuit which embodies enough to indicate how the tube can be used in a transmitting and receiving system is shown in Fig. 216. The circuit to the left of the antenna represents the transmitting circuit and is practically identical with that shown in Fig. 179 (page 307). This set was designed by the engineers of the Western Electric Company for use on aeroplanes. The plate and filament voltages are supplied by a wind-driven generator, the voltage of which is regulated by means of a thermionic valve in the manner described in Section 51. When transmitting, the switch T is thrown to the left. The several parts of this switch are shown in separate parts of the diagram for the sake of simplicity, but they are all operated from one lever. When receiving, the switch is thrown to the right. This cuts out the voltage of the generator, the plates of the detector and amplifier tubes being supplied by storage batteries. The receiving circuit to the right of the antenna is a simple circuit with inductive connection between the tubes.

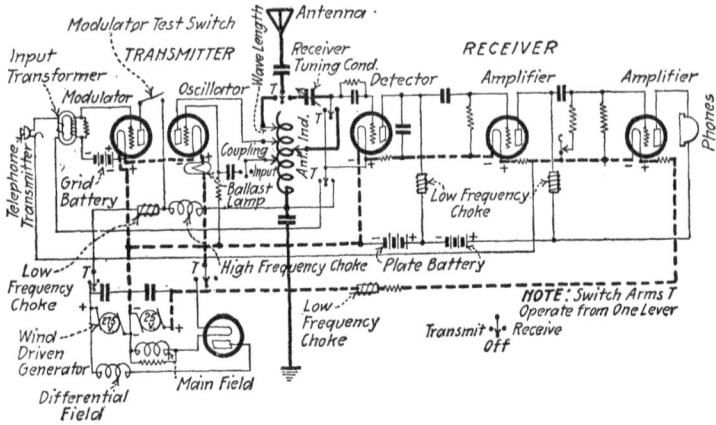

FIG. 216.

A simplified drawing of such a receiving system is shown in Fig. 217. Since this circuit has a blocking condenser C_s in the input, the grid is connected through the input circuit to the positive pole of the filament battery. C_2 is a high frequency leak, and the high resistance leak r and condenser C_3 serve to maintain the grid of the amplifier tube at a d-c. potential equal to that of the negative end of the filament. If necessary, a grid battery can be inserted to keep the grid of the amplifier negative. This circuit shows two batteries, E_b, for supplying the plate currents. The second battery is inserted when it is necessary to operate the amplifier at a higher plate voltage than the detector.

The transmitting system shown in Fig. 216 has the advantage of high efficiency, but, on the other hand, the oscillation coil of the oscillator forms part of the antenna, and therefore the frequency of the wave radiated changes with slight changes in the inductance of this coil.

A system which is not subject to this disadvantage is shown schematically in Fig. 218. The principle of this system consists in modulating a low-power radio frequency current and then amplifying this modulated current before it enters the antenna.

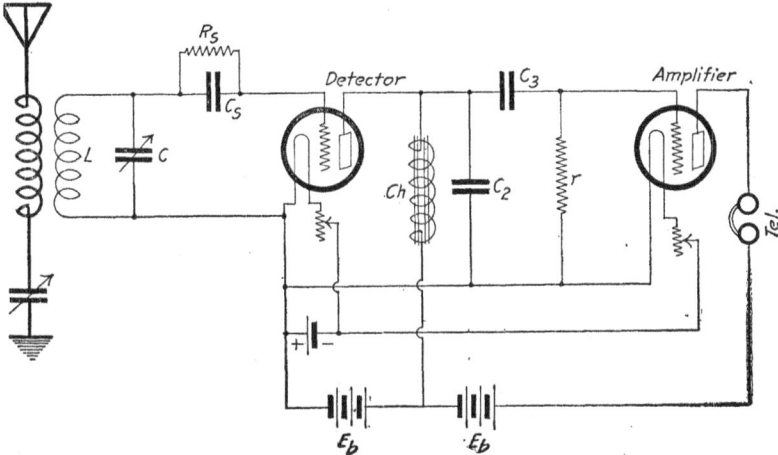

FIG. 217.

The oscillator O may be a vacuum tube oscillator giving the radio current at the desired frequency. The voltage obtained from this generator and the speech voltage obtained from the transmitter are both impressed on the modulator M. This part of the system does not have to handle large amounts of power. The oscillator tube can be a small type of tube operating on a plate voltage of

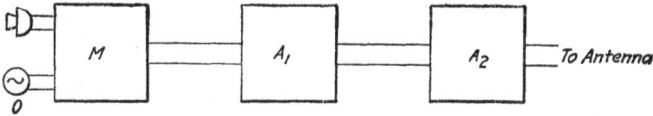

FIG. 218.

100 to 200 volts or even less, and the modulator can be a tube of about the same capacity. The speech voltage obtained from the telephone transmitter can then be impressed directly on the input of this modulator without amplification. The modulated output is amplified by an amplifier A_1, and can, if necessary, be further amplified by means of a power tube, or a set of power tubes oper-

ated in parallel.   The output from these power tubes is impressed on the antenna.   It will be evident that in such a system detuning of the antenna does not change the frequency of the radiated wave, but merely changes the energy radiated.

This system is of great advantage when using very short waves such as could be used in small portable transmitting sets, in which case the antenna is small enough to be carried in the hand.   For such short waves (about 20 to 100 meters) a change in the antenna capacity caused by the approach to objects does not appreciably change the frequency of the radiated wave but only the amount of energy radiated, because the antenna does not form part of the oscillation circuit that determines the frequency of the oscillations.   This frequency is determined by the constants of the oscillator O.

The arrangement shown in Fig. 218 embodies the system that was used in 1915 by the American Telephone & Telegraph Company and Western Electric Company in transmitting speech from Arlington, Va., U. S. A., to Paris and Honolulu.   The type of power tube used in these experiments is shown in Fig. 133, page 245. In these experiments the waves were received by the zero beat method and also by first amplifying the incoming waves and then detecting them.

113. Multiplex Telegraphy and Telephony.   It was pointed out in Section 99 that for a single telephone transmission a band of frequencies is required which covers a range of about 2000 cycles per second, and the antenna must be so tuned that it will transmit about equally well all frequencies lying within this range.   Antennæ as commonly used will usually transmit a wider range of frequencies than that required for one telephone transmission.   This is especially the case when the antenna is tuned to a high frequency, say a million per second.   The width of the transmission band of an antenna depends, of course, on its selectivity or sharpness of tuning.   If the transmission band is wider than the range of frequencies needed for a telephone transmission, it is theoretically possible to transmit more than one telephone message simultaneously from one tower by connecting to the sending antenna a number of transmitting sets, the frequencies of which are so distributed in the transmission range of the antenna that the frequency bands of their modulated waves do not overlap, and to the receiving antenna corresponding receiving sets each are tuned selectively to receive the waves from one transmitting set. These frequency bands and their separation are so small compared with the frequencies themselves, that it is very difficult to separate them at the receiving station. Besides, small changes in the frequencies of one or more of the transmitting sets, coupled to the same antenna, are likely to cause overlapping of the bands.

Instead of impressing the modulated waves from different transmitting sets of the same antenna, it is possible to modulate one high frequency wave with two or more different frequency bands. If these frequency bands lie in the audible range, such a system obviously does not allow of more than one telephone transmission. In the case of telegraph transmission, on the other hand, the required frequency band is much narrower, since here we transmit only one note, so that more than one such band lying within the audible range could be used as messages to modulate one high frequency carrier wave. Such a modulated wave can be passed through a detector at the receiving station and the original message frequencies in the output of the detector can be separated from one another by the use of appropriate filters or selective circuits; but the output circuit of the detector will contain not only the message frequencies with which the carrier wave is modulated, but also currents of double the message frequencies and currents, having frequencies equal to their sums and differences, and care must be taken to separate the original message frequencies from these.

Multiplex transmission has been realized in practice to be a. system involving what may be termed "double modulation" and "double detection." Here each of the various audio frequency currents (telephone or telegraph messages) is used to modulate one of a number of intermediate or auxiliary frequencies lying above the audible range, and the resulting modulated waves are all made to modulate a single high frequency or main carrier wave.

At the receiving end this doubly modulated main carrier wave is impressed on a detector, the output of which will contain the modulated auxiliary carriers. It will, however, also contain a number of other frequencies, containing among others the sums and differences of the modulated auxiliary carriers. These can be separated from the auxiliaries themselves by selective circuits if the frequency bands of the latter are not too close together. The modulated auxiliary carriers once separated can be passed through individual detectors, the output of which will contain the original message frequencies.

Numerous modifications of this general idea involving other special methods of modulation and detection are of course possible.

CHAPTER X

MISCELLANEOUS APPLICATIONS OF THERMIONIC TUBES

Besides the functions that have been discussed in previous chapters, there are a number of miscellaneous uses to which the thermionic vacuum tube can be applied. Some of these will now be described. The list is not intended to be complete, but merely to give an indication of what can be done with a three-electrode device. Many uses of the tube will suggest themselves to those who are acquainted with the principles of its operation.

114. The Vacuum Tube as an Electrostatic Voltmeter. Three of the properties of the tube, namely, the unilateral conductivity, amplification and small power consumption in the input circuit, make it possible to use this tube to measure small a-c. voltages by electrostatic means. Suppose, for example, that a source of small a-c. voltage is included in the grid circuit, which also contains an adjustable grid battery. The output circuit contains the plate battery and a d-c. measuring instrument. The plate and grid batteries are adjusted so that the operating point lies on the extreme lower end of the characteristic. The current in the plate circuit will then be just reduced to zero. If an alternating potential now be applied to the grid, the potential of the grid during the negative half cycle becomes more negative than that maintained by the grid battery, and the current in the plate circuit is still zero. During the positive half cycle, however, the grid becomes less negative and a current impulse is sent through the plate circuit. The result is a direct current indicated by the measuring instrument connected in the plate circuit of the tube. The amount by which the negative potential of the grid must be increased to reduce the current again to zero, measures the peak value of the alternating potential on the grid.[1]

In order to get a more sensitive device, if necessary, an amplifier could be added to the output of the tube. A circuit arrange-

[1] R. A. HEISING, U. S. patent 1232919.

ment that can be used, and which was suggested by R. H. Wilson, is shown in Fig. 219. The alternating voltage is impressed at *B*. The d-c. grid potential is supplied by means of the batteries E_c, and the potentiometer arrangement is indicated in the circuit. This arrangement makes it possible to adjust the grid potential to a value where the plate current of the first tube is just reduced to zero. The output of the voltmeter tube contains a high resistance of, say, 400,000 ohms, the grid and filament of the amplifier tube being connected to the ends of this resistance. The amplifier tube operates on the steepest part of its characteristic.

The way to operate a vacuum tube voltmeter circuit is to slide the contact on the potentiometer until the current of the first tube

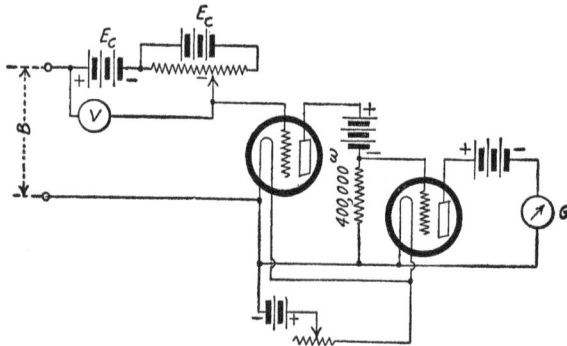

Fig. 219.

drops to zero. The corresponding grid potential is measured with the voltmeter *V*. This obtains when the current registered by the galvanometer or microammeter in the output circuit of the amplifier tube is the value given by this tube for zero potential on its grid. This can easily be tested by short-circuiting its input. The input voltage is then applied and the contact on the potentiometer again adjusted until the galvanometer *G* shows no change in the current. This grid potential is again indicated by the voltmeter *V*. The difference between this and the first reading gives the peak value of the input voltage.

It will be noticed that the source of a-c. voltage which is to be measured is connected to a circuit which can be regarded as practically an open circuit. The power consumption, due to the electrostatic capacities between the electrodes of the tube is usually very small, so that this arrangement acts practically as an electrostatic voltmeter, and because of the amplification produced by the tube, the arrangement can be used to measure very small alternating voltages. Sometimes, however, the plate current does not give a sharp intercept on the voltage axis, but tails off gradually, and the lower part of the curve may be very nearly linear. This is shown in an exaggerated way in Fig. 220. In such case it is best to operate the tube a small distance up the curve so that

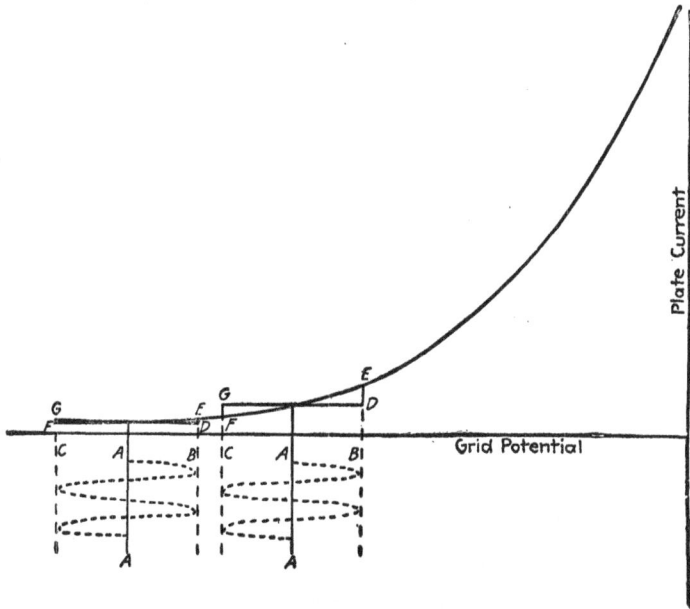

Fig. 220.

the increase in current during the positive half wave is greater than the decrease during the negative half wave. When the characteristic tails off and accuracy is desired, it may be necessary to apply a correction. Fig. 221 shows the relation between the true voltage and the voltage measured by a tube which did **not** show a sharp intercept on the grid voltage axis.

115. High Tension Voltmeter. The three-electrode tube makes it possible to measure extremely high voltages with comparative ease, by an arrangement that has been suggested to me by Dr. E. R. Stoekle. In this case the high voltage to be measured is applied between filament and plate. By means of a battery and potentiometer the grid is adjusted until the current in the plate

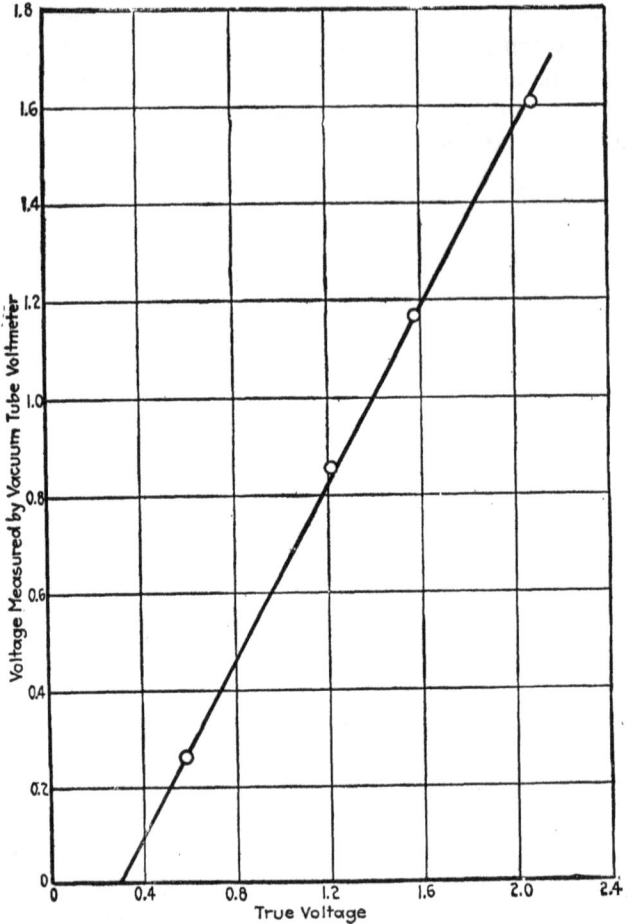

FIG. 221.

circuit is reduced to zero. Since the current through the tube is given by

$$I = f(E_p + \mu E_g),$$

it follows that the voltage to be measured is μE_g when the current is just reduced to zero. By using a tube which has a large value of

μ the necessary grid battery voltage need not be large. Thus, if $\mu = 1000$, a grid battery of 50 volts is sufficient to measure 50,000 volts. Such a tube should be designed with long side tubes as shown in Fig. 222, to prevent arcing across the glass.

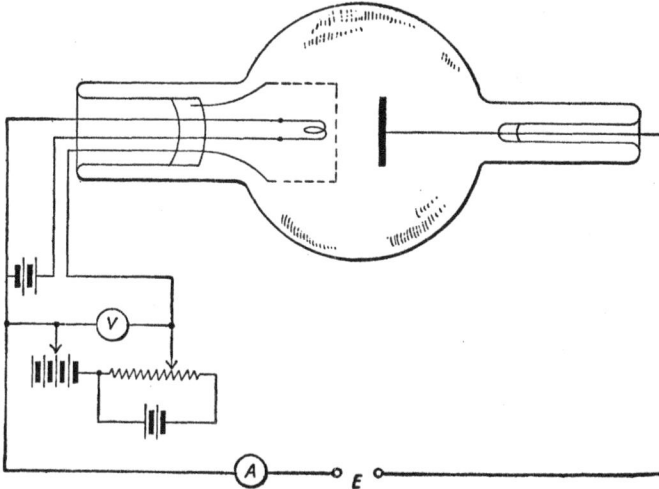

Fig. 222.

116. The Audion Voltage and Current Regulator. The follow-ing circuit arrangements, for which I am indebted to Dr. P. I. Wold, indicate how the tube can be used to control the output of a generator or other source. The generator S supplies power to

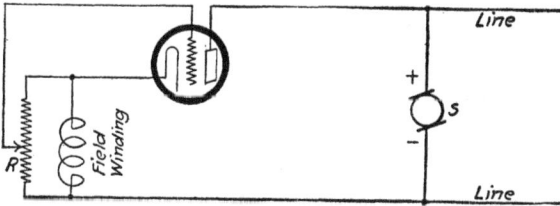

Fig. 223.

the line as shown in Fig. 223. To prevent voltage fluctuations the vacuum tube is connected in series with the field winding of the generator. The grid of the tube is connected to a convenient point on the resistance R in parallel with the field winding. Suppose, now, that the voltage of the generator tends to increase.

This increases the flow of current through the resistance R, which makes the grid more negative than before, thus increasing the resistance of the tube and, therefore, decreasing the current through the tube and field winding, because the tube and field

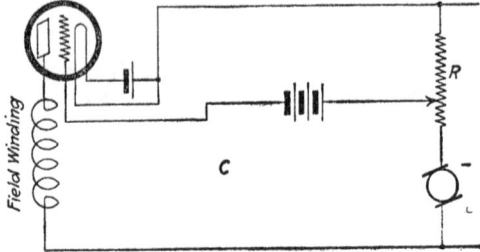

Fig. 224.

winding are in series. By this means, any tendency for the voltage to increase is counteracted. The same thing, of course, holds when the voltage tends to drop.

Fig. 224 shows an adaptation of this method of control for keeping the current output of the generator constant. It will be

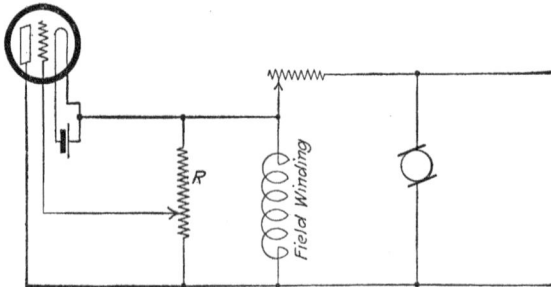

Fig. 225.

seen that if the current through the resistance R tends to increase, the grid tends to become more negative with respect to the filament, thus increasing the resistance of the tube and decreasing the current through the field winding.

Fig. 225 shows a circuit arrangement in which the tube and field winding are connected in parallel. The high resistance R to which the grid is connected is also in parallel with the field winding. In this case, when the output voltage of the generator tends to increase, the grid tends to become less negative with respect to the filament, on account of the increased voltage drop established in the resistance R. This tends to reduce the resistance of the tube, thereby decreasing the current through the field winding, which is in parallel with the tube.

117. Power-limiting Devices. It has been pointed out in Chapter VI that the thermionic valve acts not only as a rectifier but also as a power-limiting device, because, while it blocks current in one direction, the current in the other direction cannot exceed the saturation value. This is, therefore, the maximum current that can be transmitted through the tube.

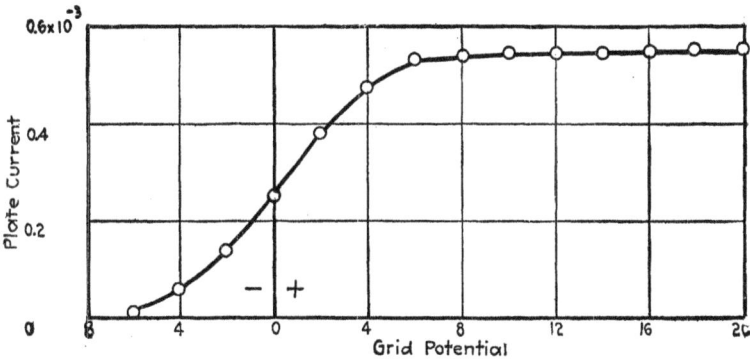

Fig. 226.

The three-electrode tube can be very well adapted for the purpose of limiting the output power. If, for example, the grid becomes sufficiently negative, the plate current is reduced to zero. If, on the other hand, the grid becomes sufficiently positive the plate current reaches a saturation value. It will be evident from the explanations given in Chapter VII, that this saturation value may be due to two causes: Firstly, the strength of the field in the space between filament and grid may become sufficiently great to pull the electrons away from the filament as fast as they are emitted. This gives rise to the ordinary saturation current. If the cathode has a smooth and pure surface, the knee of the curve, where it bends over to the saturation current, is fairly well defined. In cases of filaments having rough surfaces, on the other hand, the saturation current is approached gradually and the curve does not become quite parallel to the voltage axis. Under such conditions the device is not very suitable for power-limiting purposes, but but use can be made of a second factor which limits the current. If the external circuit contains a resistance, the potential of the plate decreases as the potential of the grid increases, on account of the voltage drop established in the external resistance in the plate circuit. A condition can then be reached where the positive potential of the grid becomes comparable with that of the plate, and in that case a large proportion of the electrons will be attracted to the grid, thus limiting the flow of electrons to the plate. This factor results in a very good curve, a sample of which is shown in Fig. 226. In obtaining this curve the voltages were so adjusted

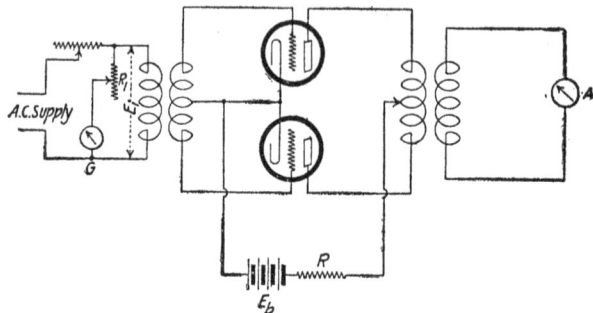

FIG. 227.

that the saturation value was obtained at a positive grid potential equal to the negative grid potential that just sufficed to reduce the current to zero, thus resulting in a curve which is nearly symmetrical with respect to the axis of zero grid potential.

Instead of using a single tube we can make use of the push-pull arrangement which was described in Chapter VII. This gives a good circuit for power-limiting purposes. The arrangement is shown in Fig. 227. The input voltage was measured by means of a thermo-couple, G and the output current was measured with the a-c. ammeter A. The result obtained is shown in Fig. 228. For low input voltages the alternating current in the output increased practically linearly with the voltage, but became limited to a value of about 3.6 milliamperes, beyond which it did not increase, although the input voltage was increased to 10 volts, as shown by the curve. The current was actually measured for input voltages up to 42 volts. At the higher voltages the current showed a tendency to decrease.

118. The Ionization Manometer. It was shown in Section 36 that if the gas pressure in a tube is so low that the mean free path of the electrons in the gas is large compared with the distance between the electrodes, the pressure is proportional to the number of positive ions formed by collision of the electrons with the residual gas molecules. This is expressed in equation (6), Chapter V, and was verified by O. E. Buckley,[1] and used by him for the construction of an ionization manometer.

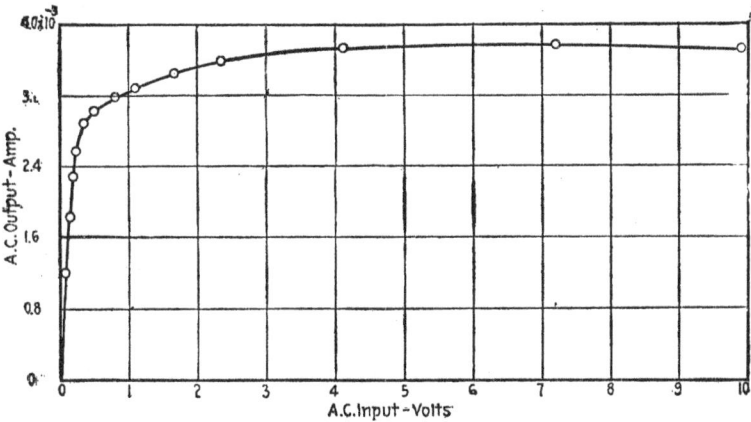

FIG. 228.

This device consists of three electrodes connected in a circuit as shown in Fig. 229. The grid is maintained at a positive potential with respect to the filament, and the plate is kept negative with respect to the negative end of the filament. The electrons emitted from the filament are attracted toward the grid, some going to the grid and some passing through the openings in the grid. Between the grid and the plate, however, is a retarding field, and, since the plate is negative with respect to the filament none of the electrons that are emitted from the filament can reach the plate; they attain their maximum speed in the neighborhood of the grid. Those that pass through the grid are retarded and finally return to the grid before they have a chance of reaching the plate. It is usually sufficient if the plate is maintained at a poten-

[1] Proc. National Academy of Science, Vol. 2, p. 683, 1916. See also DUSHMAN and FOUND for calibration of this gauge for various gases, Phys. Rev., Febr., 1920, p. 133.

tial of about 10 volts negative with respect to the negative end of the filament. Now, if there are gas molecules in the space, positive ions will be formed by collision, if the electrons move with a sufficiently high speed. The positive ions that are formed between filament and grid go to the filament, but those that are formed between grid and plate are attracted to the negative plate, thus giving a current through the galvanometer G; and this current is a measure of the number of positive ions formed. The total number of electrons flowing to the grid can be measured with the ammeter A, and must be kept constant. It is, therefore, desirable to work on the saturation part of the curve. The grid battery E_c

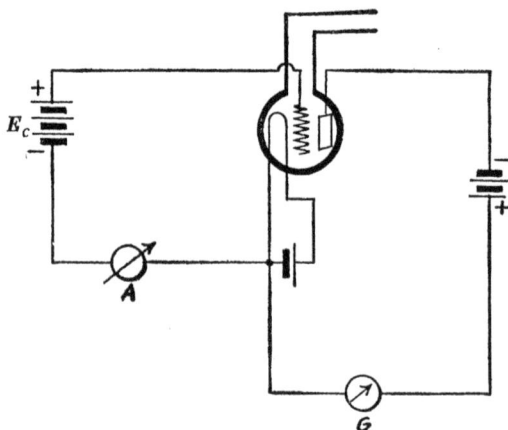

Fig. 229.

should be about 200 volts, but depends, of course, on the construction of the device.

This type of gauge has the disadvantage that gases affect the emission of electrons from the filament; that is, the saturation current is dependent on the amount and kind of gas coming in contact with the surface of the filament. The electron emission from oxide-coated filaments is not as susceptible to the influence of gas as the emission from some metallic filaments, such as tungsten, and can therefore be used to advantage in ionization manometers. If the filament is operated at a high temperature, the effect of gas on the emission becomes less. (See Chapter V.) Other means can be used for keeping the grid current constant. The grid circuit can, for example, be connected to a regulating scheme, similar to those described in Section 116.

The ionization gauge is very convenient for measuring quick changes in pressure because the positive current indicated by the galvanometer G is a direct measure of the amount of gas present in the device. On the other hand, the manometer must be calibrated separately for each kind of gas that may be encountered, since the amount of ionization produced depends on the size of the gas molecules.

The proportionality between the gas pressure and the positive current, as shown by equation 6, Chapter V., holds, in gauges of the most commonly used types, for pressures below about 1 micron. The manometer can, therefore, be calibrated against a McLeod gauge for pressures of about 1 micron, or somewhat less, where the McLeod gauge gives reliable readings. This, then gives the manometer constant K in the equation.

$$P = K\frac{I_p}{I_n}, \quad \cdots \cdots \cdots \quad (1)$$

where I_p is the positive current registered by the galvanometer G, I_n the electron current to the grid, and P the pressure of the gas. This equation can then be used to measure pressures down to very low values where the McLeod gauge is quite unreliable.

119. Heterodyne Method of Generating Currents of Very Low Frequency with the Vacuum Tube. In Section 96 a circuit was shown for the production of low frequency currents with the vacuum tube. The frequency is determined by the constants of the oscillation circuit; hence, a very low frequency requires the use of large inductances and capacities. If it is necessary to avoid the use of such large coils and condensers, use can be made of a scheme which was suggested to me by Dr. P. I. Wold. This system consists of using two vacuum tube generators to give frequencies differing by an amount equal to the frequency that it is desired to obtain. The output currents from these tubes are both impressed on the input of another tube, which operates as a modulator. The output of the modulator contains, among others, a frequency equal to the difference between the frequencies impressed on its input. (See Section 109.) Thus, if the two generators give frequencies of, say, 99 and 100 cycles per second, then the output of the modulator will contain a current having a fre-

quency of one cycle per second.   In the output of the modulator
can be inserted a filter to by-pass all frequencies higher than the
one desired.   This method, of course, requires that the frequencies
of the two generators be maintained constant to a high degree,
since a small change in either of them will cause a relatively large
change in the low frequency obtained in the output of the modulator.

120. The Thermionic Valve as a High Tension Switch.   On
very high voltage power transmission lines, it is necessary to use
especially designed switches for making and breaking the circuit.
To prevent the arcing that ordinarily would take place when
breaking a high voltage circuit, the thermionic valve could be
used in the manner suggested by Mr. J. R. Carson.   The valve
is inserted directly in the line and will transmit current in one
direction when the filament is hot.   When it is necessary to stop
the flow of current, we can, instead of directly breaking the circuit, simply cut out the filament current of the valve.   The current flowing through the valve then dies down smoothly in a period
which is short enough for ordinary work, but still large enough to
prevent arcing.   For the transmission of current in both directions,
we can, of course, insert two valves, one to transmit current in
one direction and the other in the opposite direction.

121. Devices Employing Secondary Electron Emission.   The
emission of electrons from cold electrodes under the impact of
electrons (a phenomenon which is known as secondary electron
emission or delta rays) results in a falling characteristic, as shown
by the portion ABC of Fig. 16, page 48.   The manner in which
this characteristic is obtained is explained in Section 23.   A. W.
Hull [1] has made use of this phenomenon in the construction of a
negative resistance amplifier and oscillator and has called the
device a Dynatron.   In the circuit shown in Fig. 15, the electrons
coming from the filament impinge on the grid and so emit secondary
electrons from it.   These are drawn to the plate by the positive
potential on the plate supplied by the battery E.   When the
number of secondary electrons emitted from the grid becomes large
enough in proportion to the number of electrons striking it, the
electron current flowing into the grid decreases as the potential
of the grid is increased.

There is another way in which the tube can be used to give a

[1] Proc. I.R.E., Vol. 6, p. 5–35, 1918.

negative resistance. In this case the grid is maintained at a positive potential with respect to the plate. The electrons passing through the grid and striking the plate cause the emission of secondary electrons from the plate and these are then drawn over to the positive grid. The circuit arrangement for this case is shown in Fig. 230. If r_0 is a resistance placed in the plate circuit, then a potential E_p on the plate is given by

$$E_p = E_b - r_0 I_p, \quad \ldots \ldots \quad (2)$$

where I_p is the current in the plate circuit, and E_b the voltage impressed on the plate circuit. Taking the point B of Fig. 16 as

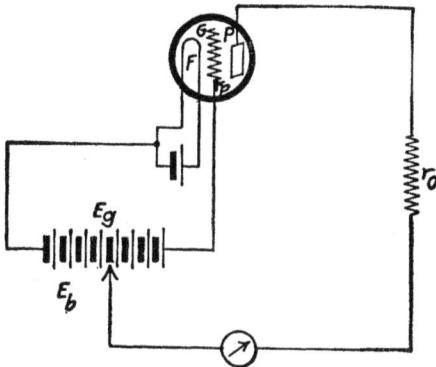

FIG. 230.

the origin of coordinates, the sloping part ABC of the characteristic can be represented by the equation

$$I_p = -mE_p = -\frac{E_p}{r_p} \quad \ldots \ldots \quad (3)$$

where r_p is the plate resistance of the tube. Substituting equation (2) into this equation we obtain,

$$I_p = \frac{E_b}{r_0 - r_p}. \quad \ldots \ldots \ldots \quad (4)$$

Differentiating I_p with respect to E_b, and multiplying by r_0, we get:

$$r_0 \frac{\partial I_p}{\partial E_b} = \frac{r_0}{r_0 - r_p} \quad \ldots \ldots \ldots \quad (5)$$

that is, a potential variation applied to the plate gives a voltage variation in the resistance r_0 which can be made very large by making r_0 nearly equal to r_p. The device thus operates as an amplifier.

Instead of using only the negative resistance characteristic of the device, connected in the manner explained above, Hull also made use of the normal amplifying property of the tube by inserting a second grid. This device he called a "Pliodynatron." By this means he has been able to obtain a voltage amplification of 1000 fold. To obtain such a high voltage amplification, however, it is necessary to make r_p and r_0 nearly equal. When this is done the device becomes unstable and needs careful adjustment and constant attention. It was, however, found possible to obtain a voltage amplification of 100 fold without trouble. It is doubtful if a device of this kind is as good as the audion, because by properly designing the audion it is easy to obtain a voltage amplification of several hundred fold, and since the audion does not possess a falling characteristic, its operation is stable no matter how high the amplification constant be made. In cases where it may be necessary to use a negative resistance device, however, the dynatron will be found to be of value and better than an arc, which also shows a negative resistance characteristic. The dynatron, for example, does not depend for its operation on the ionization of gas or vapor and is therefore more reliable.

The dynatron can be used also to produce sustained oscillations if it be connected in a circuit of the type commonly used in connection with arcs. Fig. 231, for example, shows a circuit which makes possible the production of sustained oscillations with the dynatron. It was shown in Chapter VIII that the total resistance in the output circuit will be zero if the effective resistance of the oscillation circuit L, C, r, namely $r_0 = \dfrac{L}{Cr}$, is equal to the negative resistance of the tube. Hence, by adjusting the capacity C and the inductance L so that r_0 becomes equal to the negative plate resistance of the tube, the total resistance of the output circuit is zero, and the device will produce sustained oscillations.

122. Tubes Containing More than One Grid. Various investigators have suggested using two grids instead of one for special purposes. Thus, R. A. Heising, for example, used a double grid tube as a modulator, in which the radio frequency is impressed between the filament and one grid, and the audio frequency between the filament and the other grid.

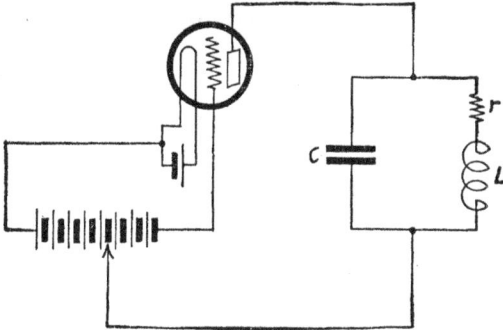

FIG. 231.

There are various circuits in which the tubes with double grids can be used. In one type of circuit, for example, the grid nearest to the anode is used as an auxiliary anode, while the grid nearest to the cathode operates as a controlling electrode. In this case,

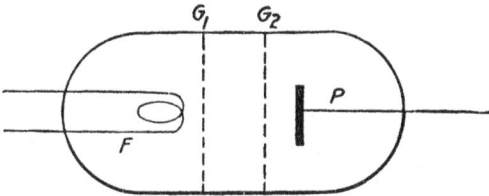

FIG. 232.

we can obtain the expression for the effective voltage as follows: Referring to Fig. 232, let us first regard grid G_1 as a cathode. Let the amplification constant of the system so formed be $\mu_2 = \dfrac{1}{\gamma_2}$; then the effective voltage between G_1 and G_2 is

$$E_2 = \gamma_2 E'_p + E_{g2}, \quad \cdots \cdots \quad (6)$$

where E'_p is the potential on the plate, and E_{g2} the potential on the grid G_2. This expression can be regarded as the effective anode potential for the system: F (cathode), G_1 (controlling electrode) and G_2P(anode). Hence, the effective voltage between F and G_1 is

$$E_1 = \gamma_1 E_2 + E_{g1}, \quad \ldots \quad (7)$$

where E_{g1} is the potential on the grid G_1. Substituting the value of E_2 given by equation (6), we get the effective voltage between F and G_1[1]:

$$E_1 = \gamma_1 \gamma_2 E'_p + \gamma_1 E_{g2} + E_{g1}, \quad \ldots \quad (8)$$

and the current is

$$I_2 = f_2(\gamma_1 \gamma_2 E'_p + \gamma_1 E_{g2} + E_{g1}). \quad \ldots \quad (9)$$

In the case of a tube containing only one grid the corresponding expression for the current is

$$I_1 = f_1(\gamma E_p + E_g), \quad \ldots \quad (10)$$

where γ is the reciprocal of the amplification constant μ.

If we make the potential of the plate in the case of the double grid tube equal to n times that of the grid G_2, we can write equation (9):

$$I_2 = f_2\left[\left(\gamma_1 \gamma_2 + \frac{\gamma_1}{n}\right)E'_p + E_{g1}\right]. \quad \ldots \quad (11)$$

This expression can be compared with that which holds for a single grid tube. Suppose that the amplification constant of the single grid tube is such that $\gamma = \gamma_1 \gamma_2$. Then we find that the ratio of the negative potentials on the controlling grid, that are necessary to reduce the current to zero in the two cases, is

$$\left(1 + \frac{1}{\gamma_2 n}\right)\frac{E'_p}{E_p}. \quad \ldots \quad (12)$$

Thus, suppose the potential of the plate is the same in both cases, and let $n = 2$, and $\gamma_2 = 0.1$. Then the intercept of the characteristic on the axis of controlling grid voltage, in the case of the two grids, is about six times as large as in the case of the tube containing only one grid. Other comparisons can be made if the form of the characteristic of the double grid tube is known.

In operating a double grid tube in the manner described above, the grid G_2 is usually sufficiently positive to draw an appreciable

[1] See also BARKHAUSEN, Jahrb. d. drahtlosen Tel. u. Tel., Vol. 14, p. 43, 1919.

number of electrons to it, and this decreases the current to the plate. In general, such tubes are not as good as audions when used for ordinary purposes. Circuits can, of course, be easily devised which enable one to make use of variations in the current flowing to the second grid, as well as that flowing to the plate or anode.

Another way in which double grid tubes can be used is to use the grid nearest to the anode as the controlling electrode, and apply a positive potential to the grid nearest to the cathode, as has, for example, been done by Langmuir. This grid should then preferably be placed close to the cathode, and the potential applied to it should be high enough to pull the electrons away from the filament as fast as they are emitted from it, thus giving the condition for the saturation current. In this case the space charge between the filament and the first grid is small. If the second grid is kept negative with respect to the first grid, the electrons passing through the first grid will be slowed down in approaching the second grid, thus increasing the space charge between the two grids, but the electrons in the space are now spread throughout a greater volume, instead of being concentrated around the filament, and hence, potential variations applied to the second grid can be expected to produce relatively large changes in current flowing to the anode. Here, again, the first grids robs the plate of current, but the circuit could be so arranged that use is made of the variation in both currents, namely, that flowing to the plate and that to the first grid.

In using such devices, where electrons impinge on a conductor which does not have the highest positive potential in the system, the effect of secondary electron emission must be taken into consideration, because if the electrons impinge with sufficient violence on a conductor, secondary electrons are emitted from it, and if there is another conductor, which is positive with respect to the first, the secondary electrons will be drawn over to this positive conductor. If the velocity with which the electrons impinge on the first conductor is large enough, so many secondary electrons can be emitted that electrons will flow out of this conductor instead of into it, thus reversing the direction of the current.

INDEX